Conceptual Wavelets In Digital Signal Processing

An In-Depth, Practical Approach for the Non-Mathematician

Conceptual Wavelets In Digital Signal Processing

An In-Depth, Practical Approach for the Non-Mathematician

D. Lee Fugal

Space and Signals Technical Publishing

San Diego, California

Conceptual Wavelets in Digital Signal Processing
An In-Depth Practical Approach for the Non-Mathematician
by D. Lee Fugal

P.O. Box 1771, San Diego, CA 91977-1771
www.ConceptualWavelets.com
www.SpaceAndSignals.com

ISBN 978-0-9821994-5-9
LCCN 2008910885

Printed in the United States of America, First Edition, First Printing, 2009

MATLAB® is a registered trademark of *The Mathworks, Inc*. Natick, MA

Cover image by D. Lee Fugal. Continuous Wavelet Transform Display of a Quadratic Chirp Signal. Wavelet used: Daubechies 10.

Cover Design by D. Casey Wright of Pix 3-D, www.pixgroup.com

Publisher's Cataloging-in-Publication ***(Provided by Quality Books, Inc.)***

Fugal, D. Lee.
Conceptual wavelets in digital signal processing : an in-depth, practical approach for the non-mathematician / D. Lee Fugal.
p. cm.
Includes bibliographical references and index.
LCCN 2008910885
ISBN-13: 978-0-9821994-5-9
ISBN-10: 0-9821994-5-7

1. Wavelets (Mathematics) 2. Signal processing--Digital techniques. I. Title.

QA403.3.F84 2009 515'.2433
QBI08-600337

I dedicate this book to my beautiful and wonderful wife, Rhea. Without your support this book would not have been possible. Thank you for your patience, encouragement, and the love that we share.

*"The main thing is the content, not the mathematics.
With mathematics one can prove anything."*

— Albert Einstein

CONTENTS

"Simplicity is the ultimate sophistication."

— Leonardo Da Vinci

"The paradox of simplicity is that making things simpler is hard work!"

—Bill Jensen
www.SimplerWork.com

PREFACE

Understanding & Harnessing Wavelet "Elephants"

It was six men of Hindustan
To learning much inclined,
Who went to see the Elephant
(Though all of them were blind)
That each by observation
Might satisfy the mind.

The first approached the Elephant
And happening to fall
Against his broad and sturdy side
At once began to bawl:
"Bless me, it seems the Elephant
Is very like a wall".

The second, feeling of his tusk,
Cried, "Ho! What have we here
So very round and smooth and sharp?
To me 'tis mighty clear
This wonder of an Elephant
Is very like a spear".

The third approached the animal,
And happening to take
The squirming trunk within his hands,
Then boldly up and spake:
"I see," quoth he, "the Elephant
Is very like a snake."

The Fourth reached out an eager hand,
And felt about the knee.
"What most this wondrous beast is like
Is mighty plain," quoth he;
"'Tis clear enough the Elephant
Is very like a tree!"

The Fifth, who chanced to touch the ear,
Said: "E'en the blindest man
Can tell what this resembles most;
Deny the fact who can,
This marvel of an Elephant
Is very like a fan!"

The Sixth no sooner had begun
About the beast to grope,
Than, seizing on the swinging tail
That fell within his scope,
"I see," quoth he, "the Elephant
Is very like a rope!"

And so these men of Hindustan
Disputed loud and long,
Each in his own opinion
Exceeding stiff and strong,
Though each was partly in the right
And all were in the wrong.

"*The Six Blind Men and the Elephant*" — A Poem by John Godfrey Saxe (1816 – 1887)

The subject of wavelets is certainly an "Elephant"—besides being *extremely powerful*, it can also become *extremely "heavy"* in terms of math. And the various methods of looking at the subject are staunchly defended by the various authors "*Exceeding stiff and strong*".

There are a lot of ways to look at the wavelet "elephant". Vector Spaces, Function Spaces, Frame Theory, Set Theory, Matrices and Transposes, Finite Elements, Continuous Time Representations (with double infinite integrals), extensions of the Fast Fourier Transform or the Short-Time Fourier Transform and even a curve on a sphere in 4-dimensional space—all these are used by various authors with the same evangelical zeal as the learned men of Indostan. None of these are really "wrong",

How This Book Differs From Other Wavelet Texts

"Conceptual Wavelets in Digital Signal Processing", however, is vastly different from other books in that we use numerous examples, figures, and demonstrations to show how to understand and use wavelets. This is a very complete and in-depth treatment of the subject, but from an intuitive, conceptual point of view. We let you look at a few key equations found in the more mathematically oriented texts—but only *after* the concepts are demonstrated and understood.* Then if you desire further study from traditional texts, this allows you to recognize these equations and understand in advance how they relate to the real world having actually seen them "in action".

It has been gratifying to present the 3-day course "Wavelets: A Conceptual, Practical Approach" at universities, corporations, and conference centers around the country for the past few years. Much of this book is "built" on these slides and improved by the comments and suggestions from the attendees. Those with little or no math background have expressed gratitude for being able to "see the elephant" enough to understand it and use it's power. Those with a strong mathematical background have expressed thanks for new insights and intuitive understanding that was not immediately evident from the equations.

One of the principle contributions of wavelets has been to bring those academic fields together to observe the "elephant" and to "satisfy the mind". It is not surprising then that you will find this particular pachyderm described in terms of *wavelets, wavelet filters, wavelet transforms, filter banks, multirate systems, matched filtering, multiresolution analysis* and so on.

This author's background is in Digital Signal Processing (Fast Fourier Transforms, Digital Filtering, etc.). and the description of the elephant is no doubt biased toward time or frequency representations of data. You will soon learn, however, that both the *power* and the *complexity* of wavelets lies in the fact that they deal with (are localized in) *both* time *and* frequency! It is especially important to understand that this dual (time/frequency) nature adds literally another dimension to wavelets! Instead of the data being shown as a function of time *or* as a function of frequency we now can look at the data simultaneously in terms of time *and* frequency (or at least effective frequency).

---

* The occasional equation, if especially relevant to the explanations on a particular page, can also be found in the footnotes on that page.

Is the extra dimension of effort in learning to use these wavelet tools worth it? I believe the answer from all the authors would be a resounding "Yes!!". They, like this author, have seen how powerful and how handy this "animal" can be for processing signals or images that have "events" (changes in amplitude, frequency or shape) that start and stop. These "non-stationary" signals—the most interesting kind—are as diverse as a human heartbeat, a telemetry pulse from a missile (friendly or not), earthquakes seismic data, and the financial patterns in the stock market.

Because of all this tremendous capability, most authors (including myself) sincerely try to prevent the student from becoming discouraged when facing this extra dimension. Thus you will see terms in the title of many wavelet books and websites such as "Gentle", "Tutorial", "Friendly", "*Really* Friendly", "A Primer", "Made Easy", and even "For Kids" (we would like to meet these child prodigies!).

This is also why authors will teach using the tools with which they are most familiar—and for most wavelet writers this is *applied mathematics* (a look inside traditional wavelet texts quickly reveals a heavy dependence on math). This book takes a very different approach in that it *doesn't* rely on proofs, theorems, lemmas, etc. etc. to try to teach *concepts* through *equations*. We emphasize informed *use* of wavelets and leave the rigorous *proofs* to scholarly texts. In the appendix, we reference some excellent traditional mathematics-based texts, articles, and websites for additional study if desired.

This author also feels strongly (no lack of evangelical zeal here!) that it is important to be sure that wavelet data is not misused or misinterpreted. As with any technology, blindly following equations out of context without understanding the concepts behind them can lead to misinformation. We will show you how to "read" the results of wavelet displays correctly. We point out common pitfalls in wavelet transforms and how to avoid them. The real-world intuitive understanding you will obtain from reading this book should allow you to take full advantage of the powerful capability of wavelets with confidence in obtaining true and meaningful results and without fear of degradation of the data.

How This Book Is Laid Out— Study Suggestions

(How do you eat an elephant?—one bite at a time).

So how do you "digest" this book? One chapter at a time. Each chapter is designed to build on the previous chapters to help you gain a conceptual understanding of wavelets.

Chapter One presents an overview of wavelets, wavelet filters and wavelet transforms and shows a little of what they can do. This should put you way ahead of the "six blind men from Indostan" by providing a good "peek at the pachyderm". The familiar FFT/DFT is first reviewed and then compared to the Continuos Wavelet Transform

(CWT). The conventional (decimated) Discrete Wavelet Transform (DWT) and the Undecimated DWT (UDWT) are introduced and compared. Examples of the capabilities of these transforms are shown, along with a short overview of the various types of wavelets

For a first "bite", you might also want to look at a short article "Wavelets: Beyond Comparison" written by the author as a staff tutorial for Applied Technology Institute. (**www.aticourses.com/ati_tutorials.htm**).

Chapter Two provides a step-by-step walk-through of the Continuous Wavelet Transform using a very simple example of 8 exam scores and a Haar wavelet. We actually construct a CWT display from this data to learn how to "read" this type of display.

Chapter Three uses the same example of 8 exam scores and the Haar wavelet filters, but provides a step-by-step walk through of the Undecimated Discrete Wavelet Transform (UDWT—a.k.a. RDWT, A' Trous, or Shift Invariant). We show how this is very similar to the Continuous Wavelet Transform from Chapter Two.

Chapter Four highlights the downasmpling and upsampling that is added to the UDWT to produce the better-known conventional (decimated) DWT. We continue with the same example of 8 exam scores and the Haar wavelet in another step-by-step walk-through. We also show some simple compression and de-noising and take our first look at some DWT displays.

Chapter Five shows how some wavelet filters ("crude" wavelets) are generated using an explicit mathematical equation. The equations, though very simple, are continuous in time and so we show how to get the discrete points needed to produce actual digital wavelet filters of varying length.

If Chapter Five can be thought of as "Filters from Wavelets", then Chapter Six can be thought of as "Wavelets from Filters". Starting with wavelet filters that have very few points (only 2 for the Haar), we learn how we can interpolate or "stretch" them to hundreds of points. Then we learn how to use these very long filters to produce filters of *any* desired length. In other words we go from "Filters to Wavelets to Filters".

In Chapter Seven we explore further and compare the 3 major types of wavelet transforms—the Continuous Wavelet Transform (CWT), the Undecimated Discrete Wavelet Transform (UDWT), and the conventional (decimated) Discrete Wavelet Transform (DWT). We also look briefly at the Wavelet Packet Transform (WPT). We examine the strengths and weaknesses of each type and show the general types of application of each. We compare and relate each type of transform to the others and show how, in each type, that we are still comparing data with the various wavelet filters.

Now somewhat familiar with the wavelet transforms, in Chapter Eight we look at what gives the "elephant" such strength and power—the Perfect Reconstruction Quadrature Mirror Filters (PRQMF). We not only discover the amazing "mirror" relationships these filters have to each other, but that they are actually factors of the relatively simple Halfband Filters—the very Heart and Soul of the wavelet "elephant" We demonstrate *orthogonality* in wavelet filters by comparing them to simple Cartesian coordinates (x-y-z). In the last section we show how the halfband filters can also be factored another way into *biorthogonal* wavelets filters. There is some mathematics (mostly at the high-school algebra level) in this chapter, but this is in line with a major goal of the book—to introduce some key equations found in conventional wavelet literature after providing an intuitive understanding of the concepts (in this case *spectral factorization*).

In Chapter Nine we introduce a few more desirable qualities of the various wavelets—some by comparing with arbitrary or "fake" wavelets to highlight these qualities. We will demonstrate such concepts as *regularity, vanishing moments,* and stretching and sliding the wavelet filters to "*match*" the hidden event in a signal (and thus determine its location, frequency, and general shape). We also talk about the adaptability of wavelets and about spending too much effort to find the "perfect" set of wavelet filters (the sport of basis hunting). We then demonstrate an even easier way to find the "magic numbers" of the filters by using the above desirable qualities to create some simple equations and then using direct substitution to solve them.

By Chapter Ten, we have learned much about the properties of the various wavelets and wavelet filters and how they can (or cannot) be used in the various wavelet transforms. We now proceed to look at the major wavelet families and how the specific properties lead to some practical applications. We first look at some *crude wavelets* such as the Mexican Hat, Morlet, Gaussian, and Meyer. We then examine the complex versions of some of these crude wavelets (and some others) such as the Complex Shannon (Sinc), Complex Frequency B-Spline, Complex Morlet and the Complex Gaussian. We proceed to the *orthogonal* wavelets such as the Haar, Daubechies Family, Symlets, Coiflets, and the Discrete Meyer. Finally, we look at the *Biorthogonal* and *Reverse Biorthogonal* wavelets. In the last section we present a table summarizing the attributes of these various wavelets.

Chapter 11 gives you a break from examining the various aspects of the elephant and lets you "hop on the Howdah" (seat affixed to an elephant's back) for a little "ride". In this chapter we present some case studies of the applications of various wavelets to some real-life problems. We show how to separate noise from the signal by matching various wavelets to *either* the noise or the signal and then modifying the Discrete Wavelet Transform (DWT} results to keep only the "clean" signal. We also revisit the Continuous Wavelet Transform (CWT) to demonstrate its power. We perform compression and denoising on the classic "Barbara" image and even show how to remove "freckles" (skin imperfections). We highlight a pathological case where the Undecimated DWT will outperform the conventional (decimated) DWT and provide suggestions for insuring the integrity of the data. In all these "non-stationary" examples we show why an FFT (or an STFT) would not work as well—or not work at all.

In Chapter 12 we learn how a conventional DWT can downsample repeatedly and still have *alias cancellation* (if done correctly). We look in more detail at the alias cancellation capability of the Perfect Reconstruction Quadrature Mirror Filters from Chapter Eight. Perfect Reconstruction and Alias Cancellation are demonstrated in both the time and the frequency domains. In the last section we look at some equations found in much of the conventional literature that describe Alias Cancellation and Perfect Reconstruction (often called "No Distortion"). We relate these equations to the concepts we have already learned in this chapter.

The final chapter, Chapter Thirteen, clarifies some additional key equations from the traditional wavelet literature by explaining the concepts behind them and/or demonstrating some alternative methods. In particular, the (continuous time) Wavelet Function and Scaling Function Dilation Equations are examined. We show how the same results are obtained by the process of repeated upsampling and convolution. We then show how this upsampling/convolution process can produce artifacts in a conventional DWT that look like the "Wavelet Function" or "Scaling Function" itself! (don't miss this one!). In the final section we explain some other terms found in much of the wavelet literature such as *forward DWT, inverse DWT, Fast Wavelet Transform, Fast Inverse Wavelet Transform,* and *Wavelet Domain* and how they relate to the intuitive concepts and terms we have learned in this book.

The Appendices are not "required reading" *per se*, but are included to provide enrichment and help. Appendix A presents additional ways to relate wavelet transforms to the more familiar Fourier Transforms and conventional DSP Filtering. Appendix B discusses Heisenberg Boxes and the Heisenberg Uncertainty Principle as applied to wavelets. Despite the Einstein-sounding name this is actually relativity – er – *relatively* simple (sorry, couldn't resist). Appendix C is a reprint of a short 7-page article "Wavelets: Beyond Comparison" by the author. This simple "Staff Tutorial" written for the Applied Technology Institute (Riva, Maryland) is also included here for your convenience Appendix D is a resource that presents the author's recommendations for wavelet books (at all levels), wavelet articles, and wavelet websites.

As you can see, the chapters build upon each other to allow you to observe and experience "this marvel of (a wavelet) elephant" It is sincerely hoped that with the concepts learned in these chapters and (if needed) some additional specific application-oriented literature and software, you will be able to harness this "mammoth" power to allow you a much more complete understanding of your data and to work with it in a more efficient and cost-effective manner.

"An investment in knowledge pays the best interest."
—Benjamin Franklin

ACKNOWLEDGMENTS

Sir Isaac Newton said “If I have seen further it is by standing on the shoulders of Giants”. While I don’t claim to be in the same league as Newton, I certainly need to acknowledge the marvelous contributions of the pioneers of wavelets such as **Meyer, Morlet, Daubechies, Coifman, Donoho, Grossman, Mallet, Wickerhauser** and others. They have brought into existence, nurtured, and helped wavelets grow from a mathematical curiosity to a powerful tool in wide use and in many disciplines.

I also wish to acknowledge personal help and specific advice on wavelets from **Amara Graps, Kurt Dobson, Michael Unser, Todd Moon** and **Behrouz Farhang-Boroujeny**. Thanks for sharing your expertise and for your patience.

I am most grateful to some established authors that have helped me in writing this book. The general layout is patterned after ‘“*The Scientist and Engineer’s Guide to Digital Signal Processing*” by **Steven W. Smith.** Thanks for all your help, Steve! **Richard G. Lyons** (“*Understanding Digital Signal Processing—Second Addition”)* provided invaluable help and guidance and even some proofreading. Your “tips and tricks” in avoiding pitfalls in writing a book were also of enormous help and I am most grateful. **David Adamy** *(“EW 101—A First Course In Electronic Warfare*” and subsequent books) is extremely busy as a sought-after lecturer around the world but still found some time to advise me on writing and publishing a book. Thanks, Dave. The title of this book was inspired by “*Conceptual Physics—Seventh Edition*” by **Paul. G. Hewitt**. Thank you, Paul, for your advice and encouragement I hope that in some small way my book teaches *Wavelets* like your wonderful book teaches *Physics*—through *comprehension* more than *computation*.

In every case, these experts have not displayed the slightest hint of arrogance or conceit at being so good at what they do. Rather they have shown—as I suppose most of the top people in any field would show—only kindness, helpfulness, courtesy, patience, and a down-to-earth attitude that belies their standing as true giants in their fields. Their attitude (and hopefully my own) is to try to *Instruct* rather than *Impress* and in so doing they end up doing both!

I was lucky enough to learn Digital Signal Processing from the late **Dr. Thomas G. Stockham**, a pioneer in DSP and inventor of the “overlap-add method” in convolution. I felt excited and honored but very inadequate when some years later I taught “his” class at the **University of Utah**. Other professors and department heads that have been supportive of my decision to write a book on wavelets, in addition to those already mentioned, include **Dr. Jon Mathews, Dr. Brian Jeffs, Dr. Neil Cotter, Dr. Paul Kuhn, Dr. Michael Rice, Dr. Ed Price, Dr. Doug Chabries, Dr. fred harris (lowercase preferred), Dr. Bob Short,** and **Dr. Jay Smith.**

I have also been fortunate in my consulting practice to meet wonderful clients, some of whom have become good friends and have provided technical support and encouragement to this project. The list includes **Lee Savage, Chuck Clark, Bob Byard, Kent Jolley, Dick Wiley, Ernie Bouchard, Scott Andrews, Forrest Stafansen,** and **Jay Linehan**.

I owe a deep debt of gratitude to **Stacey Wright** of **Pix Interactive 3-D**, **(www.PixGroup.com)** for help at the many stages of the development of the book and the website (**www.ConceptualWavelets.com**). Your expertise and professionalism, along with your advice and encouragement, had a big part in taking this book from a dream to a reality.

A special thanks goes to **Jim Jenkins** and his fine staff at **Applied Technology Institute** who have provided the promotional and logistical support that has enabled me to present the 3-day course "Wavelets: A Conceptual, Practical Approach" at various locations around the country.

The ultimate source of support is of course from my immediate and extended family. Thank you **Rhea, Sandy, Stacey, Becky, Michael, Brent, Casey, Ryan, Shannon, Isaac, Ken, Maxine, John, Gordon, Deanne,** and of course my parents **Margaret** and **Delbert** for believing in me.

The last acknowledgment is to you, the reader, for having the courage to embark on a journey that you probably have heard was difficult but that has the promise of rich rewards as you add the power of wavelet processing to your professional repertoire. John A. Shedd in 1928 wrote "A ship in harbor is safe—but that is not what ships are built for". As you leave the safe harbor of conventional Digital Signal Processing to sail upon the Wavelets, may you find the treasures you seek. Welcome Aboard!

CHAPTER 1

Preview of Wavelets, Wavelet Filters, and Wavelet Transforms

As mentioned in the Preface, wavelets are used extensively in many varied technical fields. They are usually presented in mathematical formulae, but can actually be understood in terms of simple comparisons or correlations with the signal being analyzed.

In this chapter we introduce you to wavelets and to the wavelet filters that allow us to actually use them in Digital Signal Processing (DSP). Before exploring wavelet transforms as comparisons with wavelets, we first look at some simple everyday "transforms" and show how they too are comparisons. We next show how the familiar discrete Fourier transform (DFT) can also be thought of as comparisons with sinusoids. (In practice we use the speedy fast Fourier transform (FFT) algorithm to implement DFTs. To avoid confusion with the discrete wavelet transforms soon to be explored, we will use the term fast Fourier transform or FFT to represent the discrete Fourier transform.)*

Time signals that are simple waves of constant frequencies can be processed in a straightforward manner with ordinary FFT methods. Real-world signals, however, often have frequency content that can change over time or have pulses, anomalies, or other "events" at certain specific times. They can be intermittent, transient, or noisy. This type of signal can tell us where something is located on the planet, the health of a human heart, the position and velocity of a "blip" on a RADAR screen, stock market behavior, or the location of underground oil deposits. For these signals, we will usually do better with wavelets.

Jargon Alert: Signals (or noise) that stay at a constant frequency are called "stationary signals" in wavelet terminology. Signals that can change over time are called "non-stationary".

One final thought before beginning: Wavelets deal simultaneously with both time and frequency and require some effort to master. However their powerful capabilities in achieving this feat make them well worth the effort. This conceptual method makes

* MATLAB also uses the term "FFT" rather than "DFT" to compute the discrete Fourier transform.

learning them possible without advanced math skills and gives you a gut-level comprehension in the bargain.

The goal of this preview chapter is to introduce you to some new concepts, show you some basic diagrams, familiarize you with the jargon, and give you a preliminary feel for what's going on. Please don't be discouraged if everything is not obvious at first glance.

The next few short chapters will walk you step-by-step through the main concepts and the later chapters should answer most of the remaining questions. You should then be prepared to correctly and confidently use wavelets and to better understand the more advanced math-based texts and papers after you have seen wavelets "in action".

1.1 What is a Wavelet?

A *wavelet* is a waveform of limited duration that has an average value of zero. Unlike sinusoids that theoretically extend from minus to plus infinity, wavelets have a beginning and an end. Figure 1.1–1 shows a representation of a continuous sinusoid and a so-called "continuous" wavelet (a Daubechies 20 wavelet is depicted here).

Sinusoids are smooth and predictable and are good at describing constant-frequency (stationary) signals. Wavelets are irregular, of limited duration, and often non-symmetrical. They are better at describing anomalies, pulses, and other events that start and stop within the signal.

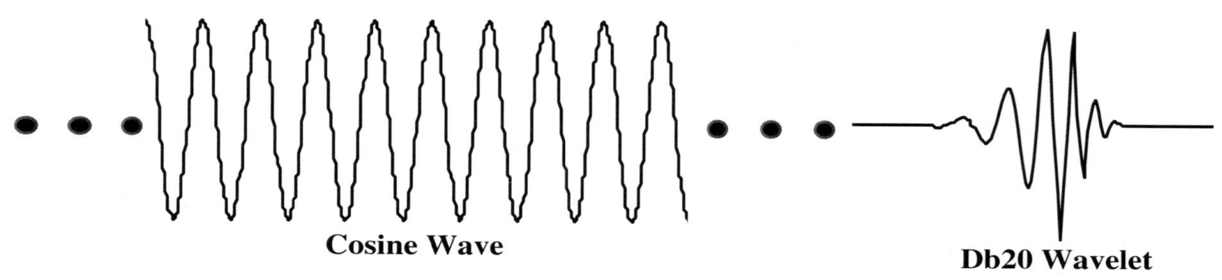

Figure 1.1–1 A portion of an infinitely long sinusoid (a cosine wave is shown here) and a finite length wavelet. Notice the sinusoid has an easily discernible frequency while the wavelet has a *pseudo frequency* in that the frequency varies slightly over the length of the wavelet.

Figure 1.1–2 shows how wavelets can be stretched or "*scaled*" to the same frequency as the anomaly, pulse, or other event. Notice that as the wavelet is stretched it has a lower frequency. Wavelets can also be *shifted* in time to line up with the event. Knowing *how much* the wavelet was stretched and shifted to line up (correlate) with the event gives us information as to the time and frequency of the event.

Jargon Alert:* In DSP *scaling* usually means changing the amplitude of a signal or waveform. In *wavelet* terminology, however, the term *scaling* means stretching or shrinking the wavelet in time. Thus the term *scaling* usually has reference to the *frequency* (or more precisely *pseudo frequency*) of the wavelet. The term *dilation* is also used to describe either stretching *or* shrinking the wavelet in time (despite the dictionary definition).

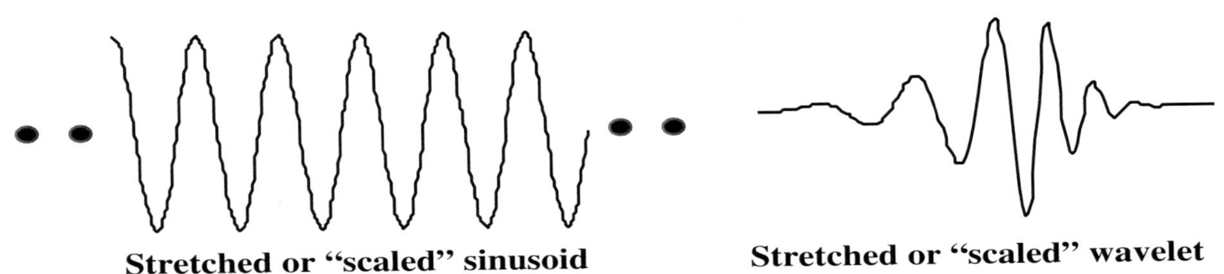

Figure 1.1–2 The infinitely long sinusoid is stretched (or *scaled* in wavelet terminology) and is now a lower frequency. The Db20 wavelet is also stretched (scaled) and its *pseudo frequency* (average frequency) is also lower.

1.2 What is a Wavelet Filter and how is it different from a Wavelet?

Wavelets are a child of the digital age. Some wavelets are defined by a mathematical expression and are drawn as continuous and infinite. These are called *crude wavelets.* However to use them with our digital signal, they must first be converted to *wavelet filters* having a finite number of discrete points. In other words, we evaluate the *crude wavelet equation* at the desired points in time (usually equispaced) to create the filter values at those times.

* By this definition "dilated pupils" can mean eyes constricted to pinhole openings. "When I use a word it means just what I choose it to mean—neither more nor less."—Humpty Dumpty in *"Through the Looking Glass"* by Lewis Carroll.

***Jargon Alert*: "Crude" wavelets are generated from an explicit mathematical equation.**

Figure 1.2–1 shows the whimsically-named Mexican Hat *crude* wavelet that looks like the side view of a sombrero. The mathematical expression for this particular wavelet as a function of time (t) is given by

$$mexh(t) = \left\{2/(\sqrt{3}\pi^{-1/4})\right\}(1-t^2)e^{-t^2/2}$$

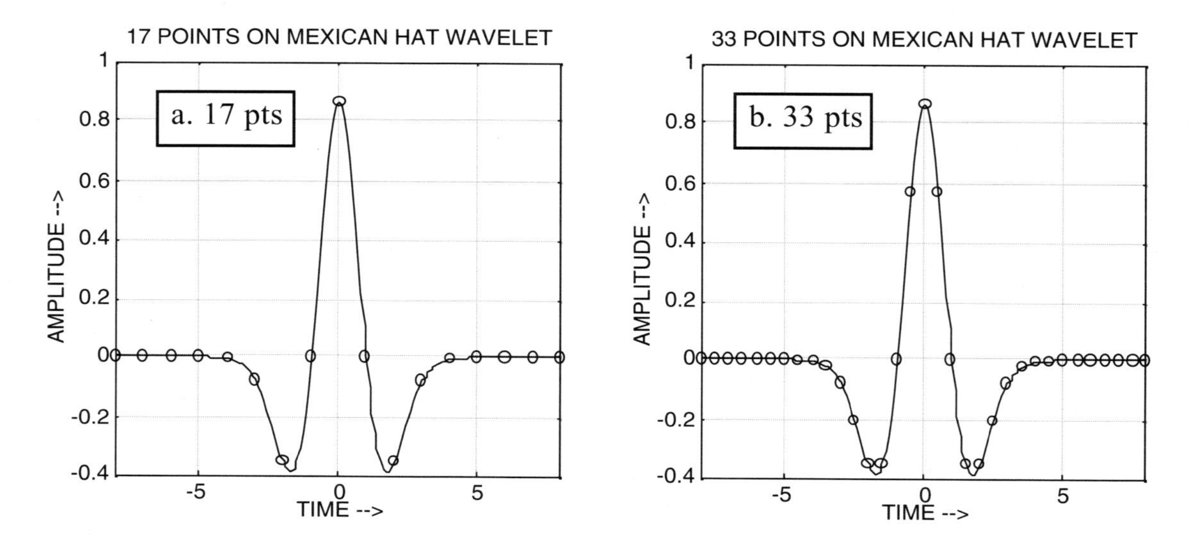

Figure 1.2–1 "Crude" Mexican Hat Wavelet with 17 points (a) then 33 wavelet filter points superimposed on the continuous *(crude)* representation (a then b). Although the defining equation describes an infinite, continuous waveform, by using equispaced discrete points we have created discrete, finite-length filters ready for use with digital computers.

Other wavelets *start out* as filters having as little as 2 points. Then an approximation or *estimation* of a continuous wavelet (for depictions) is *built* by interpolating and extrapolating more points. For these wavelets, there really is no true *continuous* form, only an estimation built from the original filter points. Figure 1.2–2 shows the 4 original filter points (plus 2 zeros at the same spacing) of the Daubechies 4 (Db4) wavelet superimposed on a 768 point estimation of a *"continuous wavelet"* built from these points.*

* We will demonstrate later how we go from 6 points to 12, 24, etc. to 768. MATLAB uses 768 points as a suitable approximation (estimation) of a "continuous" Db4 wavelet.

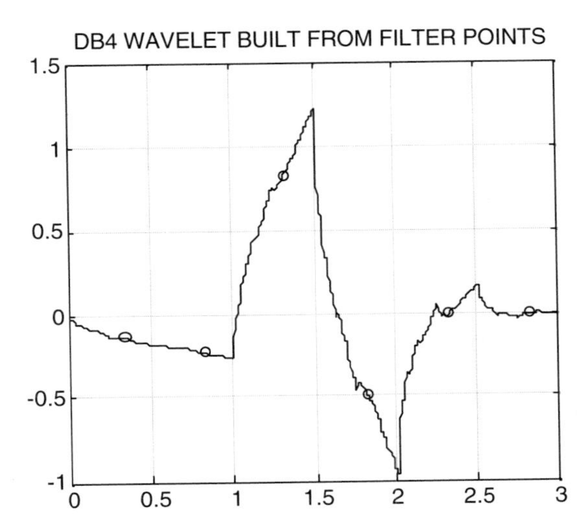

Figure 1.2–2 768 point estimation of a "continuous" Daubechies 4 wavelet built from 6 equispaced filter points (the 4 original filter points and 2 end zeros) superimposed on the graph.

Some wavelets have symmetry (valuable in human vision perception) such as the Biorthogonal wavelet pairs. Shannon or "Sinc" wavelets can find events with specific frequencies. (These are similar to the "sin(x)/x" sinc function filters found in traditional DSP.) Haar wavelets (the shortest) are good for edge detection and reconstructing binary pulses. Coiflets wavelets are good for data with self-similarities (fractals) such as financial trends. Some of the wavelet families are shown below in Figure 1.2–3.

You can even create your own wavelets, if needed. However there is "an embarrassment of riches" in the many wavelets that are already out there and ready to go. With their ability to stretch and shift, wavelets are extremely adaptable. You can usually get by very nicely with choosing a less-than-perfect wavelet. The only "wrong" choice is to avoid wavelets entirely due to an abundant selection.

As you can see (Fig. 1.2–3), wavelets come in various shapes and sizes. By stretching and shifting ("*dilating* and *translating*") the wavelet we can "match" it to the hidden event and thus discover its frequency and location in time. In addition, a particular wavelet shape may match the event unusually well (when stretched and shifted appropriately). This then tells us also about the *shape* of the *event* (It probably looks like the wavelet to obtain such a good match or *correlation*.) For example, the Haar wavelet would match an abrupt discontinuity while the Db20 would match a *chirp* signal (see the first and fourth wavelets in Fig. 1.2–3).

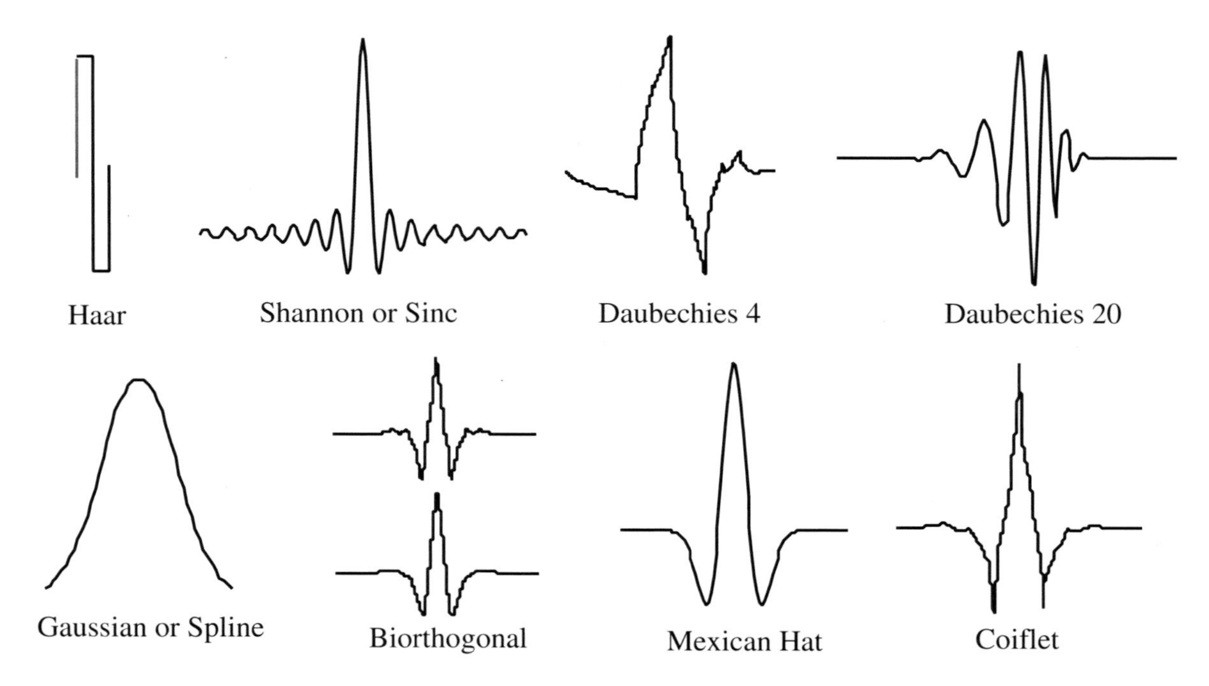

Figure 1.2–3 Examples of types of wavelets. Note 2 wavelets for the Biorthogonal. The Shannon, Gaussian, and Mexican Hat are "crude" wavelets that are defined by an explicit mathematical expression (and whose wavelet filters are obtained from evaluating that expression at specific points in time). The rest are estimations of a "continuous" wavelet built up from the original filter points.

***Jargon Alert*: Shifting or sliding is often referred to as *"translating"* in wavelet terminology.**

1.3 The value of Transforms and Examples of Everyday Use

Perhaps the easiest way to understand wavelet transforms is to first look at some transforms and other concepts we are already familiar with.

The purpose of any transform is to make our job easier, not just to see if we can do it. Suppose, for example, you were asked to *quickly* take the year 1999 and double it. Rather than do direct multiplication you would probably do a home-made *"millennial transform"* in your head something like 1999 = 2000 – 1. Then after transforming you would multiply by 2 to obtain 4000 – 2.

You would then take an "*inverse millennial transform*" of 4000 – 2 = 3998 for the correct answer. You would have described the years in terms of millennia (2000 – 1, 4000 – 2). In other words you *compared* years with millennia.

Another even more common example is when you ask a dieter how the program is working out. They will usually tell you their weight loss, but not their current weight. This in spite of the fact they have been doing daily forward and inverse transforms between the bathroom scale reading and the "brag value* that they share with the world. Here they would be describing their progress in terms of weight loss (instead of bulk weight).

A more advanced example is of course the fast Fourier transform or FFT which allows us to see signals in the "frequency domain". Fig. 1.3–1 shows us the constituent sinusoids of different frequencies (spectrum) that make up the signal. In other words, we are *correlating* (comparing) the signal with these various sinusoids and describing the signal in terms of its frequencies).

***Jargon Alert*: The use of the FFT is now so commonplace that it's results are referred to as the *"frequency domain"* of a signal. ("*Time domain*" is simply the original amplitude vs. time plot of a signal.)**

Thus we can say that in the FFT we are comparing and describing the signal in terms of sinusoids of different frequencies or "*stretched*" sinusoids (to use wavelet terminology). In the wavelet transforms we will be comparing and describing the signal in terms of *stretched* and *shifted* wavelets.

The FFT also allows us to manipulate the transformed data and then do an inverse FFT (*IFFT*) for custom filtering such as eliminating constant frequency noise. For signals with embedded events (the most interesting kind!) the FFT tells us the frequency of the event but not the time that it occurred.

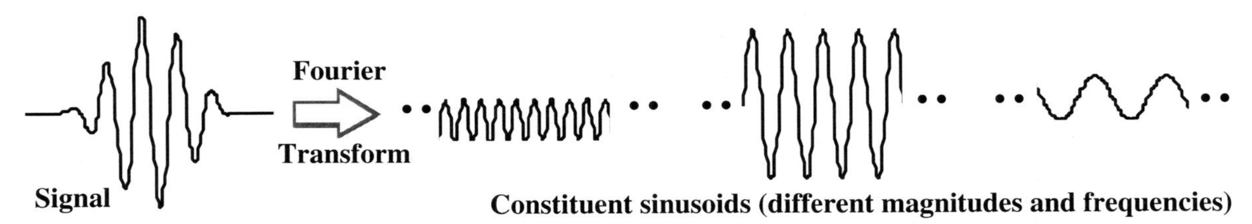

Figure 1.3–1 The signal can be transformed into a number of sinusoids of various sizes and frequencies. When added together (inverse), these sinusoids reconstruct the original signal.

* Apologies to Sir Lawrence Bragg, a Physics and Signal Processing pioneer.

1.4 Short-Time Transforms, Sheet Music, and a first look at Wavelet Transforms

A possible solution to providing both time *and* frequency information about an embedded event might be to divide the total time interval into several shorter time intervals and then take the FFT for each interval. This *time-windowing* method would narrow down the time to that of the interval where the event was found. This *short-time Fourier transform* (STFT) method has been around since 1946 and is still in wide use today.

While the STFT gives us a compromise of sorts between time and frequency information, the accuracy is limited by the size and shape of the window. For example, using many time intervals would give good time resolution but the very short time of each window would not give us good frequency resolution, especially for lower frequency signals.

Longer time intervals for each window would allow us better frequency resolution, but with these fewer, longer windows we would suffer in the time resolution (i.e. with very few windows we would have very few times to associate with the event). Longer time intervals are also not needed for high frequency signals.

Wavelet transforms allow us variable-size windows. We can use long time intervals for more precise low-frequency information and shorter intervals (giving us more precise time information) for the higher frequencies.

We are actually already familiar with this concept. Ordinary sheet music is an everyday example of displaying both time and frequency information—and it happens to be set up very similar to a wavelet transform display.

Besides demonstrating the concept of longer time for lower frequencies and shorter times for higher frequencies, sheet music even has a logarithmic vertical frequency scale (each octave is twice the frequency of the octave below it). Musicians know that low notes take longer to form in a musical instrument. (Engineers know it takes a longer *time* to examine a low *frequency* signal.) This is why the Piccolo solo from John Philip Sousa's "Stars & Stripes Forever" (Fig. 1.4–1) can't be played on a Tuba.[*]

[*] There are in fact recordings of tuba players that can press the big valves fast enough to play all the notes in the piccolo solo, but some of the notes themselves do not have enough time to form in the horn and are not heard. DSP engineers would refer to this as *insufficient integration time.*

Figure 1.4–1 Portion of the piccolo solo from John Philip Sousa's "Stars and Stripes Forever"

Figure 1.4–2 shows octaves of the note "C" on sheet music (left). A wavelet display is very much like the configuration to the right. If we use a base 2 log scale so we have increasing *frequency* in powers of 2 as the y axis we can see a remarkable comparison between the sheet music and the wavelet display. (We will see more of this *octave* or *power of 2* behavior a little later in the *discrete wavelet transforms.*)

If you think about it, sheet music has another dimension besides discrete time and frequency. The volume or *magnitude* of each note is indicated by discrete indicators such as *ff* for *fortissimo* or very loud.* Sheet music even describes the time increments. For example, Tempo 60 indicates 60 beats per minute or 1 beat per second.

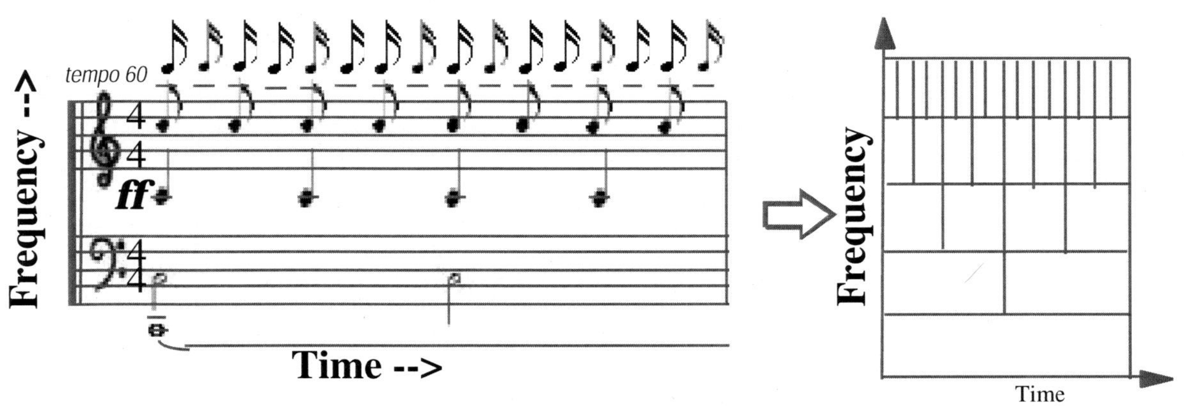

Figure 1.4–2 Comparison of 5 octaves of the note "C" on sheet music to a (vertically inverted) wavelet transform time/frequency diagram. Note the powers of 2 in both time (linear scale) and frequency (log scale) and that higher frequencies require less time for adequate resolution. Note how magnitude is indicated by *fortissimo* (*ff*) on sheet music. Magnitude on the time/frequency diagram at right is indicated by the brightness or color. Note that both representations indicate discrete frequencies, discrete times, and discrete magnitudes.

---

* *Fortissimo* is Italian for very loud, *pianissimo* or *pp* indicates very soft. The other intensity notations (*f, mf, mp, p, etc.*) indicate discrete levels of loudness. The modern piano or *pianoforte* got its name from its ability to play notes either *piano* (soft) or *forte* (loud).

In most (but not all) texts, the wavelet display at the right of Figure 1.4–2 is drawn inverted ("flipped" vertically). In other words the high frequencies are on the bottom and the lower frequencies are on top. The y axis on most wavelet displays shows increasing *scale* (stretching of the wavelet) rather than increasing frequency.

Figure 1.4–3 shows a "sneak preview" of a wavelet transform display (c) and how it compares to an ordinary time-domain (a) and frequency-domain (b) display.

Imagine if a musical composer had to "de-compose"* by changing a single note. If he used a musical form similar to (a) he would have to change *all* the notes at a particular time (or "beat"). If he used a form similar to (b) he would have to change *all* the notes of a particular frequency (or "pitch"). Using a form similar to (c), as is sheet music, he can change only one note. This is how denoising or compression is accomplished using wavelet technology. An unwanted or unneeded portion of data (a computational "wrong note" if you will) can be easily identified and then changed or deleted without appreciable degradation of the signal or image.

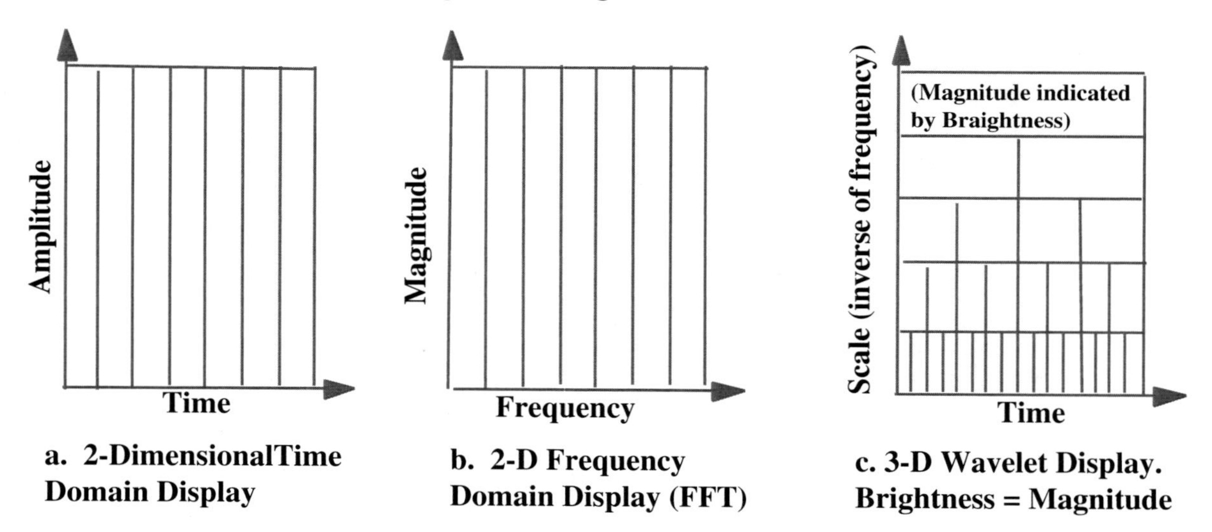

Figure 1.4–3 Time domain, frequency domain, and "wavelet domain" display. Note that the wavelet display (c) incorporates both time *and* frequency. Note the similarity to sheet music except that this display (c) is inverted with increasing stretching or *scale* (inverse of frequency) as the vertical or y-axis.

* "*Decomposing*" is a term actually used in the *discrete wavelet transforms* we will soon discuss. Beethoven is even now decomposing. (You knew that was coming, right?)

1.5 Example of the Fast Fourier Transform (FFT) with an Embedded Pulse Signal

In this example we start with a point-by-point comparison of a time-domain pulse signal (A) with a high frequency *sinusoid* of constant frequency (B) as shown in Figure 1.5–1. We obtain a single "goodness" value from this comparison (a *correlation value*) which indicates how much of that particular sinusoid is found in our original pulse signal.

We can observe that the pulse has 5 cycles in 1/4 of a second. This means that it has a frequency of 20 cycles in one second or *20 Hz.*

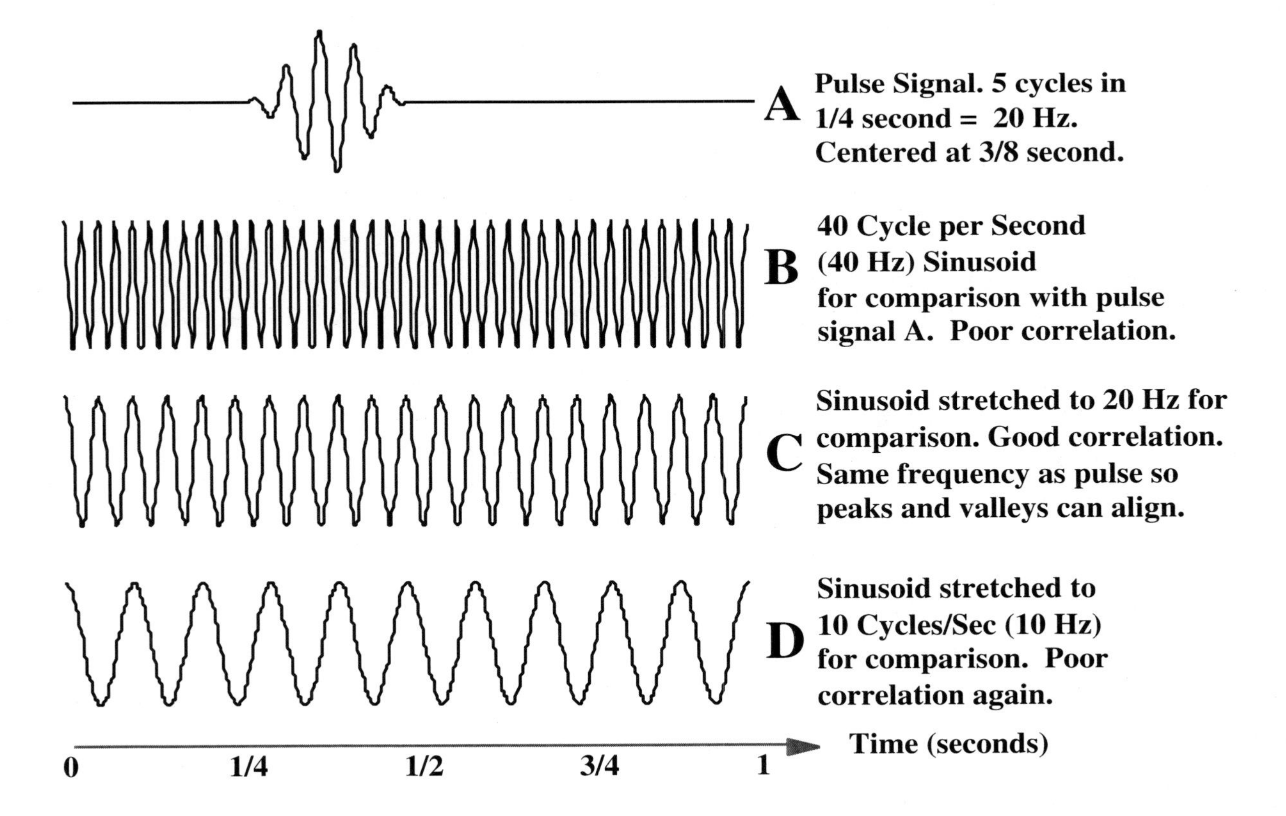

Figure 1.5–1 FFT-type comparison of Pulse Signal with several stretched sinusoids. The pulse (A) has 5 discernible "peaks" (local maxima) and 5 discernible "valleys" (local minima). These peaks and valleys will line up best with those of sinusoid C. (We discuss shifts in time or *phase shifts* to better align the pulse and sinusoid in a later chapter).

The first comparison sinusoid (B) has twice the frequency of the pulse or 40 Hz. Even in the time interval where the signal is non-zero (the pulse) it doesn't seem intuitively like the comparison would be very good. (A small mathematical correlation value bears this out.)

By lowering the frequency of (B) from 40 to 20 Hz (waveform C) we are effectively "stretching in time" (*scaling*) the sinusoid (B) by 2 so it now has only 20 cycles in 1 second. We compare (C) point-by-point again over the 1 second interval with the pulse (A). This time the correlation of the pulse (A) with the comparison sinusoid (C) is very good. The peaks and valleys of (C) and of the pulse portion of (A) align in time (or can be easily phase-shifted to align) and thus we obtain a large correlation value.

This same diagram (Fig. 1.5–1) shows us one more comparison of the pulse with (A). This is our original sinusoid (B) stretched by 4 so it has only 10 cycles in the 1 second interval. The correlation is poor once again because the peaks and valleys of (A) and (D) no longer line up. We could continue stretching until the sinusoid becomes a straight line having zero frequency or "DC" (named for the zero frequency of Direct Current) but all these comparisons will be increasingly poor.

An actual FFT compares many "stretched" sinusoids ("*analysis signals*") to the original pulse rather than just the 3 shown in Figure 1.5–1. The best correlation is found when the comparison sinusoid frequency exactly matches that of the pulse signal. Figure 1.5–2 shows the first part of an actual FFT of our pulse signal (A).

The locations of our sample comparison sinusoids (B, C, and D) are indicated by the large dots. (The spectrum of our pulse signal is shown by the solid curve.) Again, the FFT tells us correctly that the pulse has primarily a frequency of 20 Hz, but does *not* tell us where the pulse is located in time.*

* The generalized mathematical equation for the DFT (implemented by the FFT algorithm) is a shortcut for indicating the real cosines and imaginary sines ($\mathbf{X_k}$) that make up the signal ($\mathbf{x_n}$). (We showed *cosines only* in the above example to simplify and visually portray the process.)

$$X_k = \sum x_n \cos(2\pi nk / N) - j\sum x_n \sin(2\pi nk / N)$$

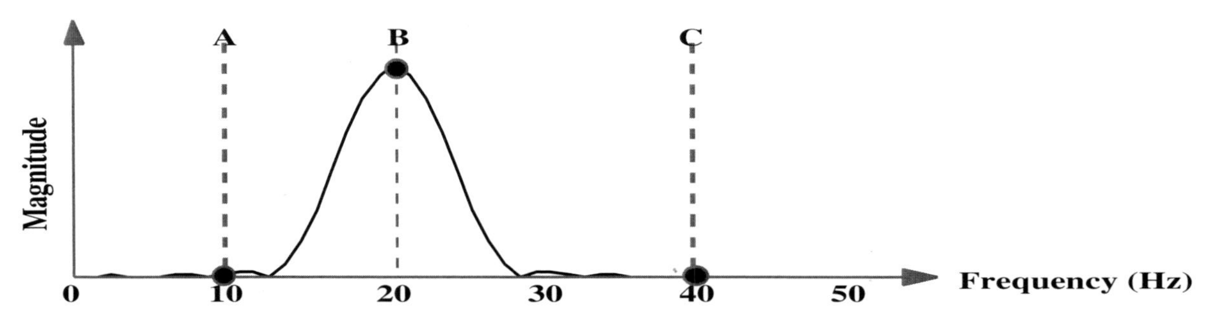

Figure 1.5–2 Actual FFT plot of the above pulse signal with the three comparison sinusoids.

1.6 Examples using the Continuous Wavelet Transform

Wavelet transforms are exciting because they too are comparisons, but instead of correlating with various stretched, constant frequency sinusoid *waves* they use smaller or shorter waveforms ("*wave–lets*") that can start and stop. In other words, the *fast Fourier transform* relates the signal to *sinusoids* while the *wavelet transforms* relate signals to *wavelets*. In the real world of digital computers, *wavelet transforms* relate our discrete, finite (digital) *signal* to the discrete, finite, *wavelet filters*.

Fig. 1.6–1 shows us some of the constituent wavelets that have been shifted and stretched (from the mother wavelet) that make up the signal. In other words, we are *correlating* (comparing) the signal with these various shifted, stretched wavelets. An actual wavelet transform compares many stretched and shifted wavelets ("*analysis wavelets*") to the original pulse rather than just these few shown in Figure 1.6–1.

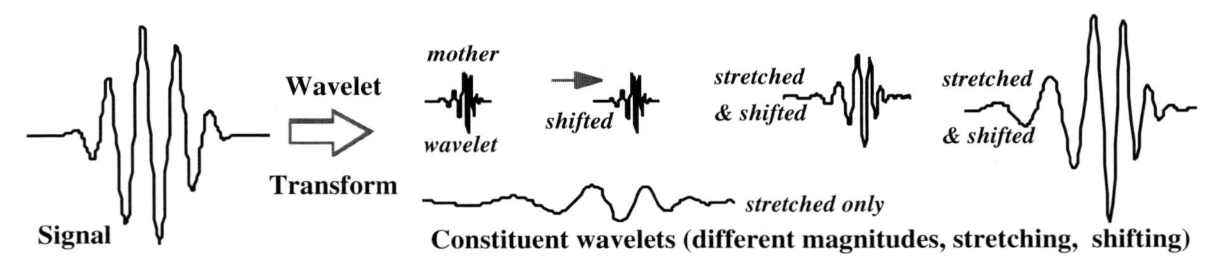

Figure 1.6–1 The signal can be transformed into a number of wavelets of various stretching, shifting, and magnitude. When added together these wavelets reconstruct the original signal.

Figure 1.6–2 demonstrates the stretching and shifting process for the continuous wavelet transform. Instead of sinusoids for our comparisons, we will use wavelets. Waveform (B) shows a Daubechies 20 (Db20) wavelet about 1/8 second long that starts at the beginning (t = 0) and effectively ends well before 1/4 second. The zero values are extended to the full 1 second. The point-by-point comparison* with our pulse signal (A) will be very poor and we will obtain a very small correlation value.

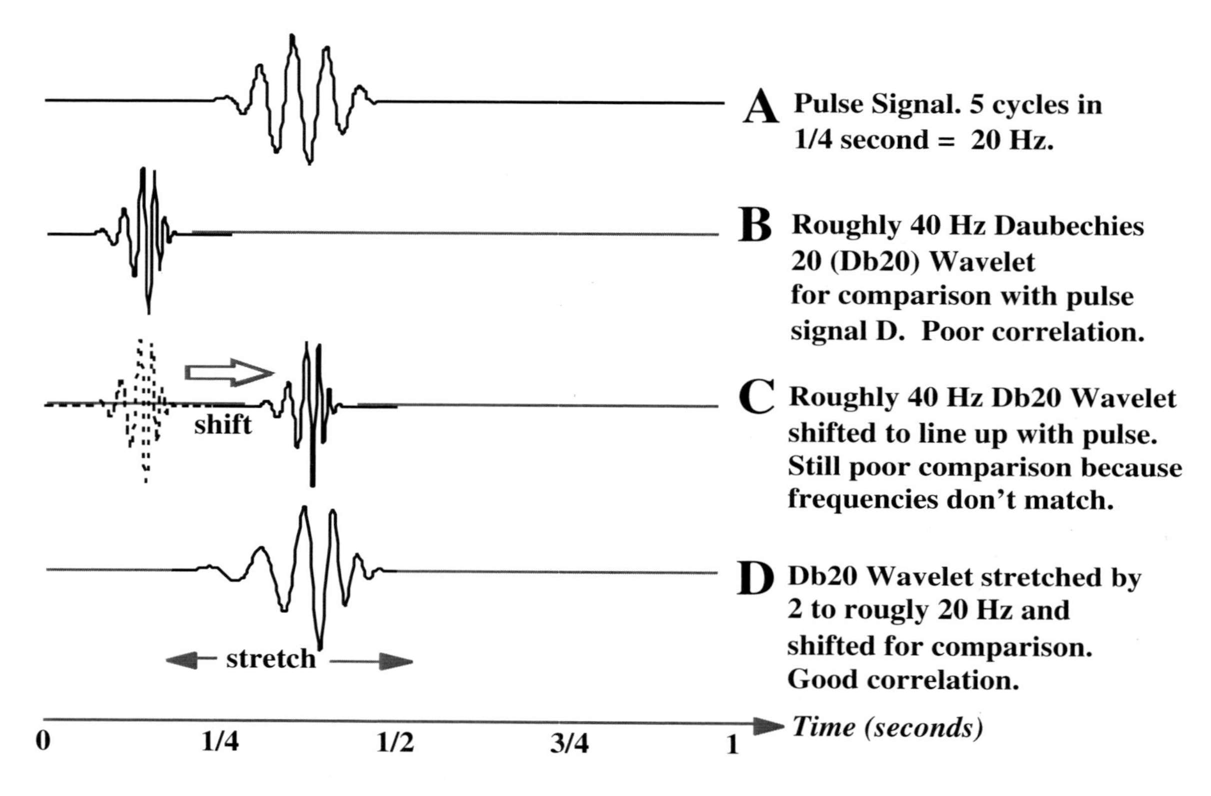

Figure 1.6–2 CWT-type comparison of pulse signal with several stretched and shifted wavelets. If the energy of the wavelet and the signal are both unity, these the comparisons are *correlation coefficients*. Note: Knowing how much (B) was stretched and shifted to match (A) tells us the location and approximate frequency of the pulse. Also, a good match with this particular wavelet tells us that the pulse looks a lot like the wavelet (sinusoidal in this case).

* The pulse and the wavelets are drawn here as continuous functions. In DSP we would have a finite number of data points for the signal and we would be comparing these point-by-point with the finite number of values of the Db20 wavelet filter.

In the previous FFT discussion we proceeded directly to stretching. In the wavelet transforms here we first shift the unstretched basic or *mother* wavelet slightly to the right and perform *another* comparison of the signal with this new waveform to get *another* correlation value. We continue to shift and when the Db20 wavelet is in the position shown in (C) we get a little better comparison than with (B), but still very poor because (C) and (A) are different frequencies.

***Jargon Alert*: The unstretched wavelet is often referred to as the "*mother wavelet*". The Db20 wavelet filter we are using here starts out as 20 points long (hence the name) but can be stretched to many more points. (A counterpart lowpass filter used in the upcoming *discrete wavelet transform* is often called a "*father wavelet*". Honest!)**[*]

After we have continued shifting the wavelet all the way to the end of the 1 second time interval, we start over with a slightly stretched wavelet at the beginning and repeatedly shift to the right to obtain *another full set* of these correlation values.

Waveform (D) shows the Db20 wavelet stretched to where the frequency (pseudo-frequency—ref. Fig. 1.1–1) is roughly the same as the pulse (A) and shifted to the right until the peaks and valleys line up fairly well. At these particular amounts of shifting and stretching we should obtain a very good comparison and a large correlation value. Further *shifting* to the right, however, even at this same stretching will yield increasingly poor correlations. Further *stretching* doesn't help at all because even when lined up, the pulse and the over-stretched wavelet won't be the same frequency.

In the CWT we have one correlation value for every shift of every stretched wavelet.[†] To show the correlation values (quality of the "match") for all these stretches and shifts, we use a 3-D display. Figure 1.6–3 shows a Continuous Wavelet Transform (CWT) display for the pulse signal (A) in our example.

[*] Mathematically speaking, we replace the infinitely oscillating sinusoid basis functions in the FFT with a set of locally oscillating basis functions which are stretched and shifted versions of the fundamental, real-valued bandpass *mother wavelet.* When correctly combined with stretched and shifted versions of the fundamental, real-valued lowpass *father wavelet* they form an orthonormal basis expansion for signals.

[†] The generalized equation for the CWT (shown below) is a shortcut that shows that the correlation coefficients depend on both the stretching and the shifting of the wavelet, ψ, to match the signal ($\mathbf{x}_n$ here) as we have just seen. The equation shows that when the "dilated and translated" wavelet matches the signal the summation will produce a large correlation value.

$$C(stretching, shifting) = \sum x_n \psi\,(stretching, shifting)$$

The bright spots indicate where the peaks and valleys of the stretched and shifted wavelet align best with the peaks and valleys of the embedded pulse (dark when no alignment, dimmer where only *some* peaks and valleys line up, but brightest where *all* the peaks and valleys align). In this simple example, stretching the wavelet by a factor of 2 from 40 to 20 Hz (stretching the filter from the original 20 points to 40 points) and shifting it 3/8 second in time gave the best correlation and agrees with what we knew *a priori* or "up front" about the pulse (pulse centered at 3/8 second, pulse frequency 20 Hz).

We chose the Db20 wavelet because it looks a little like the pulse signal. If we didn't know *a priori* what the event looked like we could try several wavelets (easily switched in software) to see which produced a CWT display with the brightest spots (indicating best correlation). This would tell us something about the shape of the event.

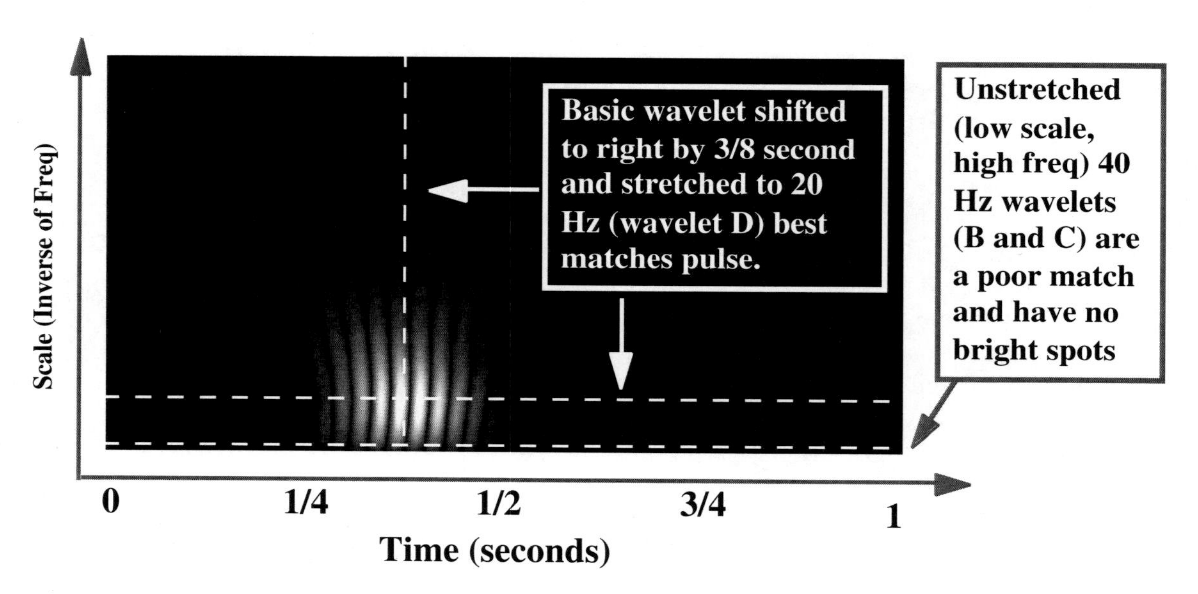

Figure 1.6-3: Actual CWT display of the above example indicating the time and frequency of the Pulse Signal. Shifting or *translation* of the wavelet (filter) in time is the x or horizontal axis, stretching or *dilation* of the wavelet (the inverse of its pseudo frequency) is the y or vertical axis, and the "goodness" of the correlation of the wavelet (at each x-y point) with the signal (pulse) is indicated by brightness. The fainter bands indicate where some of the peaks and valleys line up while the center of the brightest band (in the cross-hairs) shows the best "match" or correlation.

For the simple tutorial example above we could have just visually discerned the location and frequency of the pulse (A). The next example is a little more representative of wavelets in the real world where location and frequency are not visible to the naked eye. Wavelets can be used to analyze *local* events as we will now see.

We construct a 300 point slowly varying sine wave signal and add a tiny "glitch" or discontinuity (in slope) at time = 180 as shown in Figure 1.6–4 (a). We would not notice the glitch unless we were looking at the closeup (b). Using a conventional fast Fourier transform (FFT) on the signal shows its frequency components (Fig. 1.6–5). The low frequency of the sine wave is easy to notice, but the small glitch cannot be seen.

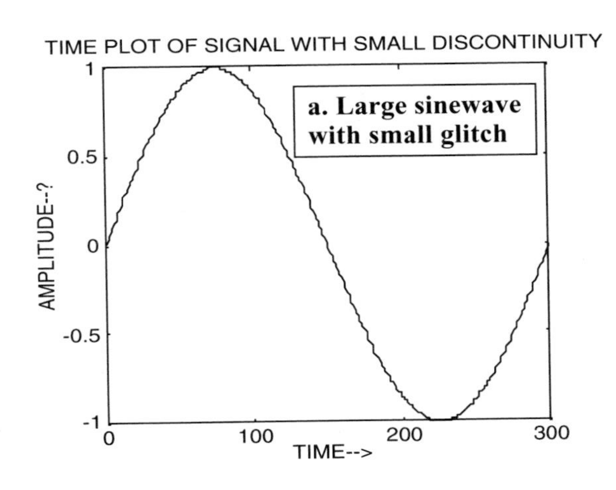

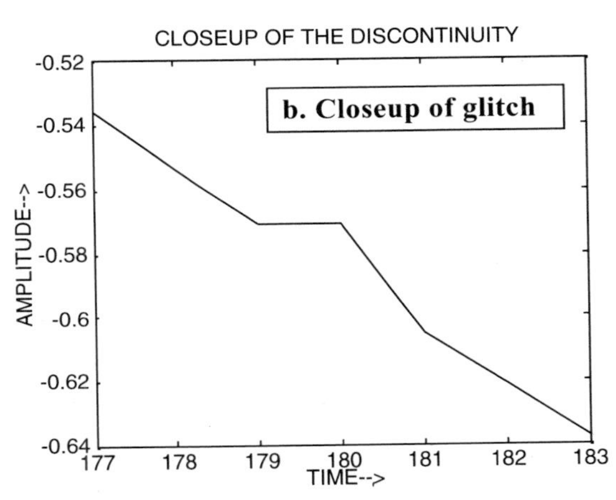

Figure 1.6–4 Very small discontinuity at time = 180 (a) cannot be seen without a closeup (b).

While neither the (full-length) time domain or frequency domain display tell us much about the glitch, the CWT wavelet display (Fig. 1.6–6) clearly shows a vertical line at time = 180 and at low scales. (The wavelet has very little stretching at low scales, indicating that the glitch was very short.) The CWT also compares well to the large oscillating sine wave which hides the glitch. At these higher scales the wavelet has been stretched (to a lower frequency) and thus "finds" the peak and the valley of the sine wave to be at time = 75 and 225 (see Fig. 1.6–4). For this short discontinuity we used a short 4-point Db4 wavelet (as shown) for best comparison.

Note that if we were to vertically invert this display with the lower frequencies shown on the bottom, we would see many similarities to the sheet music notation described earlier.

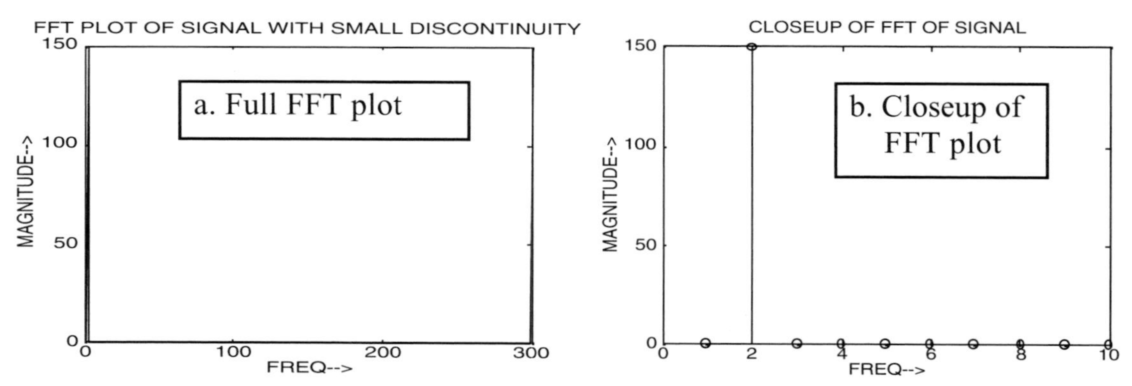

Figure 1.6-5 FFT magnitude plot (a) clearly indicates the presence of a large low-frequency sinusoid. A closeup of the FFT (b) further defines the sine wave in frequency, but does not help to find the glitch. Note: Even if the glitch were large enough to show a noticeable frequency component in the FFT, this would still not indicate the *time* of the glitch-event.

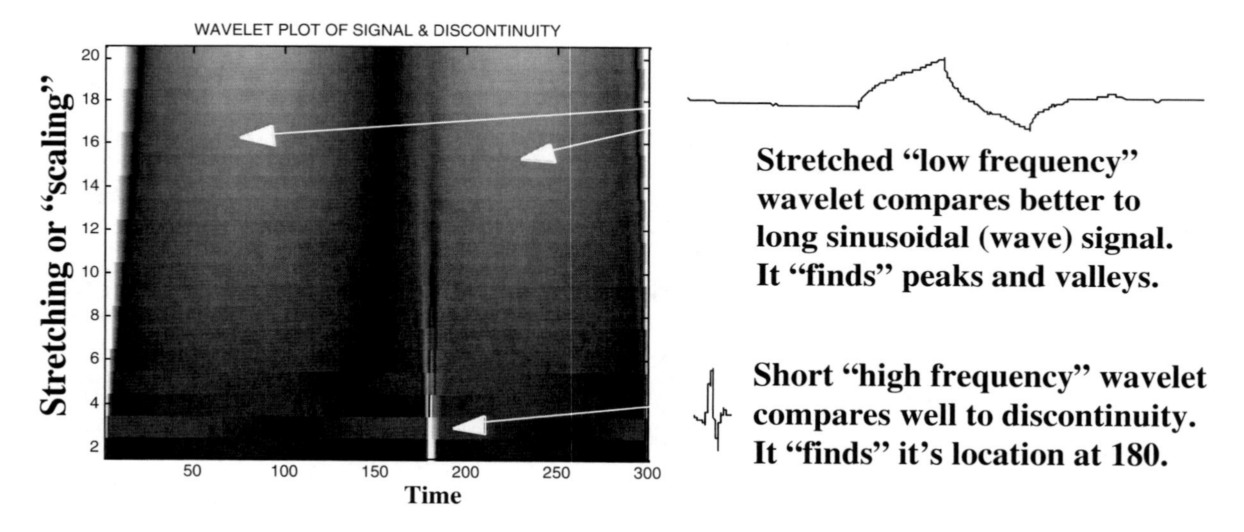

Figure 1.6-6 CWT display of result of correlation of signal with various scales (stretching) of the Daubechies 4 (Db4) wavelet. The short mother wavelet (filter) at scale = 2 is only 4 points long (the "continuous" estimation of the Db4 is drawn). This short filter compares well with the short glitch at time 180. The stretched wavelet (filter) at scale = 20 (top) is about 50 points long and compares better to the large 300 point sinusoid of the main signal than to the glitch.*).

* We show only to scale = 20 here. A CWT display with much larger scale values would show the best correlation with the sinusoid to be at about scale = 150. The Db4 wavelet filter is stretched to 300 points at scale = 150, and best fits the 300 points of the sine-wave signal. However the glitch at time = 180 would not be so easily discernible on such a large scale display ant thus it is not shown· We can also adequately locate the peak and valley of the sine wave at 75 and 225 using just this abbreviated-scale CWT plot—compare Fig 1.6–3 (a).

1.7 A First Glance at the Undecimated Discrete Wavelet Transform (UDWT)

Besides acting as a "microscope" to find hidden events in our data as we have just seen in the *continuous* wavelet transform (CWT) display, *Discrete wavelet transforms* (DWTs) can also separate the data into various frequency components, as does the FFT. We already know that the FFT is used extensively to remove unwanted noise that is prevalent throughout an entire signal such as a 60 Hz hum.

Unlike the FFT, however, the *discrete wavelet transform* allows us to remove frequency components at *specific times* in the data. This allows us a powerful capability to throw out the "bad" and keep the "good" part of the data for denoising or compression. Discrete wavelet transforms also incorporate easily computed *inverse* transforms (IDWTs) that allow us to reconstruct the signal after we have identified and removed noise or superfluous data.

A fair question before proceeding is "What is *continuous* about the *continuous wavelet transform* in our world of digital computers that works with discrete data? Aren't *all* these transforms "discrete"? What then differentiates these *discrete* wavelet transforms from the so-called *continuous* ones?"

The answer is that although all wavelet transforms in DSP *are* technically *discrete**, the so-called *continuous* wavelet transform (CWT) differs in how it stretches and shifts. The term "continuous" in a CWT indicates all possible integer factors of shifting and stretching (e.g. by 2, 3, 4, 5, etc.) rather than a mathematically continuous function. By contrast, we will see that *discrete* wavelet transforms stretch and shift by powers of 2. Another difference is that the *continuous* wavelet transform uses only the one wavelet filter while the *discrete* wavelet transform uses 3 additional filters as we will soon discover. We will now look at the 2 best known and most utilized of the DWTs—the *conventional discrete wavelet transform* (DWT) and the *undecimated discrete wavelet transform* (UDWT).

***Jargon Alert*: Stretching or shifting by powers of 2 is often referred to as *"dyadic"*. For example, *dyadic dilation* means stretching (or shrinking) by factors of 2 (e.g. 2, 4, 8, 16 etc).**

---

* As is the *discrete Fourier transform* (implemented by the FFT algorithm).

The *undecimated discrete wavelet transform* (we'll explain why it's called "*undecimated*" in a moment) is not as well known as the *conventional* discrete wavelet transform. However it is simpler to understand than the conventional DWT, compares better with the *continuous wavelet transform* we have just studied and is similar enough to the DWT to provide a clear learning "bridge". UDWTs also don't have the "aliasing" problems we will soon encounter and discuss in the conventional DWT.

Figure 1.7–1, (a) shows the simplest UDWT. The first thing you will notice on this signal flow diagram is that it has 4 filters. This is called a *filter bank* for this reason. (We will run into this type of figure a lot during the book, but it's not necessary to completely understand it at this point in the preview.) These 4 filters are closely related (complimentary) as we will see later.

The left half of the UDWT is called the *decomposition* or *analysis* portion and comprises the *forward* transform. The right half is called the *reconstruction* or *synthesis* portion and comprises the *inverse* transform.* The vertical bar separating the 2 halves represents the area where we can add more complexity (and capability), but we proceed by ignoring the bar for now.

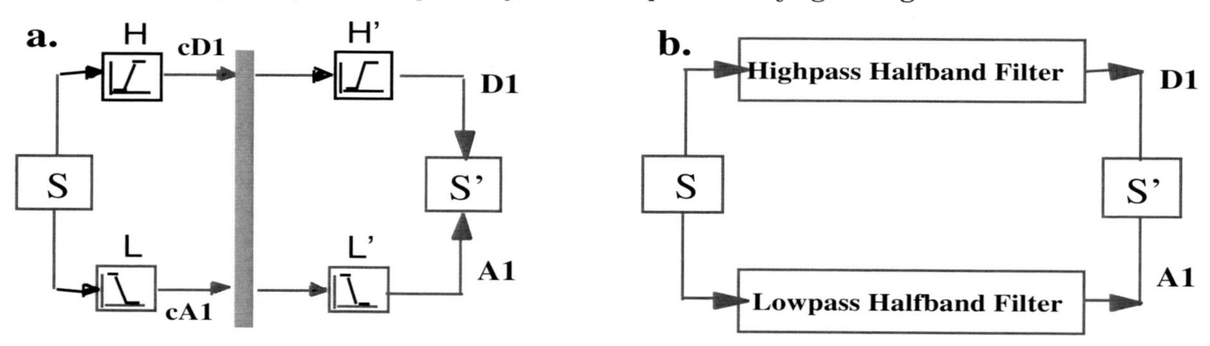

Figure 1.7–1 Single-level undecimated discrete wavelet transform (UDWT) filter bank shown at left (a). The *forward transform* or *analysis* part is the half to the left of the vertical bar and is usually referred to as the *decomposition portion*. The *inverse transform* or *synthesis* part is the half to the right of the vertical bar and is called the *reconstruction* portion. The bar itself is where additional levels of decomposition and reconstruction can be inserted, producing higher level UDWTs. If the data is left unchanged (no activity in the vertical bar) the functional equivalent of this single-level UDWT is shown in the right diagram (b).

---

* The terms UDWT and IUDWT are occasionally used as labels for the forward and inverse transforms. Usually, however, the term UDWT refers to both halves. The discrete Fourier transform (DFT) and the functionally-equivalent *fast Fourier transform* (FFT) also use the terms *analysis* and *synthesis* to describe their left and right (forward and inverse) halves (FFT and IFFT).

***Jargon Alert*: "*Decomposition*" in wavelet terminology means splitting the signal into 2 parts using a highpass and a lowpass filter. Each of the 2 parts themselves can be *decomposed* further (split into more parts) using more filters. "*Reconstruction*" means using filters to combine the parts. "*Perfect reconstruction*" means that that the signal at the end is the same as the original signal (except for a possible delay and a constant of multiplication).**

On the upper path of Fig. 1.7–1, (a) the signal, **S**, is first filtered by **H** (*highpass decomposition filter*) to produce the coefficients **cD1**. At this point we can do further decomposition (analysis) for compression or denoising, but for now we will proceed directly to reconstruct (synthesize) the signal. **cD1** is next filtered by **H'** (*highpass reconstruction filter*) to produce the *Details* (**D1**). The same signal is also filtered on the lower path by the *lowpass decomposition filter* **L** to produce the coefficients **cA1** and then by the *lowpass reconstruction filter* **L'** to produce the *Approximation* (**A1**).

***Jargon Alert*: "*Approximation*" in Wavelets is the smoothed signal after all the lowpass filtering. "*Details*" are the residual noise after all the highpass filtering. cA1 designates the *Approximation Coefficients* and cD1 designates the *Details Coefficients*. These coefficients can be broken down (decomposed) into further coefficients in higher level systems (depicted by the vertical bar in the center of the diagrams).**

H and **H'** together produce a highpass *halfband filter* while **L** and **L'** produce a lowpass *halfband filter* as seen in Fig. 1.7–1, (b). These 4 *wavelet filters are* non-ideal filters and there is some overlap as depicted below in the frequency allocation diagram (Figure 1.7–2).

***Jargon Alert*: "*Halfband filters*" split the frequency into a lowpass and a highpass "half" (A1 and D1 here), usually with some overlap. We refer to all 4 filters as *wavelet filters*, but some texts refer to the 2 lowpass filters as *scaling function filters* and the 2 highpass as *wavelet filters*. Some call only H' the *wavelet filter*. Again, *Caveat Emptor*.**

Fig. 1.7–1 showed a *single-level* UDWT. Fig. 1.7–3 shows a *2-level* UDWT. (Note the additional decomposition and reconstruction as **cA1** is split into **cD2** and **cA2**.) Multi-level UDWTs allow us to stretch the filters, similar to what we did in the CWT, except that it is done dyadically (i.e. by factors of 2). The stretching is done by *upsampling by 2* (e.g. "$\mathbf{H}_{up}$") and then filtering (a common method of interpolation in DSP). With this further decomposing and reconstruction we can split the signal into more frequency bands (Fig. 1.7–4).

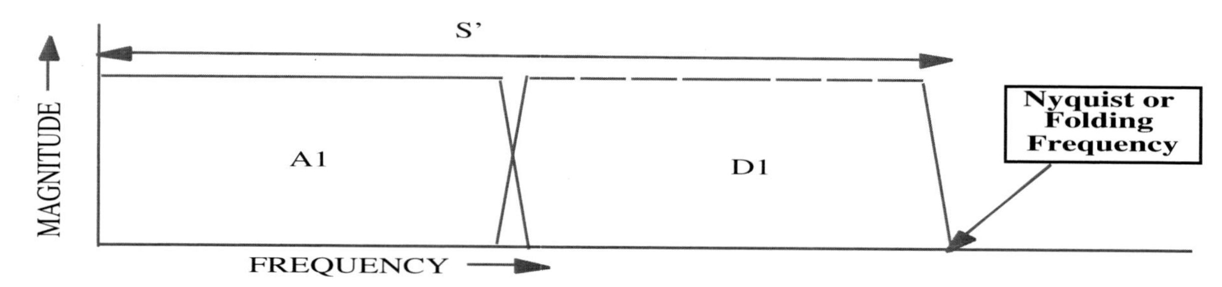

Figure 1.7–2 Frequency allocation after a single-level UDWT*. The diagram is illustrative only and the actual shape depends on the wavelet filters. Note overlap from non-ideal filtering. When the Details and Approximations are added together they *reconstruct* **S'** (**D1** + **A1** = **S'**) which is identical to the original signal, **S**, except for a delay and usually a constant of multiplication. For a very simple denoising, we could just discard these high frequencies in **D1** (for whatever time period in the signal we choose) and **A1** by itself would be a rudimentary "denoised" signal.

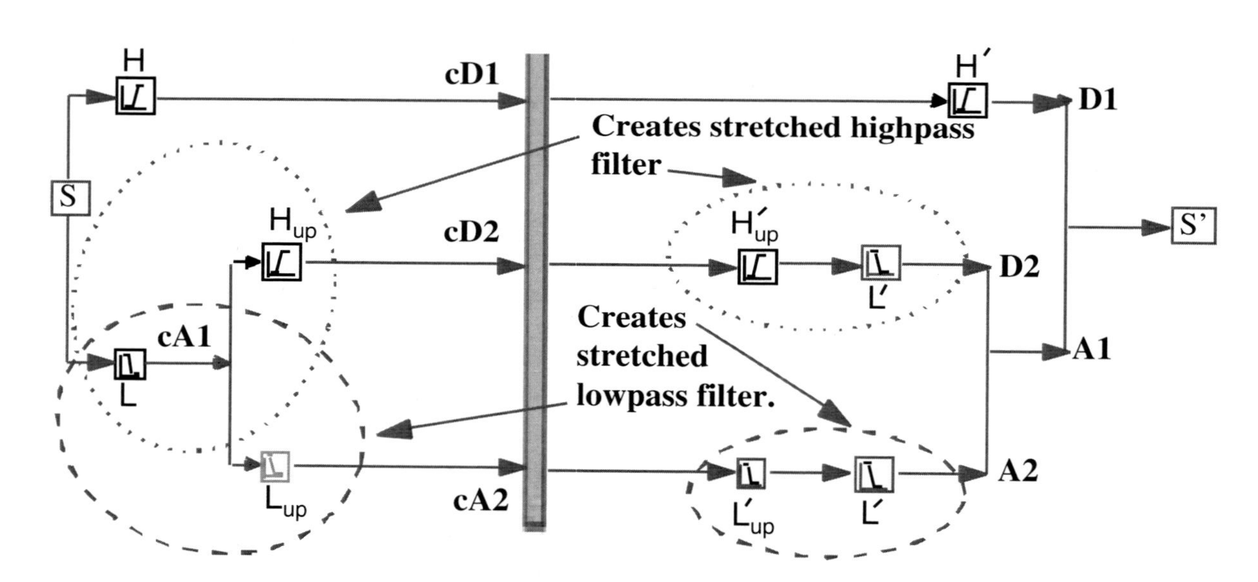

Figure 1.7–3 A 2-level UDWT. The signal, **S**, is split into **cD1** and **cA1**. We then split **cA1** into **cD2** and **cA2**. The final signal, **S'**, is now reconstructed by combining **A1** and **D1**. Since **A1** is obtained by combining **D2** and **A2**, **S'** = **A1** + **D1** = **A2** + **D2** + **D1**. We could do some denoising or compression at this point. If there was nothing of interest in **D2**, for example, we could zero it out and would have **S'** = **A2** + **0** + **D1** = **A2** + **D1**. Notice that if we set the coefficients **cD2** to zero that this would also cause **D2** to be zero (filtering of zeros still produces zeros). We will discuss multi-level UDWTs in detail later.

* The Nyquist or "folding" frequency is the highest possible frequency without aliasing (discussed shortly).

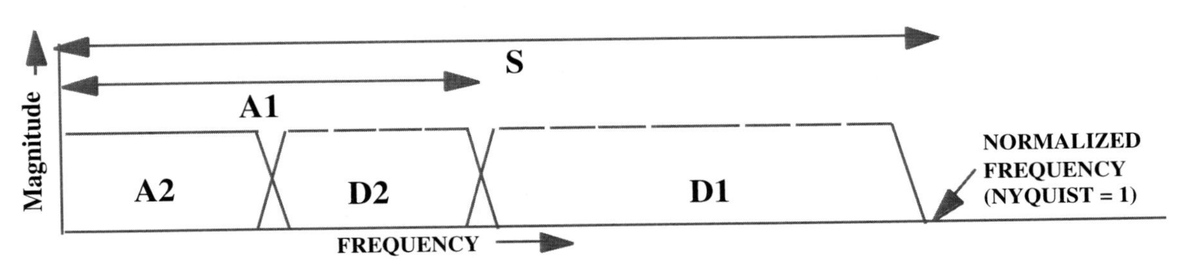

Figure 1.7–4 Frequency allocation after a 2-level UDWT. Note that the A1 is now split into 2 sub-bands. This allows us better flexibility in denoising or compression.

> ***Jargon Alert: "Upsampling by 2"*** **means placing zeros between the existing data points. For example, A time-sequence of the numbers [6, 5, 4, 3] would become with upsampling by 2**
>
> `[6, 0, 5, 0, 4, 0, 3].`
>
> **or in some cases**
>
> `[0, 6, 0, 5, 0, 4, 0, 3, 0]`
>
> **with a leading and/or a trailing zero (more on interpolation later).**

The UDWT (sometimes referred to as the *redundant* DWT or RDWT) with it's method of inserting zeros as part of the stretching of the filters is thus also called the *"A' Trous"* method which is French for *"with holes"* (zeros).

A 4-level UDWT with more stretched filters splits the signal into 5 frequency sub-bands (Figure 1.7–5).

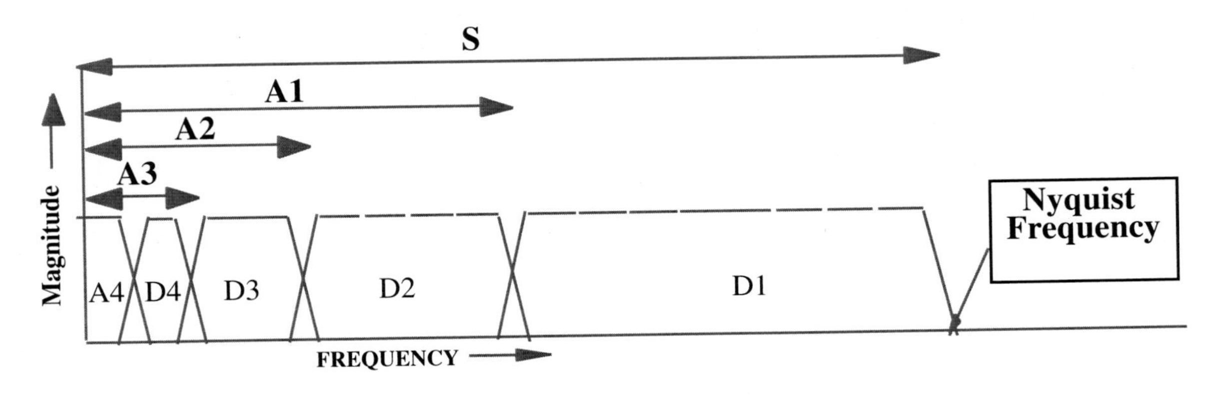

Figure 1.7–5 Frequency allocation in a 4-level UDWT. Note that **S** is split into **A1** and **D1**, **A1** is split into **A2** and **D2**, **A2** is split into **A3** and **D3**, and finally **A3** is split into **A4** and **D4**.

The 4-level UDWT signal flow diagram is not shown in this preview because of its size and complexity, but it functions very similar to the 2-level UDWT except that the filters are stretched not only by 2, but also by 4 and 8 to give us these additional sub-bands. Don't worry if you don't understand these multi-level systems in this preview. We'll talk more about them later. They are presented here mainly to show you what they look like and how they have stretched filters similar to the CWT. Hang in there.

1.8 A First Glance at the conventional Discrete Wavelet Transform (DWT)

We stretched the wavelet continuously (by integer steps) in the CWT and dyadically (by factors of 2) in the UDWT. In the conventional DWT we shrink the *signal* instead (dyadically) and compare it to the unchanged wavelet filters. We do this through "*downsampling by 2*".

***Jargon Alert*: "*Downsampling by 2*" means discarding every other signal sample. For example a sequence of numbers (signal) [5 4 3 2] becomes [5 3] (or [4 2] depending on where you start). This is also referred to in wavelet terminology as "*decimation by 2*" (in spite of the dictionary definition for the prefix "Deci").**

A single-level conventional DWT is shown in Figure 1.8–1 with the *decomposition* or *analysis* portion on the left and the *reconstruction* or *synthesis* portion on the right half*. Downsampling and upsampling by 2 is indicated by the arrows in the circles. For example, if we downsampled then immediately upsampled [5 4 3 2] we would first have [5 3] and then [5 0 3 0].

Further decomposition and reconstruction (a higher level DWT) is done in the vertical bar separating the 2 halves. The single-level *DWT* shown here is the same as the single level *UDWT* except that in discarding every other point, we have to deal with *aliasing*. We must also be concerned with *shift invariance* (do we throw away the odd or the even values?—it matters!).†

* As with the UDWT, the term DWT usually refers to both the left half or *forward* DWT and the right half or *inverse* DWT (occasionally called the IDWT).

† The UDWT has neither of these problems. We will discuss aliasing and shift invariance more in later chapters.

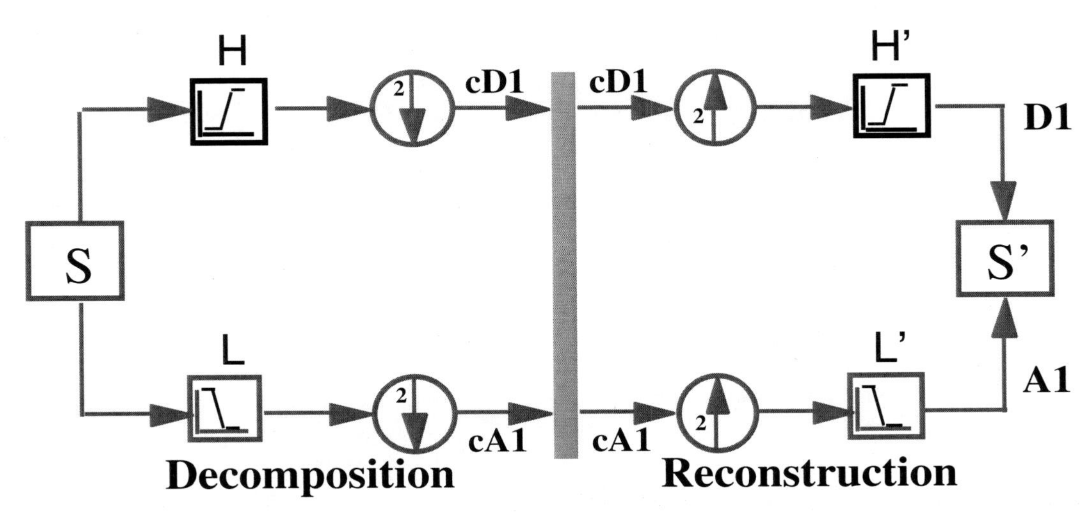

Figure 1.8–1 Single-level conventional DWT. Similar to the single-level UDWT with the same filters (H, H', L, and L') but with upsampling and downsampling. With no activity in the vertical bar, the coefficients **cD1** and **cA1** will be unchanged between the end of decomposition and the start of reconstruction. Notice that with downsampling the coefficients **cD1** and **cA1** are about half the size as those in the Undecimated DWT (UDWT).

> ***Jargon Alert: "Aliasing"* means 2 or more signals have the same sample values. One pathological example of aliasing caused by downsampling by 2 would be a high frequency oscillating time signal:**
>
> `[1 -1  1 -1  1 -1  1 -1  1 -1  1 ...`
>
> **If we downsample by 2 we have left over**
>
> `[1     1     1     1     1     1 ...`
>
> **which is a DC (zero frequency) signal. This is obviously *not* the high frequency signal we started with but an "alias" instead.**

With the potential for aliasing problems because of downsampling we would not expect to be able to perfectly reconstruct the signal as we did in the UDWT. One of the remarkable qualities of DWTs is that with the right wavelet filters (H, H', L and L') we *can* perfectly reconstruct, even with aliasing! The stringent requirements on the wavelet filters to be able to cancel out aliasing is part of why they often look so strange (as we saw in Figure 1.2–3).

***Jargon Alert*: Filters in these filter banks that are able to cancel out the effects of aliasing (if used correctly) are called "*Perfect Reconstruction Quadrature Mirror Filters*" or PRQMFs".**

As with the UDWT, we can denoise our signal by discarding portions of the frequency spectrum—as long as we are careful not to discard vital parts of the alias cancellation capability. Correct and careful downsampling also aids with compression of the signal. With downsampling, **cD1** and **cA1** are only about half the size as in the UDWT. So compared to the conventional DWT, the UDWT is "redundant". This is why it's often called a *redundant* DWT.

Multi-level conventional DWTs produce the same frequency sub-bands as the multi-level UDWTs we saw earlier (if the aliasing is correctly dealt with). Figure 1.8–2 shows a 2-level conventional DWT. The frequency allocation is the same as the 2-level UDWT (see Fig. 1.7–2). Notice that we use the same 4 wavelet filters (**H**, **H'**, **L**, and **L'**) repeatedly in a conventional DWT.

It's usually the high-frequencies that comprise the noise in a signal, thus we decompose the lower frequencies in these multi-level transforms. Figure 1.8–2 shows **cA1** split into **cA2** and **cD2** but **cD1** is not split further. We can, of course split these Details further if we want to. This is done using a *wavelet packet transform* and we will look at these in later chapters.

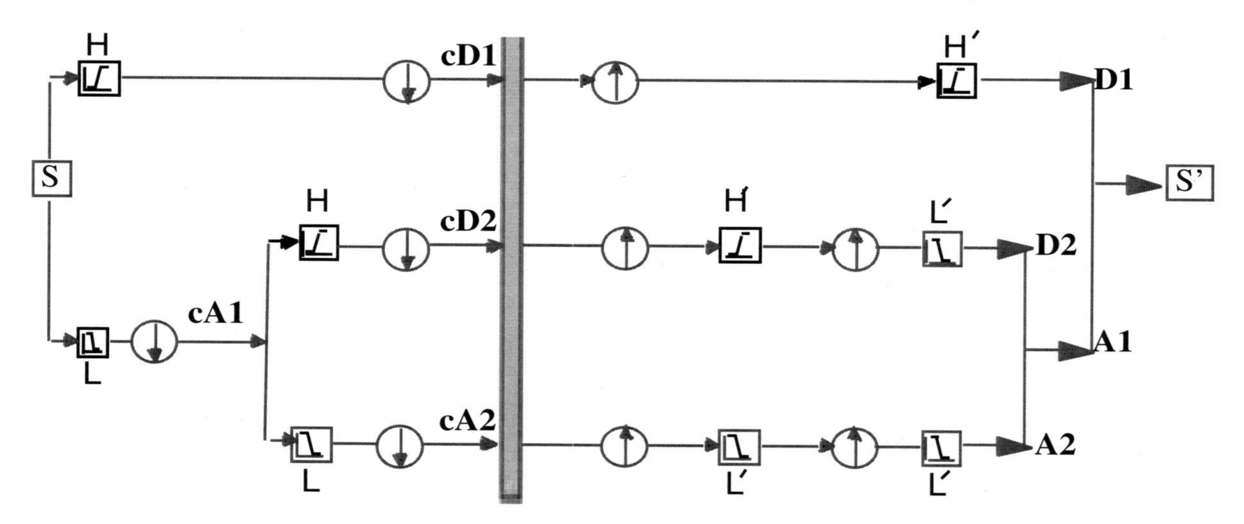

Figure 1.8–2 2-level conventional DWT. Instead of *stretching the filters* as in the UDWT (and CWT), we "shrink" the signal through downsampling and use the same 4 filters (**H**, **H'**, **L**, and **L'**) throughout. Note that with downsampling **cD2** and **cA2** are about 1/4 the size as those in the UDWT.

1.9 Examples of use of the conventional DWT

As mentioned, an important advantage of a wavelet transform is that, unlike an FFT, we can *threshold* the wavelet coefficients for only part of the time.

Jargon Alert: **To *"threshold"* (used as a verb here) means to disallow all numbers that are either greater or less (depending on the application) than a specified value or threshold (used now as a noun).**

We will use a seven-level DWT for this next example. Instead of simply **A1** and **D1** as we saw in Figure 1.8–1, we would have further decomposition of **A1** into **A2** and **D2**, then **A2** into **A3** and **D3**, and so on until **A6** is decomposed into **A7** and **D7**. The frequency allocation for a conventional DWT (assuming no aliasing problems) is the same as that for the UDWT. For example, see figure 1.7–5 for the allocation by a 4-level DWT (or 4-level UDWT).

Suppose we had a binary signal that had a great deal of noise added which changed frequency as time progressed (e. g. "chirp" noise). Using a 7-level DWT the noise would appear at different times in the different frequency sub-bands (**D1**, **D2**, **D3**, **D4**, **D5**, **D6**, **D7** and **A7**). We could automatically threshold out the noise at the appropriate times in the frequency sub-bands and keep the "good" signal data.

A portion of the original noiseless binary signal is shown in Figure 1.9–1. The values alternate between plus and minus one (a Polar Non-Return to Zero or PNRZ signal). We next bury the signal in chirp noise that is 10,000 times as great (80 dB). Looking at the signal buried in noise (Figure 1.9–2) we see only the huge noise in the time domain (a). Using an FFT on the noisy signal we see only the frequencies of the noise (b).

However, using a conventional DWT with a time-dependant automatic threshold for the various frequency sub-bands, we are able to reconstruct the binary signal (see Fig. 1.9–3) from the "scraps" left over after the chirp segments were thresholded out at the appropriate times. (More details on how this was done will be given later).

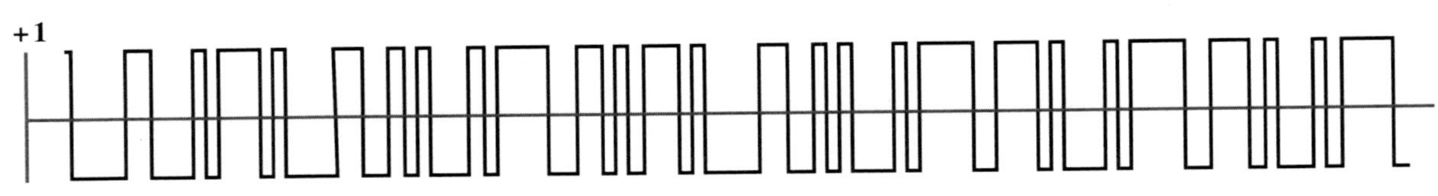

Figure 1.9–1 Portion of original binary signal. Values alternate between plus and minus one.

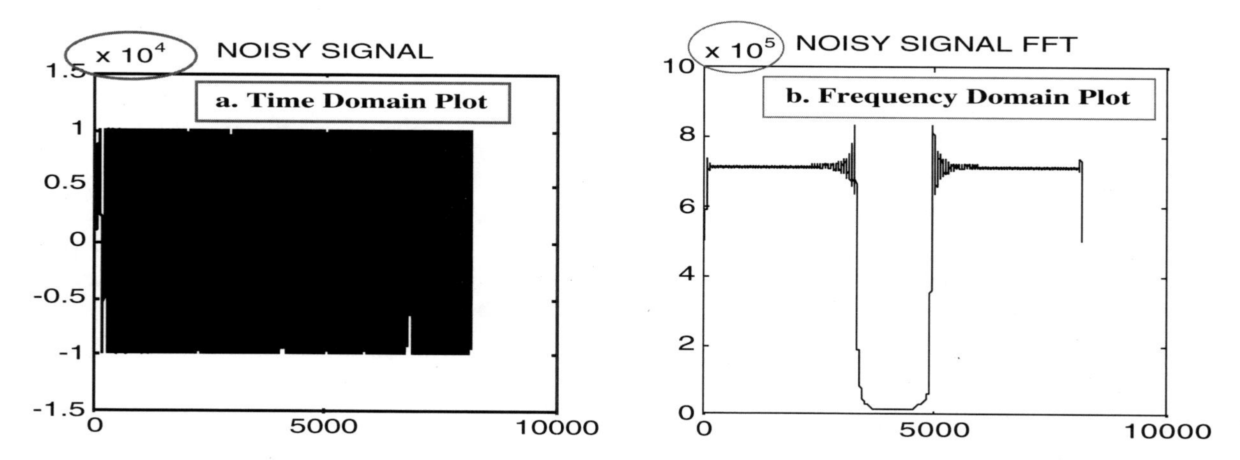

Figure 1.9–2 Signal buried in 10,000 times chirp noise is undetectable in either the amplitude vs. time plot (a) or the magnitude vs. frequency (FFT) plot (b).

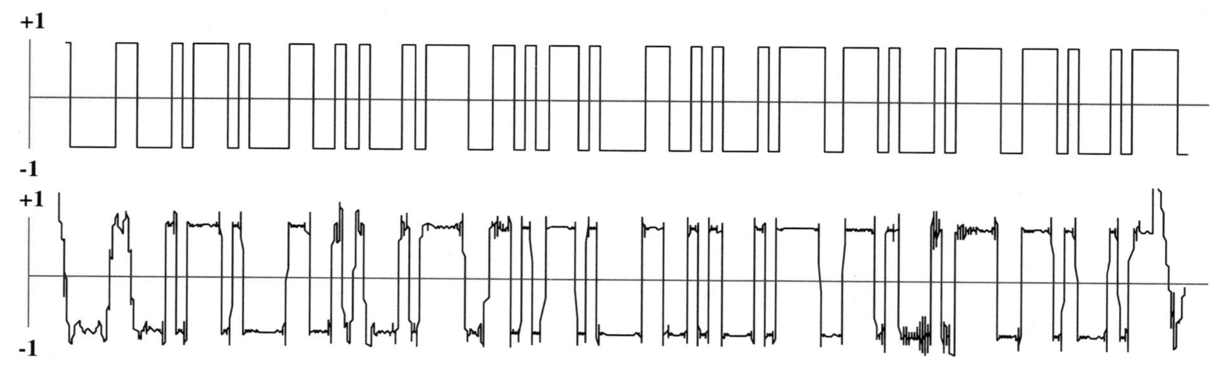

Figure 1.9–3 Successful use of discrete wavelet transform. Portion of denoised signal using time-specific thresholding with a 7-level conventional DWT is shown at bottom. Original binary signal. is redrawn at top for comparison. The final result is not a perfect reconstruction of the original, but close enough to discern the binary values.

Modern JPEG compression also uses wavelets. Figure 1.9–4 shows JPEG image compression. The image on the right was compressed by a ratio of 91:1 using a conventional DWT with a Biorthogonal 9/7* set of wavelets.

* As will be explained further in later chapters, the Biorthogonal 7/9 filters have 7 points in **H** and **L'** and 9 points in **H'** and **L**. This particular set of wavelet filters is referred to in MATLAB as "Bior4.4"

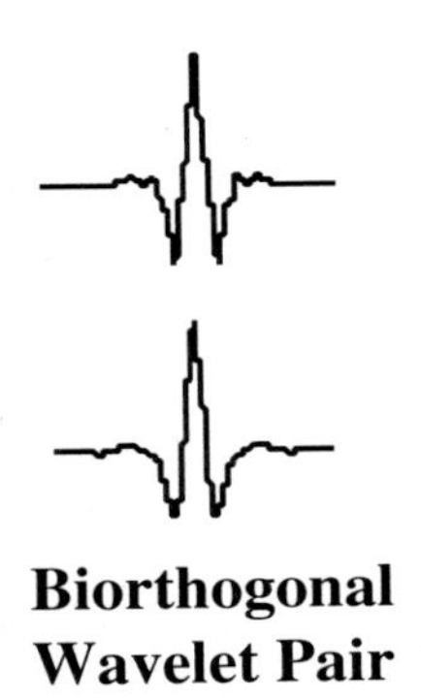

Biorthogonal Wavelet Pair

Figure 1.9–4 JPEG image compression of 91:1 achieved with a conventional DWT using a Biorthogonal 9/7 set of symmetrical wavelets.

1.10 Summary

In this preview chapter we introduced wavelets by drawing them as continuous functions, but told how they are actually implemented in a digital computer as discrete, short *wavelet filters*. We showed how some filters come from a mathematical expression for a continuous wavelet (crude wavelets) while other wavelets start out as filters with just a few points and then are built into a suitable estimation of a continuous wavelet. We then looked at various types of wavelets and their uses.

We next looked at transforms we use everyday and the (hopefully) familiar FFT and showed how they can be thought of as *comparisons* (correlations). We saw that the FFT has the shortcoming of not being able to determine the *time* of an embedded event. We discussed *short-time Fourier transforms* and then introduced the concept of wavelet transforms by comparing them to ordinary sheet music. We compared the fast Fourier transform (FFT) to the *continuous wavelet transform* (CWT) using an embedded pulse signal as an example. We next showcased the ability of a CWT to identify the time of occurrence of an embedded glitch, it's frequency, and it's general shape.

We moved on to the *undecimated discrete wavelet transform* (UDWT) and showed how it is similar in many ways to the CWT but uses all 4 wavelet filters rather than just one. We also noted that the stretching is done only by *factors of 2* (dyadically) in the UDWT rather than by every possible integer value as in the so-called "continuous" wavelet transform.

We continued building our understanding from FFT to CWT to UDWT by next moving on to the *conventional* DWT. The DWT is similar to the UDWT but introduces downsampling and thus potential aliasing problems. We mentioned special filters that (if used correctly) can cancel out the effects of aliasing! We showed two examples of uses of the DWT in signal denoising in a severe environment and in image compression (JPEG). In the next few short chapters we will do a step-by-step walk through of these various transforms.

We stress again that this preview is intended to give the reader a feel for how wavelets and wavelet transforms work. The next 12 chapters and the appendices will provide much more information and facilitate a real-world understanding and applications of these amazing tools.

Another option for understanding wavelets is to attend one of the open seminars by Mr. Fugal. The comments from attendees have been very favorable and are one reason why a book and website were developed. In fact, all the chapters in the book (including this one) are written using the completed seminar slides as the basis. Contact D. Lee Fugal at (toll-free) 877-845-6459 for information on the next open seminar. Private seminars are also available for your company or organization.

Be sure to visit our website at **www.conceptualwavelets.com** *for more information, downloads, updates, color slides, additional case studies, corrections, and FAQs.*

You can also selectively solidify your understanding by using the consulting services of Mr. Fugal to clarify or expand on specific sections. You are also welcome to contact him for comments and suggestions, a short specific question, or for general advice. He can be reached during business hours at the above number or at l.fugal@ieee.org.

There is much more to discover than can be presented in this short preview. The time spent, however, in learning, understanding and correctly using wavelets for these "non-stationary" signals with anomalies at specific times or changing frequencies (the fascinating, real-world kind!) will be repaid handsomely.

CHAPTER 2

The Continuous Wavelet Transform (CWT) Step-by-Step

In Chapter One we showed a preview of wavelets, wavelet filters, and wavelet transforms and compared them to some conventional methods such as the FFT and STFT. In section 1.6 we demonstrated the principles behind the continuous wavelet transform or CWT and produced some CWT displays.

In this chapter we will walk through, step-by-step, the process used to construct a CWT and how to generate data for a CWT display that matches MATLABs Wavelet Toolbox cwt utility. We will use a very simple, understandable "signal" and the simplest of the wavelets, the Haar. Our goal in this chapter will be to understand the basic mechanics of the CWT and how a simple CWT display is generated. We will defer to later chapters demonstrations of the power of this tool.

2.1 Simple Scenario: Comparing Exam Scores using the Haar Wavelet

Let's begin with a simple (albeit contrived) example of some exam scores at a university. Rather than using a mid-term and final exam, some instructors (including the author) prefer to assign a grade using a series of equally weighted exams. We will assume 8 exams throughout the term.

In this hypothetical example the student does fairly well the first half of the term then neglects his or her studies for the last half. Thus the exam scores for the term were **80%, 80%, 80%, 80%, 0%, 0%, 0%, and 0%**[*]

We can tell the average of all the scores (**40%**) and when the scores "tanked" after the 4th exam just by looking. Knowing the answer in advance, however, is a good way to learn and to verify the wavelet transforms. Then we can use them with confidence on real-world data where we can't simply "eyeball" the final values.

---

[*] The author actually had a student who got engaged to be married in the middle of the term and the test scores were fairly close to this hypothetical example!

We will now walk through the CWT process step by step using the simplest of the wavelet filters for this example. We begin by comparing the humble Haar wavelet filter, **[1 –1]**, with the data as shown in Figure 2.1–1.

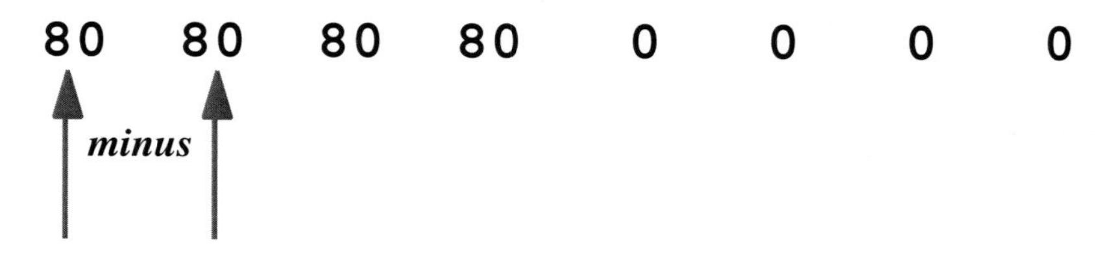

Figure 2.1–1 Comparison of the Haar wavelet filter [1 –1] with the first 2 exam scores. **80 minus 80** = **zero.**

Comparing the first 2 points with the wavelet filter we obtain **80** – **80** = **0**. For this very simple highpass filter we can say there was no change in the first 2 exam scores.

As demonstrated in the preview chapter, we shall now "shift" in time by one as shown below in Figure 2.1–2

We still have a zero value after this first shift. If we shift the wavelet filter once again to the right (not shown) we will also have **80** – **80** = **0**. However, if we shift once more as shown below in Figure 2.1–3 we will have **80** – **0** = **80**.

This is significant in that this wavelet process of comparison (correlating) and shifting has just indicated a large change between the 4th and 5th exam. We have "found the discontinuity".

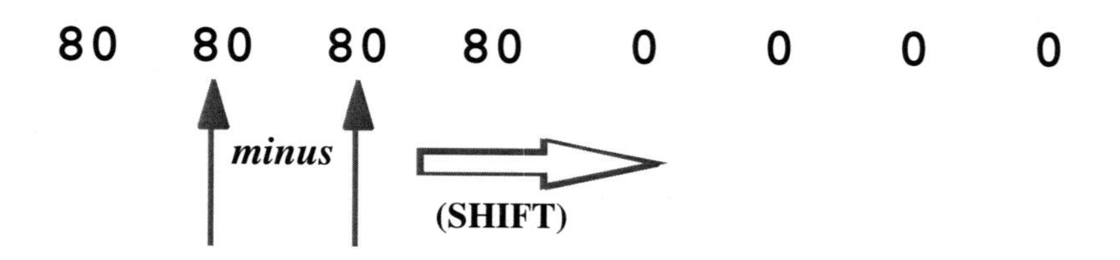

Figure 2.1–2 Comparison of the Haar wavelet filter [1 –1] with the next 2 exam scores. Again, **80 minus 80** = **zero**.

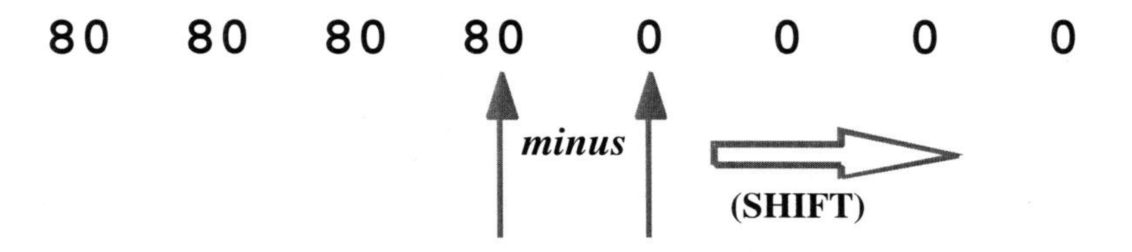

Figure 2.1–3 Comparison of the Haar wavelet filter **[1 –1]** with the 4th and 5th exam scores. We now have a non-zero value with **80 minus 0** = **80**.

We shift once more to the right (in time) and have the comparison shown in Figure 2.1–4. For this shift and for all further shifts of this 2-point wavelet filter (**[1 –1]**) we will be back to zero differences with **0** – **0** = **0.**

Thus if we were to list all the comparisons we have done so far with this Haar wavelet filter we would have the series of 7 numbers:

```
[0, 0, 0, 80, 0, 0, 0]
```

This is almost identical to what the wavelet software packages such as the MATLAB Wavelet Toolbox does. The only difference is that MATLAB has 2 additional comparisons as shown in Figure 2.1–5.

We start comparing a little sooner and end a little later as shown. To do this we assume zeros on both ends of the exam scores and we end up with a series of 9 numbers:

```
[-80, 0, 0, 0, 80, 0, 0, 0, 0]
```

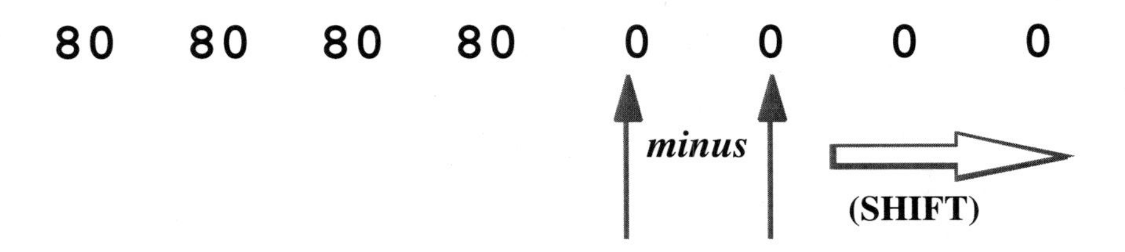

Figure 2.1–4 Comparison of the Haar wavelet filter **[1 –1]** with the 5th and 6th exam scores. The value for this and further shifts will be **0** – **0** = **0**.

Figure 2.1–5 shows us how the 2 “end” numbers (–**80** and **0**) are obtained.

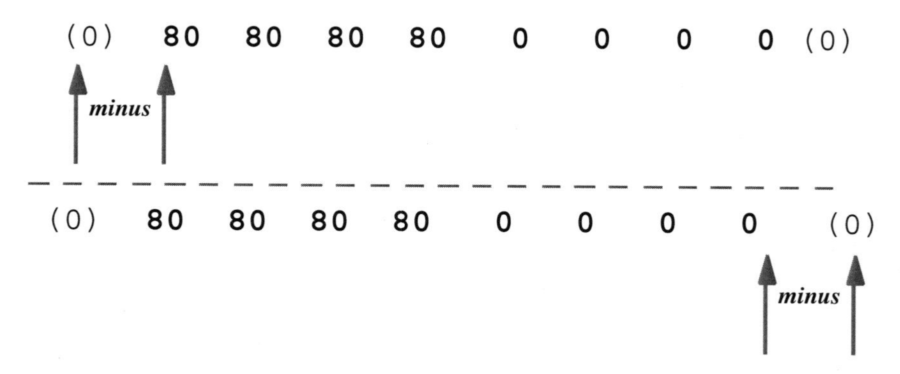

Figure 2.1–5 Comparison of the Haar wavelet filter **[1 –1]** starting one data point earlier and ending 1 data point later. Note that with zeros on both ends we would not obtain any further information from starting even earlier or ending any later than shown.

2.2 Above Comparison Process seen as simple Correlation or Convolution

This probably looks familiar as a *correlation* of the 2-point Haar wavelet filter **[1 –1]** with the exam scores "signal". We could also obtain the exact same results by performing a *convolution* of the flipped-in-time filter **[–1 1]** with the signal. Here is the MATLAB result convolving the time-reversed filter with the same 8 exam scores.

```
conv( [-1 1],[80 80 80 80 0 0 0 0] ) =

-80 0  0  0  80  0  0  0  0
```

which is identical to the results we got by hand in the last section.(2.1)*

Notice that the convolution of the 2-point filter and the 8 point signal gives us 9 points **(8+2–1)**. This is the familiar "**L+M–1**" result from filtering where **L** is the length of the signal, and **M** is the length of the filter.

*We could also have used MATLAB's cross correlation routine ***xcorr*** with the wavelet filter **(xcorr([1 -1], [80 80 80 80 0 0 0 0]))** to obtain the same sequence of numbers. However ***xcorr*** zero-pads the 2-point filter to 8 points and we are left with extra zeros. The MATLAB convolution routine ***conv*** is not only easier to work with but also leads us directly into next chapter.

In a CWT display we usually wish to keep the total number of points of the correlation (or convolution with time-reversed filters) to the length of the original signal. MATLAB, for example, has a routine ***wkeep*** that keeps only the center points. For this scenario it keeps only the middle or "left middle" 8 points of the 9-point correlation. The MATLAB code is shown here:

```
wkeep( [-80 0 0 0 80 0 0 0 0], 8 )

      = [-80 0 0 0 80 0 0 0]
```

If we were to look at the magnitude or absolute value of these numbers we would see indicators that the scores had not only a discontinuity or "drop" starting at the 5th exam but also another discontinuity or "jump" at the very first (an artifact in going from knowing nothing of the subject at the beginning of class to an 80% score on the first exam).

For the 2 points (**1** and **–1**) we used in comparing or correlating the exams we have the above 8 CWT values. For the Haar wavelet filters this is called **scale** = **a** = **2**. We recall from Chapter One that the next step in the continuous wavelet transform was to "stretch" (in time) the wavelet filter to 3 points (**scale** = **a** = **3**) and perform another correlation.

This start of the correlation or comparison process is shown below in Figure 2.2–1. The filter has now been stretched from **[1 –1]** to **[1 0 –1]**. Notice that for the middle value of the stretched filter we use the average of **1** and **–1** or **zero**. Again we have zeros outside the range of the 8 exams.

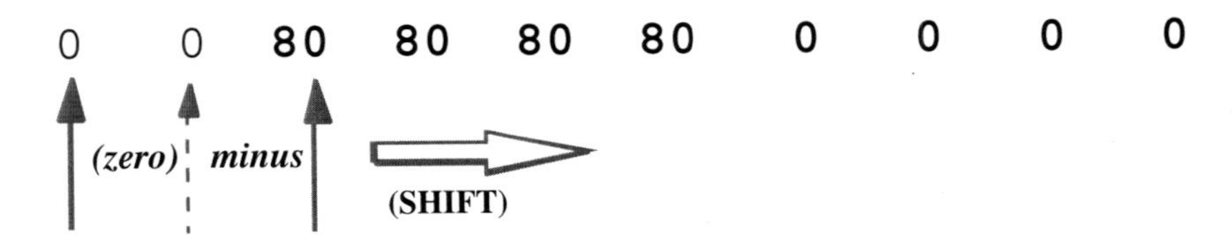

Figure 2.2–1 Comparison of the *stretched* Haar wavelet filter **[1 0 –1]** with the same 8 exam values. The first value we would have from this correlation (before shifting in time) would be **zero plus zero minus 80** = **–80.**

As depicted in Figure 2.2–1, the first value will be **1** x **0** + **0** x **0** – **1** x **80** = **–80**. As we shift and compare again, the next value will also be **1** x **0** + **0** x **80** – **1** x **80** = **–80**. Shifting once again we would have **1** x **80** + **0** x **80** – **1** x **80** = **0**. Thus the first 3 of the 10 values will be **[–80, –80, 0]**. We "cheat" at this point and use MATLABs ***conv*** routing, remembering to time-reverse the stretched 3-point filter. We see the –**80**, –**80**, and **0** values plus the other 7 here.

```
oonv( [-1 0 1], [80 80 80 80 0 0 0 0] )

= -80 -80   0   0  80  80   0   0   0   0
```

Using ***wkeep*** to keep the center 8 values we have

```
wkeep( [-80 -80 0 0 80 80 0 0 0 0], 8 )

= -80   0   0  80  80   0   0   0
```

For **scale** = **a** = **4** we have the original 2-point Haar wavelet filter **[1 –1]** now stretched to the four points **[1 1 –1 –1]**. Figure 2.2–2 shows this dyadically stretched (stretched by a factor of 2) wavelet ready to correlate with the signal. We can see from the figure that as we slide and compare the first few values will be –**80**, –**160**, –**80**, and **0**.

As before, we can use the convolution/time reversal shortcut. We set

```
exams = [80 80 80 80 0 0 0 0]
```

and let ***temp*** be the result of the convolution. We then have

```
temp = conv(exams, [-1 -1  1  1]

= -80 -160 -80 0 80 160 80 0 0 0 0
```

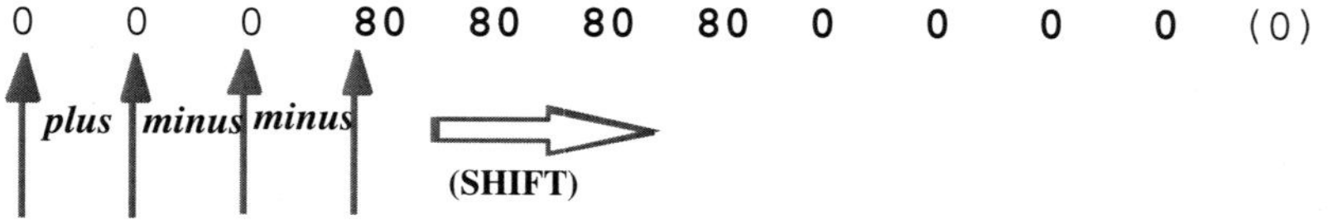

0 0 0 80 80 80 80 0 0 0 0 (0)

plus minus minus

(SHIFT)

Figure 2.2–2 Comparison of the *stretched* Haar wavelet filter **[1 1 –1 –1]** with the exam values. The first value we would have from this correlation (before shifting in time) would be **0** + **0** – **0** – **80** = –**80**.

Keeping the center-left 8 values we have for **scale** = **a** = **4** the values

```
wkeep(temp, 8) = -160 -80 0 80 160 80 0 0
```

Continuing the process with the 5-point stretched wavelet we have

```
wkeep( conv(exams,  [-1 -1  0  1  1]), 8)

= -160  -80   80  160  160   80    0    0
```

We can repeat the process as long as we need. In this case we will stop at **scale** = **a** = **10** for a Haar wavelet stretched to 10 points. We have

```
temp = conv(exams,[-1 -1 -1 -1 -1  1  1  1  1  1] )

= -80 -160 -240 -320 -320 -160 0 160 320 320 240 160
   80  0 0 0 0
```

and again keeping only the left-center 8 values we have

```
wkeep(temp,8) = -320 -160 0 160 320 320 240 160
```

We can now make an array for these scales from **a** = **2** to **a** = **10** and from that array duplicate an actual Haar CWT display for the exams. To complete our set, however, we need to determine the values for **scale** = **a** = **1**. Similar to a 3-point filter with the center point being the average of 1 and –1 or zero, this single-point filter is also the same average and is simply zero. Thus a correlation (or convolution for that matter) of zero with the exams would be simply a set of 8 zeros. Thus for **scale** = **a** = **1** we have

```
0  0  0  0  0  0  0  0
```

2.3 CWT Display of the Exam Scores using the Haar Wavelet Filter.

Placing the 10 sets of values in an array with the 10 point filter at the top and the 1 point filter at the bottom we have the values as shown below in Table 2.3–1:

(scale = 10)	-320	-160	0	160	320	320	240	160
(scale = 9)	-240	-80	80	240	320	240	160	80
(scale = 8)	-320	-160	0	160	320	240	160	80
(scale = 7)	-240	-80	80	240	240	160	80	0
(scale = 6)	-240	-160	0	160	240	160	80	0
(scale = 5)	-160	-80	80	160	160	80	0	0
(scale = 4)	-160	-80	0	80	160	80	0	0
(scale = 3)	-80	0	0	80	80	0	0	0
(scale = 2)	-80	0	0	0	80	0	0	0
(scale = 1)	0	0	0	0	0	0	0	0

Table 2.3–1 Raw correlation of exam data with various stretched Haar wavelet filters as they go from 10 points (**[1 1 1 1 1 –1 –1 –1 –1 –1]**) at **scale** = **a** = **10** down to the trivial single point (**[0]**) at **scale** = **1**.

Notice that the negative numbers also indicate a strong correlation of the signal with the wavelet filter. The MATLAB Wavelet Toolbox plots these values in a single 3-D graph using color or brightness as the magnitude (positive or negative amplitude). In other words, it changes the negative numbers to positive.

We need to make one other adjustment before graphing these numbers. Larger scales mean longer filters. In order to compensate we divide each row by the square root of the scale to "level the playing field" (correct energy representation).

For example at **scale** = **4** we divide by **sqrt(4)** = **2** and change *amplitude* to *magnitude*. So from

```
-160   -80    0    80   160    80    0    0
```

we obtain

```
80    40    0    40    80    40    0    0
```

These adjusted values are now shown for **scale** = **a** in table 2.3–2 below. Note the 2 underlined numbers that are maxima for the entire array.

(a =10)	101.2	50.6	0.0	50.6	101.2	101.2	75.9	50.6
(a = 9)	80.0	26.7	26.7	80.0	106.7	80.0	53.3	26.7
(a = 8)	*113.1*	56.6	0.0	56.6	*113.1*	84.9	56.6	28.3
(a = 7)	90.7	30.2	30.2	90.7	90.7	60.5	30.2	0.0
(a = 6)	98.0	65.3	0.0	65.3	98.0	65.3	32.7	0.0
(a = 5)	71.6	35.8	35.8	71.6	71.6	35.8	0.0	0.0
(a = 4)	80.0	40.0	0.0	40.0	80.0	40.0	0.0	0.0
(a = 3)	46.2	0.0	0.0	46.2	46.2	0.0	0.0	0.0
(a = 2)	56.6	0.0	0.0	0.0	56.6	0.0	0.0	0.0
(a = 1)	0.0	0.0	0.0	0.0	0.0	0.0	0.0	0.0

Table 2.3–2 Correlation of exam data with Haar wavelets from 10 points down to 1 point after adjusting for the length of wavelets (dividing by square root of length) and using magnitude of correlations (replacing negatives with positives).

These now correspond to the MATLAB Wavelet Toolbox CWT display (and most other CWT display software). Figure 2.3–1 shows the MATLAB display of the exam data for scales 1 through 10. The MATLAB command to do this can be written in a single line as

```
c = cwt([80 80 80 80 0 0 0 0],1:1:10,'haar','plot')
```

We can specify colors but use a gray scale here for publication. The Wavelet Toolbox Display assigns the brightest color to the highest value and the darkest to the lowest value. Table 2.3–2 showed the best correlation to be the first and fifth values of **scale** = **8**. This is seen also here in Figure 2.3–1 as the 2 brightest values on the display. As explained in the last section, the center 8 values are shown for each scale.

We can see intuitively why this **scale** = **8** gives the best “match” by looking at Figure 2.3–2 below.

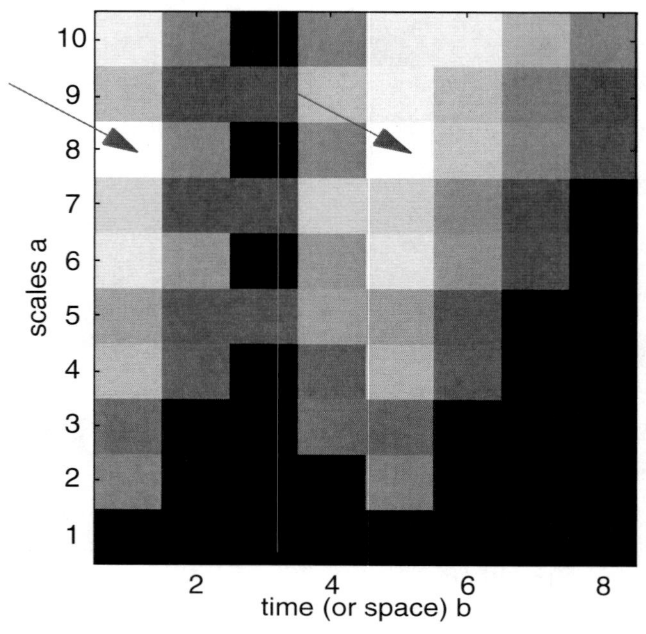

Figure 2.3–1 MATLAB Wavelet Toolbox display of the Continuous Wavelet Transform of the 8 exam scores using the Haar wavelet. Maximum values are at the brightest points (arrows).

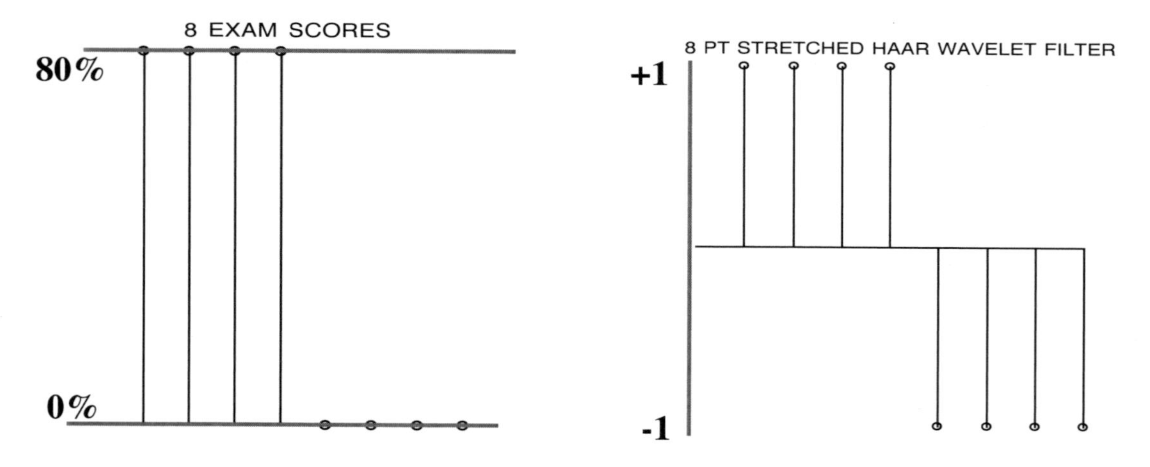

Figure 2.3–2 Similarities shown between the exam scores (left) and the 8-point stretched Haar wavelet.

We can see from Figure 2.3–2 that there will be a strong negative correlation when the four –1 values of the stretched Haar filter line up with the 4 non-zero test scores and a strong positive correlation 4 points when the four +1 values line up. These strong correlations are indicated by the 2 arrows in Figure 2.3–1.

As a further check, we compare the values from MATLABs ***cwt*** routine just prior to plotting to our own values. For **scale** = **8** we came up with the following 8 values ready to be plotted:

```
113.1   56.6    0.0   56.6 113.1   84.9   56.6   28.3
```

After the single-line MATLAB instruction to perform the CWT and display the results

```
c = cwt([80 80 80 80 0 0 0 0],1:1:10,'haar','plot')
```

the array for the 10 scales (**1:1:10** in the *cwt* instruction) is found in the variable "**c**". Looking at the values for **scale** = **8** we have:

```
c(8,:) = -113.1371  -56.5685    0.0000   56.5685
          113.1371   84.8528   56.5685   28.2843
```

which, when rounded off to 1 decimal point and taking the absolute value, matches our values exactly.

With our step-by-step walk-through of the CWT, we have thus successfully duplicated the results of the commercial MATLAB Wavelet Toolbox software.

2.4 Summary

In this chapter we performed a walk-through of the *continuous wavelet transform* and it's associated CWT display. We used a hypothetical example of 8 university exams as the "signal" and then *correlated* these scores point-by-point with the basic 2-point Haar wavelet filter (**[1 –1]**). We also showed how this is the same as *convolving* them with the time-reversed version of the filter (**[–1 1]**).

We repeated this process for stretched filters using the 3-point **[1 0 –1]** filter at **scale** = **3** and continuing to stretch the filter to 10 points

```
[1  1  1  1  1 -1 -1 -1 -1 -1]
```

at **scale** = **10**. Keeping the center 8 values (same length as the signal), we were able to produce an array of correlation values for scales ranging from 1 to 10. After normalizing the values and looking at the magnitudes we were able to see the strong correlations.

A single-line MATLAB instruction produces a display of the CWT of the exam scores with the Haar wavelet. We can see directly that the display matches the results we obtained by our walk-through of the CWT. As a further check, we looked at the results from the MATLAB *cwt* routine for **scale** = **8** and saw an exact match of the magnitudes with our own values.

We now move on to a walk-through of the simplest of the discrete wavelet transforms—the *undecimated discrete wavelet transform* or UDWT.

CHAPTER 3

The Undecimated Discrete Wavelet Transform (UDWT) Step-by-Step

In the previous chapter (Ch. 2) we did a walk-through of the continuous wavelet transform or CWT. We are now ready to look at the discrete wavelet transforms.

The 2 best known and most utilized of these are the conventional (decimated) discrete wavelet transform or DWT and the undecimated discrete wavelet transform or UDWT. A quick note about terminology is in order here. The conventional DWT is actually more complicated than the UDWT and an argument could be made that it should be called the decimated or downsampled discrete wavelet transform, leaving the shorter name to the simpler UDWT discussed in this chapter. However, the more complicated form is better known and we thus follow convention calling it the DWT and requiring the simpler form to add the descriptor "undecimated". To change things around would be similar to ordering "caffeinated" coffee and letting the word "coffee" stand for Decaf.[*]

We now proceed in this chapter to do a step-by-step walk-through of the UDWT. We will use the same simple "signal" of 8 exam scores and the same basic Haar wavelet as in the last chapter. We use some MATLAB commands such as **conv** *(for simple convolution) here to expedite the walk-through, but the results of any of these commands are easily verifiable by hand.*

3.1 Single-Level Undecimated Discrete Wavelet Transform (UDWT) of Exam Data

Figure 3.1–1 shows the flow diagram for a single-level UDWT. It splits the signal, **S**, into a hiighpass (upper) portion and a lowpass (lower) portion. As mentioned in the preview chapter (Ch. 1), **H** and **H'** combine to form a highpass halfband filter while **L** and **L'** combine to form a lowpass halfband filter.

This single-level UDWT doesn't really demonstrate the full capabilities of the UDWT and we don't do any compression or denoising between the left and

[*] The author is old enough to remember when a "regular" television meant black and white. These days the default is color and the simpler black and white TV must be specifically designated.

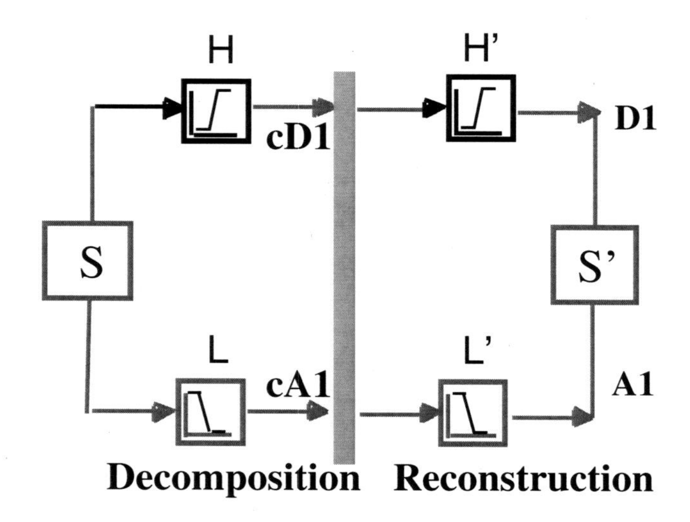

Figure 3.1–1 Signal flow diagram for the single-level *undecimated discrete wavelet transform* (UDWT). **H** and **H'** are the highpass filters and **L** and **L'** are the lowpass filters. **D1** designates the *details* while **A1** designates the *approximation*. The *coefficients* of the Details and the Approximation are designated as ***cD1*** and ***cA1*** respectively.

right halves of this transform for now. However, learning is our primary goal here and thus we do a walk-through to discover the basic mechanics of the UDWT. Specifically, we want to see if we can reconstruct the original signal, **S**, at the end of the process (**S'**).

We use the same "signal" of 8 university exam scores as in the last chapter:

```
S = [80 80 80 80  0  0  0  0]
```

We also use the Haar wavelet filters. In the last chapter we used one of these filters (and its stretched versions) with the CWT. We called it the Haar wavelet filter **[1 –1]**. There are actually *four* Haar wavelet filters to consider as we move on to discrete wavelet transforms. They are

```
H or highpass decomposition filter ("hid")     = [-1  1]
H' or highpass reconstruction filter ("hir")   = [ 1 -1]
{This is the filter we used in the CWT)
L or lowpass decomposition filter ("lod")      = [ 1  1]
L' or lowpass reconstruction filter ("lor")    = [ 1  1]
```

Notice on the upper path that the signal, **S**, is first filtered or convolved with **H**. This is exactly what we did in the CWT. You'll recall that we said convolving with **[–1 1]** is the same as correlating with **[1 –1]** (**H'**).

Let's proceed step-by-step along the top path of the UDWT flow diagram (Fig. 3.1–1). We first have the convolution of **S** = **[80 80 80 80 0 0 0 0]** with **H** = **[–1 1]** to produce **cD1**. Using MATLAB shorthand we have

```
cD1 = conv( S, [-1  1] ) = -80  0  0  0  80  0  0  0  0
```

which is identical to the CWT values at scale = 2. (section 2.2). We next convolve the above 9 numbers of **cD1** with **H'** to obtain the *details* (**D1**).*

```
d1 = conv(cD1, [1 -1) = -80  80  0  0  80  -80  0  0  0  0
```

Moving to the lower path on the flow diagram we have the same exam scores "signal" (S) convolved with **L** = **[1 1]** to produce **cA1**. We have

```
cA1 = conv(S,[1 1]) = 80 160 160 160 80 0 0 0 0
```

At the end of the lower (lowpass) path we convolve **cA1** with **L'** = **[1 1]** and we have for the *Approximation* (**A1**)

```
a1 = conv(cA1,[1 1]) = 80 240 320 320 240 80 0 0 0 0
```

We now combine **A1** and **D1** to see how well we have reconstructed the original exam scores:

```
A1 =   80 240 320 320 240  80   0   0   0   0
                          plus
D1 = -80   80   0   0  80 -80   0   0   0   0
   =    0 320 320 320 320   0   0   0   0   0
   =    sprime (S')
```

The original exam scores were

```
80%  80%   80%  80%   0%  0%  0%  0%
```

so we have **S'** = **S** to within a constant of multiplication of 4 and a delay of 1

* As to the terminology *Details* and *Approximation*, highpass filtering usually produces the noise or the "finer details" of a signal. Lowpass filtering usually produces the smoothed result or an "approximation" of the signal

Using ***wkeep*** to keep the center 8 points removes the delay and then dividing by 4 accounts for the constant of multiplication. We have once again our original exam scores of **[80 80 80 80 0 0 0 0]**.

Notice that we could remove the constant of multiplication in **D1** and **A1** *before* combining them. Dividing by 4 and keeping only the center 8 values we would have

```
D1 (adjusted) = 20   0   0  20 -20   0   0   0
A1 (adjusted) = 60  80  80  60  20   0   0   0
```

Which, when combined, would give us our 8 original exam scores. Still another way would be, as some texts do, to have the dividing "up front" on the highpass and lowpass filters. We would then have

```
H  or HP decomposition filter (hid)   = [-0.5  0.5]
H' or HP reconstruction filter (hir) = [ 0.5 -0.5]
L  or LP decomposition filter (lod)   = [ 0.5  0.5]
L' or LP reconstruction filter (lor) = [ 0.5  0.5]
```

This would produce

```
S' =  0  80  80  80  80  0  0  0  0  0
```

and, after keeping the middle 8 values, our original exam scores.

Different texts use different multiplication factors for these filters. Some use **±0.5** as we just demonstrated, some use **±1/sqrt(2) = ±0.7071**. We usually use **±1** for simplicity and divide later. Again, *Caveat Emptor*. ("Buyer beware").

3.2 Frequency Allocation of a Single-Level UDWT

In section 1.7 we mentioned that the filters **H** and **H'** combine to produce a highpass halfband filter and that **L** and **L'** combine to produce a lowpass halfband filter (Figure 3.2–1). We now perform a step-by-step verification to demonstrate that this is true.

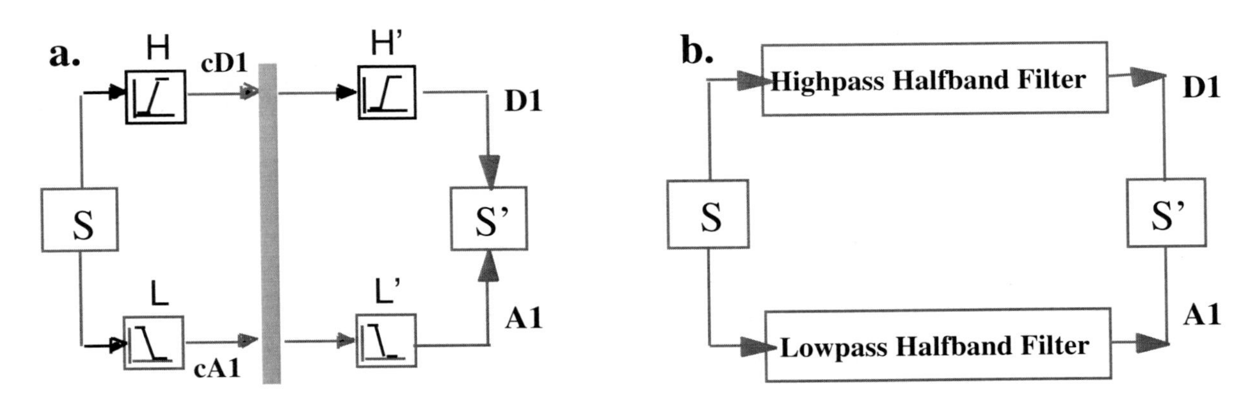

Figure 3.2–1 Single-level Undecimated Discrete Wavelet Transform (UDWT) filter bank shown at left (a). The functional equivalent of this single-level UDWT is shown in the right diagram (b).

We saw in the last section that filtering the signal, **S**, by **H** and then by **H'** means convolving **S** with **H** to produce the *Details coefficients* **cD1** and then convolving **cD1** with **H'** to produce the *Details*, **D1**. Since the order of convolution doesn't matter, we can first convolve **H** with **H'** and then convolve the result with **S**.

For the simple Haar filters we have

```
conv(H, H') = conv([-1 1], [1 1]) = [-1  2 -1] = Php
```

(a result easily verifiable by hand). We designate this highpass halfband filter result as Php to correspond to nomenclature in many wavelet texts.

Similarly for the lowpass (lower) path we have

```
conv(L, L') = conv([1  1], [1  1]) = [1  2  1] = Plp
```

If we were to add Php to Php we would have **[0 4 0]**. This is what we saw in the last section—a delay of one and a constant of multiplication of 4.

We now look at the frequency characteristics of Php and Plp as shown in Figure 3.2–2 **a** and **b**. Notice that the highpass and lowpass filters (a and b) are vertical and horizontal mirror images of the other. In each frequency graph we can see perfect symmetry around the superimposed dotted lines. Notice also that the phase (lower 2 graphs) is linear.

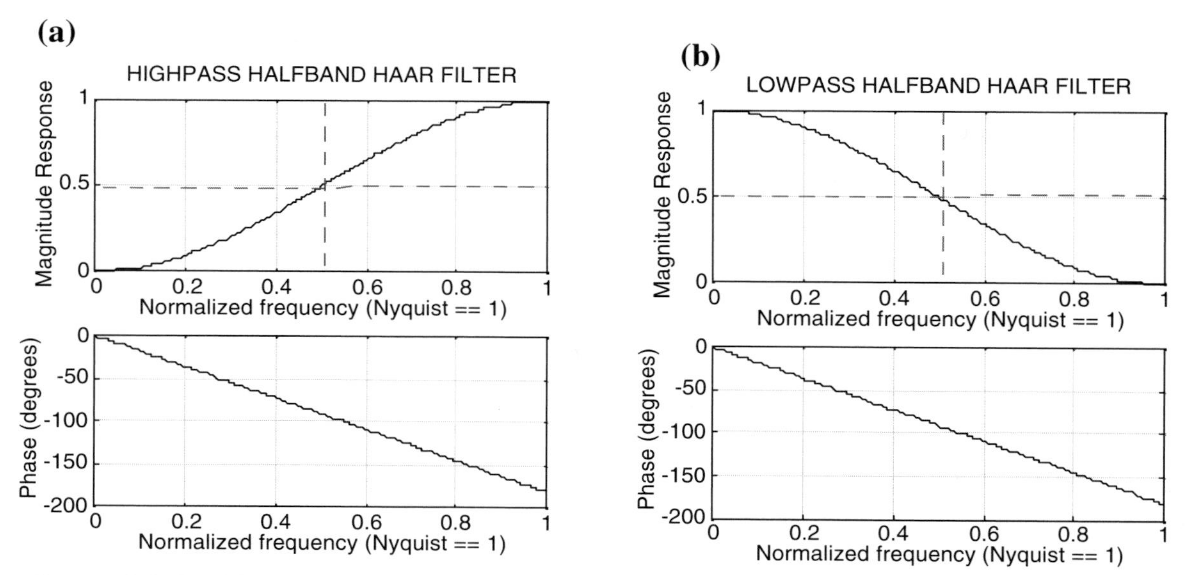

Figure 3.2–2 Frequency characteristics of the halfband highpass and lowpass filters.

We have learned that the short 2-point Haar filters have great time resolution and are suited for tasks such as edge detection. Their frequency characteristics are less impressive as we see very short stopbands and passbands and a huge transition band that spans almost from zero to Nyquist.

Nevertheless, the Haar filters do produce these perfectly valid halfband filters and we can see (Fig. 3.2–2, top graphs) that at any given frequency the sum of (a) and (b) will be 1.0. In other words, adding the filters produces a constant magnitude as the 2 "halves" make an all-pass filter.

So it's no surprize that as we process **S** on the highpass path and on the lowpass path that the high frequency and low frequency "halves" add to reconstruct the original signal.

Halfband filters are often depicted as shown in Figure 3.2–3. This shows the general form, but it must be remembered that the overlap area (transition bands) for some filters such as the Haar can occupy almost the entire range of frequency.

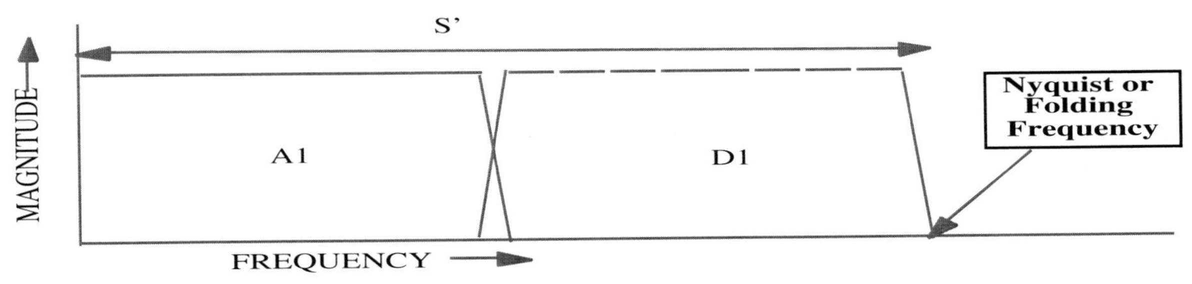

Figure 3.2–3 Frequency allocation after a single-level UDWT. The diagram is illustrative only and the actual shape depends on the wavelet filters. Note overlap from non-ideal filtering.

In denoising or compression we might wish to manipulate the coefficients (**cD1**) or the final Details themselves (**D1**) before reconstructing. Manipulating the coefficients (**cD1** or **cA1**) is done in the vertical (gray) bar shown in Figure 3.1–1. We could, for example, do a very rudimentary denoising by simply setting the details **D1** to zero. **S'** would then consist only of the *approximation* or smoothed value of the signal. We would have more flexibility, however, with the signal being decomposed into more frequency subbands. This is usually done with a multi-level UDWT.

3.3 Multi-Level Undecimated Discrete Wavelet Transform (UDWT)

We looked at the multi-level UDWT briefly in section 1.7. We will look a little closer here and even closer in upcoming chapters. Figure 3.3–1 shows a signal flow diagram of a 2-level UDWT using the basic Haar filters **H**, **H'**, **L** and **L'** along with upsampled-by-2 versions *(**Hup**, **H'up**, **Lup** and **L'up**)*. We are now ready to begin the walk-through by following the flow diagram. We use the same "exams" signal

```
S = [80 80 80 80 0 0 0 0]
```

Rather than a complete walk-through as we did in the single-level UDWT, we can "save some steps" (and learn some valuable concepts). We look first at the filters **L** and **Hup** in the oval (Fig. 3.3–1).

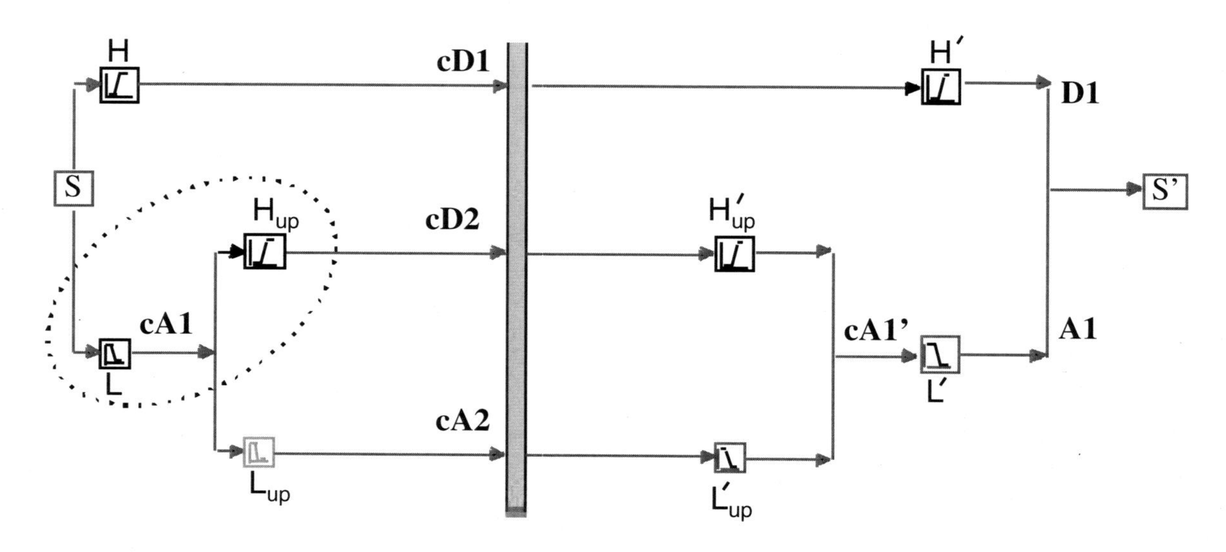

Figure 3.3–1 Flow diagram for 2-level UDWT. **Hup**, **H'up**, **Lup** and **L'up** are **H**, **H'**, **L** and **L'** upsampled by 2. Filters in the dotted line oval are examples of how a stretched filter is produced.

The convolution of **L** = **[1 1]** and **Hup** = **[-1 0 1]*** is

```
conv([1 1],[-1 0 1]) = [-1 -1  1  1]
```

which is the filter **H** = **[-1 1]** stretched by a factor of 2. In other words we are convolving the signal *on this path in the flow diagram* with a stretched filter just like we did in the CWT of the last chapter with **scale** = **a** = **4**. The only difference being that in the CWT we used filters of length 2, 3, 4, 5, etc. while in the UDWT we use *dyadic* (factors of 2) lengths 2, 4, 8, etc. It is easily verifiable by software or by hand that the *level 2 Details coefficients*, **cD2**, on Figure 3.3–1 can be obtained by convolving the signal with **L** and then convolving that result by **Hup** ***or*** by convolving the signal directly with the stretched filter.

On the bottom path of Figure 3.3–1 we can convolve **L** and **Lup** to produce

```
conv([1 1], [1 0 1]) = [1  1  1  1]
```

* We can see here why upsampling or placing zeros between the filter values is referred to in French as *'A Trous* or "with holes". The UDWT is thus called the *'A Trous Transform* in some texts and papers.

which is **L** = **[1 1]** stretched by a factor of 2. The *level 2 Approximation coefficients*, **cA2**, is obtained by filtering **S** by this stretched version.

Similarly to the CWT, we are again comparing (*correlating*) the signal with stretched filters by *convolving* the signal with time-reversed versions.

For example correlating **S** with **[1 1 –1 –1]** is the same as convolving **S** with **[–1 –1 1 1]**. (For **[1 1 1 1]** the time-reversed version is obviously unchanged).

Let us next look at Figure 3.3–2 and how **cA1** becomes **cA1'**. We see that the signal flow within the dotted oval is very similar to a single-level UDWT. The only difference being that the 4 filters are upsampled versions of the originals. Looking at the top path of this embedded UDWT (with cD2) we see that we have a convolution of **Hup** and **H'up**. This is given by

```
conv([-1 0 1],[1 0 -1]) = [-1     0     2     0     -1]
```

Similarly for the bottom path (with **cA2**) we have

```
conv([1 0  1],[1 0  1]) = [ 1     0     2     0      1]
```

Adding these 2 paths together we have

```
cA1' = [0    0    4    0    0]
```

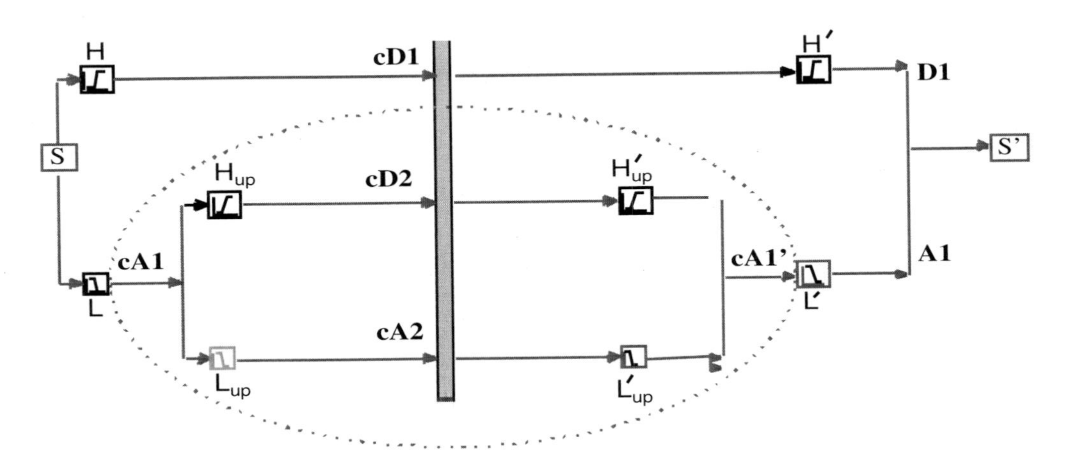

Figure 3.3–2 Flow diagram for 2-level UDWT with the embedded single-level UDWT (with upsampled filters) highlighted by the dotted oval.

In other words **cA1'** will perfectly construct **cA1** to within a delay and a constant of multiplication. If we remove the delay and divide by 4 we have **cA1'** = **cA1**. This is almost identical to what we found in the single-level UDWT, the only difference being a delay of 2 points rather than one.

The MATLAB Wavelet Toolbox software does the removal of delays by keeping the middle points and uses the "pre-divided" filters to keep the constant of multiplication at 1.0. Thus in MATLAB software **cA1'** = **cA1**.

This is an important result in that for our walk-through we can treat this 2-level UDWT much as we did the single level. In other words with no changes to any of the components Figure 3.3–2 simplifies to Figure 3.1–1 and we have already performed a walk-through of this single-level UDWT.

Note that in between the forward transform in the left half (decomposition) and inverse transform in the right half (reconstruction) we have *Details coefficients* **cD1**, **cD2**, and **cD3**. Notice that each is approximately the length of the signal (plus a few extra points from filtering) so we have a little over 3 times as much data as we started with. This is why the UDWT is sometimes referred to as the *redundant* DWT or RDWT.*

Our goal so far has been to demonstrate *perfect reconstruction* so we have left these coefficients untouched. However, we could have adjusted them for denoising or compression. We can also work with the final Approximation and Details directly. We re-draw the 2-level UDWT as shown below in Figure 3.3–3. We can see that

A1 = D2 + A2

Thus

S' = D1 + A1 = D1 + D2 + A2

* The term *redundant* does not mean superfluous or unneeded. The conventional *decimated* DWT must deal with aliasing problems caused by the removal of some of the data. In addition, the CWT with its numerous integer-stretched scales has far more data to keep track of than this "redundant" DWT.

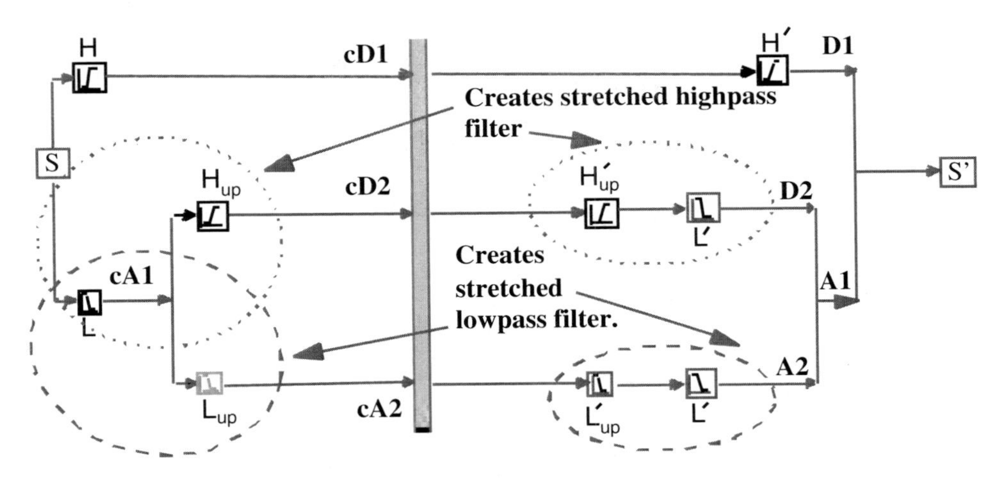

Figure 3.3–3 2-level UDWT signal flow diagram re-drawn to show how final Approximations (**A1**. **A2**) and Details (**D1**, **D2**) combine.

3.4 Frequency Allocation of a Multiple-Level UDWT

The top path in Figure 3.3–3 above is identical to the single-level case. **H** and **H'** combine to produce the highpass halfband filter **[–1 2 –1]** as was shown in section 3.2 and graphed in Figure 3.2–5 (a). Let's now look at the middle and bottom paths in the 2-level UDWT.

We saw in the last section that on the bottom path

L = **[1 1]** and **Lup** = **[1 0 1]** combine to produce the stretched filter **[1 1 1 1]** Later on the bottom path **L'** = **[1 1]** and **L'up** = **[1 0 1]** combine to produce **[1 1 1 1]** (**L** = **L'** and **Lup** = **L'up**). The combination of these 2 stretched filters is

```
conv([1 1 1 1], [1 1 1 1]) = 1 2 3 4 3 2 1
```

or we could verify this step-by step with a "convolution of convolutions"

```
conv( conv([1 1],[1,0,1]), conv([1 1],[1,0,1]) )
```

and have the same result (**[1 2 3 4 3 2 1]**).

On the "middle" path in Fig. 3.3–3 **L** = **[1 1]** and **Hup** = **[–1 0 1]** combine to form the stretched filter **[–1 –1 1 1]**. Later on the middle path **L'** = **[1 1]** and

H'up = [1 0 −1] combine to produce **[1 1 −1 −1]**. Thus the stretched-by-2 filters on the middle path combine to produce

```
conv([-1 -1  1  1],[1  1 -1 -1]) = -1 -2 1 4 1 -2 -1
```

If we add this result to that of the lower path we have

```
-1 -2  1  4  1 -2 -1
          +
 1  2  3  4  3  2  1
          =
 0  0  4  8  4  0  0
```

If we remove the delay (2) and constant of multiplication (4.0) we have the same lowpass halfband filter **[1 2 1]** that we had in the single-level case. In other words, the lowpass halfband filter in the single-level case is split into two halves itself. As shown in Figure 3.4–1, the bottom path at **A2** produces a *lowpass quarter-band filter* while the middle path at **D2** produces a *bandpass quarter-band filter*. These 2 filters combine to produce the *lowpass halfband filter* at **A1**. When combined with the highpass halfband filter at **D1**, we have the original signal, **S**, back.

The frequency transition bands from these non-ideal filters may have considerable overlap. As mentioned before, Haar filters are better used for detecting short events in the time domain than for efficient frequency allocation. There are other wavelet filters, however, that do have short transition bands and *are* well suited for frequency allocation. We used the Haar filters in this chapter for ease of explanation and, even with sloppy overlap characteristics, they do work together to perfectly reconstruct the original signal.

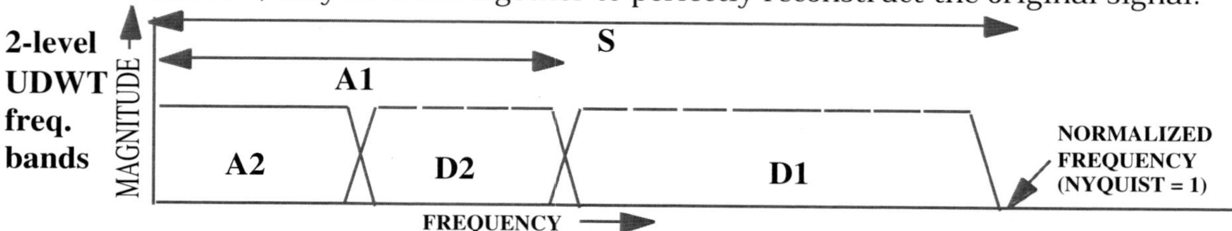

Figure 3.4–1 Frequency allocation in a 2-level UDWT. Level 2 Approximations and Details **A2** + **D2** combine to cover the lower half of the frequencies while level 1 Details **D1** covers the upper half.

We can begin to see here the utility and flexibility of the UDWT, especially for the multiple-level forms. Remember that, unlike the FFT, we can adjust Details and Approximations for any desired part of the total time. This gives us great power to work with embedded pulses or other short-time events in the signal.

Figure 3.4–2 shows the frequency allocation for a 4-level UDWT Notice that the frequency band is divided into 5 sub-bands. We can work with any of these sub-bands (for any desired length of time) for compression or denoising.

The flow diagram for a 4-level UDWT is not drawn but is similar to the 2-level UDWT in that it has "nested" single-level UDWTs. In other words for perfect reconstruction the 4-level reduces to a 3-level, the 3-level reduces to a 2-level, and the 2-level (as we saw in section 3.3) reduces to a single level.

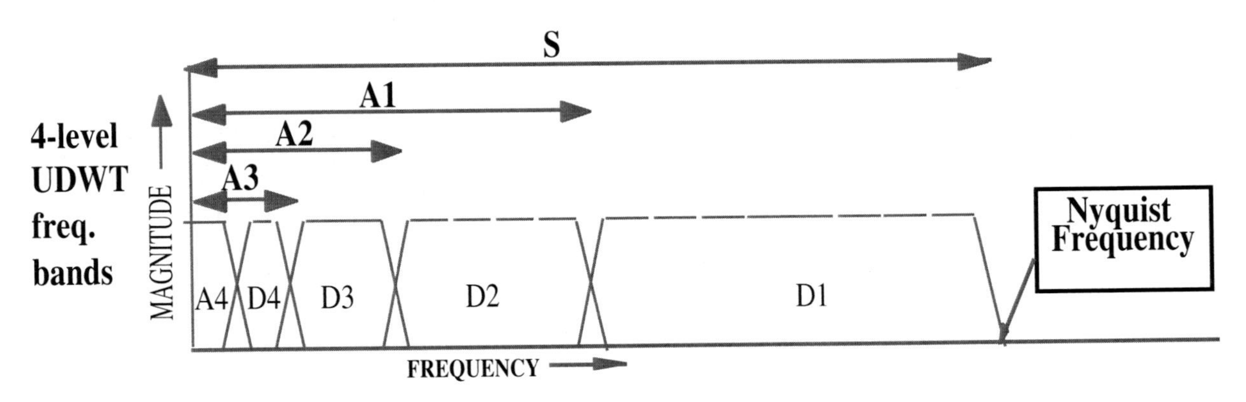

Figure 3.4–2 Frequency allocation for a 4-level UDWT. Note the various ways the signal, **S**, can be broken down. **S = A1 + D1**, **S = A2 + D2 + D1**, **S = A3 + D3 + D2 + D1**, **S = A4 + D4 + D3 + D2 + D1**.

3.5 The Haar UDWT as a Moving Averager

Before leaving the UDWT, there is one other concept we should address. DSP students may be familiar with a simple form of lowpass filtering called a *moving averager,* sometimes referred to as a *Block Averager.*[*]

In a nutshell, the moving averager does filtering by taking the average of several points then shifting forward in time by a single point and averaging the same number of points. This process tends to "average out" any spikes and thus acts as a lowpass filter.

Using our 8 exams scores again we could construct a 2-point moving averager as shown in Figure 3.5–1. Here we are taking 2 points at a time. We start with the first 2 exam scores, add them together, and divide by 2 to produce

(80 + 80)/2 = 80

or an average of 80% on the first 2 exams. We then shift the 2 arrows forward in time by 1 and average the 2nd and 3rd exam, again obtaining an 80% average. We shift again to obtain yet another 80% average. The next shift is (Fig. 3.5–1) shows the average of the 4th and 5th exam which is

(80 + 0)/2 = 40

or an average of 40% for those 2 exams. Further shifts produce averages of zero.

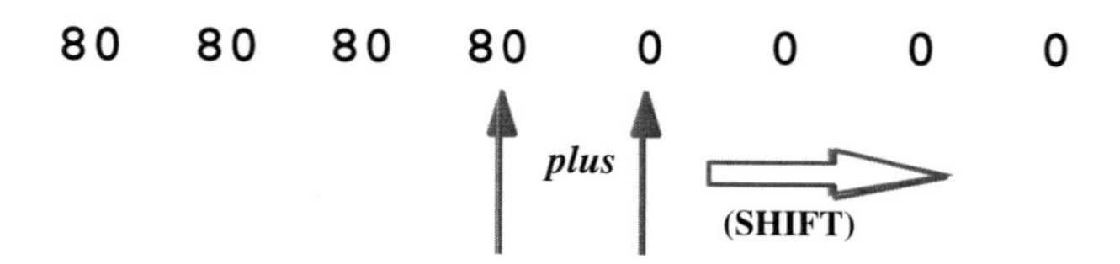

Figure 3.5–1 Averaging 2 points at a time to produce a value then shifting to average the next 2 points. The process of adding the 4th and 5th exam scores together is depicted here. We must remember to also divide by the number of points in this *moving averager* (2 points) to produce the correct average of 40% (**(80 – 0)/2 = 40**).

[*] If this is familiar, it will help you to understand the concepts of the UDWT. If this is not familiar, this section may be (1) skipped entirely or (2) studied to add the Moving Averager to your repertoire of filtering techniques.

We recognize this process as almost identical to the single-level UDWT on the bottom path (see Fig. 3.1–1). The bottom path of the 2-level UDWT (Fig. 3.3–1) is also a moving averager, but it uses 4 points in each average because ***L*** and ***Lup*** act together as the stretched filter **[1 1 1 1]**.

This 4 point block averager adds together 4 successive points and then divides by the length of the block (4) to obtain the average. In Figure 3.5–2 we show the process where the moving averager has shifted in time from the first point to the fifth point.

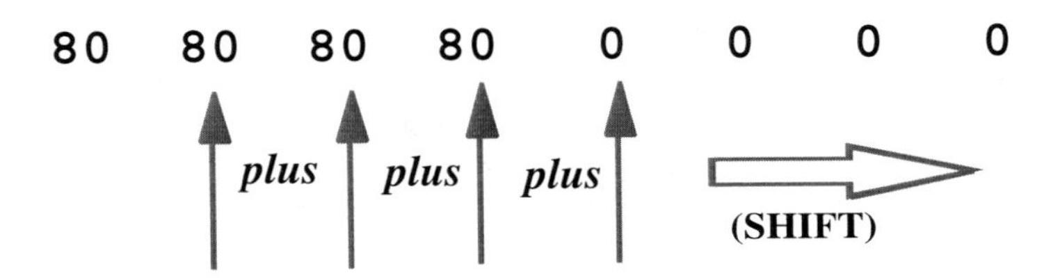

Figure 3.5–2 Averaging 4 points at a time to produce a value then shifting to average the next 4 points. The process of adding the 2nd through 5th exam scores together is depicted here. After we divide by the number of points in this *moving averager* (4) we have the correct average of 60% for these 4 exams (**(80 + 80 + 80 + 0)/4 = 60**).

3.6 Summary

In this chapter we first looked at the single-level *undecimated discrete wavelet transform* or UDWT . We saw that it uses the same filter as the CWT of the last chapter plus three additional filters. We learned that **H** and **H'** together comprise a highpass halfband filter and that **L** and **L'** comprise a low-pass halfband filter. We looked at the frequency characteristics of these half-band filters and saw perfect symmetry (even with the huge transition bands of these simple Haar filters).

We completed our step-by-step walk-through of the single-level UDWT. We saw that the final Details and Approximation (**D1** and **A1**) combine to perfectly reproduce the original signal to within a delay and a constant of multiplication. We also explored the option of using pre-divided filters and keeping only the center points as is done in the MATLAB software. This way we can set the constant of multiplication to 1.0 and remove the delay thus having true perfect reconstruction.

We introduced the coefficients for the Details and Approximation (**cD1** and **cA1**) and learned that we can modify these coefficients (or the final values **D1** and **A1**) for compression or denoising.*

We moved on to the 2-level UDWT and saw not only the basic 2-point Haar filters, but also some upsampled filters with zeros (or "holes" in French) inserted between the values. We observed that these filters combine to produce stretched versions somewhat similar to those we saw in the CWT of the previous chapter (except the stretching is done by factors of 2). We also saw how the 2 level UDWT can be thought of as a single-level UDWT with another single-level UDWT (with the upsampled filters) embedded or nested on the lowpass path.

We saw that the frequency allocation of multi-level UDWT gives us more sub-bands to work with for compression or denoising and that we have the flexibility to work with different time periods within specific sub-bands.

We concluded our discussion of the UDWT by comparing it to a *moving averager*—a handy DSP filtering technique. We now move on to the more complicated *conventional (decimated) discrete wavelet transform*, usually referred to as simply the DWT.

* After all, the underlying reason for these wavelet transforms is so we can better perform data processing tasks such as denoising and compression—not just to see if we can reconstruct the original signal.

CHAPTER 4

The Conventional (Decimated) DWT Step-by-Step

We first made our acquaintance with the conventional (decimated) Discrete Wavelet Transform or DWT in the preview chapter (Ch. 1). In the last chapter (Ch. 3) we completed our walk-through of the simpler UDWT. We now proceed in this chapter to walk-through of the slightly more complicated conventional Discrete Wavelet Transform (DWT). We will again use the same simple "signal" of exam scores and the same basic Haar wavelet.

We will find many similarities to the UDWT, but a few important differences that must be taken into account when using this tool to prevent corruption of your data.

4.1 Single-Level (Decimated) Discrete Wavelet Transform (DWT) of Exam Data

In the CWT and in the UDWT of Chapters 2 and 3, respectively, we were comparing (correlating) the signal with the stretched wavelet filters. We learned that this was the same as *convolving* the signal with the *reconstruction* filters—which happen to be time-reversed versions of the *decomposition* filters (the filters are set up this way as part of the *perfect reconstruction* process). For example, **H'** is the time reversed version of **H**. We will soon see that this relationship holds true for the conventional DWT as well.

We learned earlier that the CWT uses 1 filter and all its stretched versions (2 points, 3 points, 4 points, 5 points, etc.). We also learned that the single-level UDWT uses 4 filters (**H**, **H'**, **L**, and **L'**) and these are *dyadically* stretched (by factors of 2 to 4 points, 8 points, 16 points, etc.) in the higher level UDWTs.

As discussed briefly in the preview, instead of dyadically *stretching the filters,* the conventional (decimated) DWT dyadically *shrinks the signal* instead.

The signal flow diagram for a single-level conventional DWT is shown in Figure 4.1–1. It looks much the same as the UDWT except for downsampling by

2 and upsampling by 2 as shown in the circles with arrows inside. The downsampling by 2 is often refereed to as *decimation by 2* (ignoring the dictionary definition of the prefix *deci*).

Notice that with the downsampling, the Details Coefficients (**cD1**) and the Approximations Coefficients (**cA1**) are each about half the length of the original signal, **S**. We can now see even better why the terms *undecimated* and *redundant* are sometimes applied to the simpler UDWT version we studied in the last chapter.

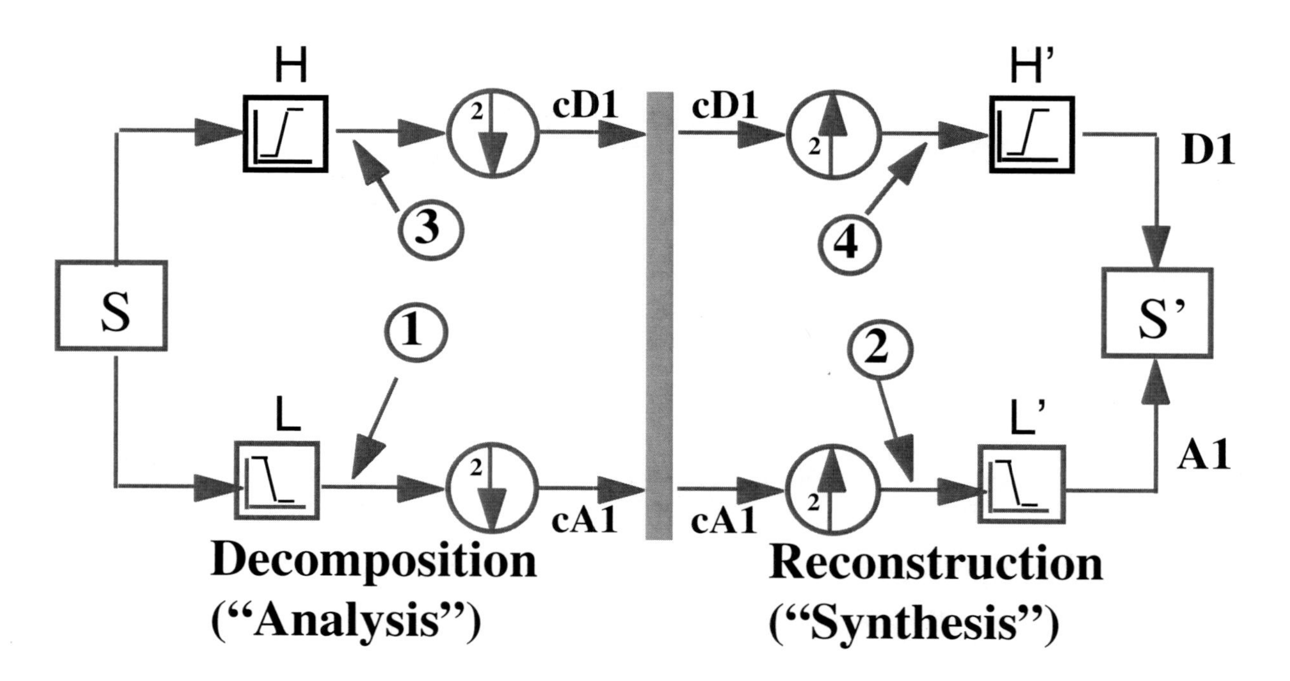

Figure 4.1–1 Single-level conventional DWT signal flow diagram. Note the many similarities to the single-level UDWT (Figure 3.1–1). H, H', L, and L' are the same. The main difference is the downsampling by 2 on both paths in the Decomposition (left) half and the upsampling by 2 on both paths in the reconstruction (right) half. The 4 arrows are to indicate checkpoints as we follow the flow step-by-step.

We have to be careful using the DWT instead of the simpler UDWT for 2 main reasons:

1. Downsampling by 2 in the DWT can produce *aliasing* (throwing away half the samples can lead to false signals).

2. This transform is not *shift-invariant* (sometimes called *time invariant).* In other words, how do you chose *which half* of the samples to throw away in the conventional DWT? Throwing away all the odd samples leaves an entirely different set of data than throwing away all the even samples. As a pathological example, suppose we had a cosine signal

```
S = 1 0 -1 0 1 0 -1 0 . . .
```

If we throw away all the even samples we have the sinusoidal signal

$S_{downsampled}$ = 1 -1 1 -1 . . .

While if we throw away all the odd samples we have the constant zero signal

$S_{downsampled}$ = 0 0 0 0 . . .

This is why the simpler (and safer*) UDWT of the last chapter is also referred to in some texts as a *shift-invariant* transform (in addition to the designations *redundant, undecimated* and *A' Trous* we have already mentioned).

To see how (and how well) this method of shrinking the signal works we now begin a step-by-step walk-through of the conventional (decimated) DWT. We will shorten the designation to *DWT* from here on.

We begin again with our "signal" as the 8 exam scores:

```
S = 80 80 80 80 0 0 0 0
```

At the #1 arrow checkpoint in Figure 4.1–1 we have convolved the signal with L = [1 1]

and the results are the same as for the UDWT (so far):

```
conv(S,[1 1]) = 80 160 160 160 80 0 0 0 0
```

But downsampling by 2 (keeping the even values) gives

```
cA1 = dyaddown(conv(s,[1  1]) = 160 160 0 0
```

Notice that when we divide by 2 (now instead of later) this gives us the averages for 4 sets of 2 of the exam scores:

```
80%  80%     80%  80%      0%  0%     0%  0%
```

* Having neither of these concerns, the simpler UDWT is a safer method in many cases. If you have the resources to store extra data, the UDWT should be considered, as a backup or at least as a sanity check.

```
cA1/2 = [80 80 0 0]
```

Notice also that the 40% average value from the 4th and 5th exams is not found—a reminder that the conventional DWT is not *shift-invariant.*

Using the "pre-divided" averages we have for the upsampled values at the #2 arrow checkpoint of Figure 4.1–1

```
dyadup(cA1/2) = 0 80 0 80 0  0  0  0  0
```

And for **A1** we have (keeping the middle 8 values)

```
A1 = wkeep(conv(dyadup(cA1/2),[1  1]), 8)

              =

80 80 80 80 0 0 0 0
```

Notice that the entire bottom (lowpass) path required a division by 2, rather than by 4 as in the UDWT.

On the upper (highpass) path at arrow #3 the results are the same as for the UDWT:

```
conv(S,[-1 1]) = -80 0 0 0 80 0 0 0 0
```

Downsampling by 2 (again keeping the even values) gives

```
cD1 = dyaddown(conv(s,[-1  1])) = 0 0 0 0
```

Notice that now we have the *differences* in the 4 sets of 2 exam scores:

```
80%  80%     80%  80%      0%  0%     0%  0%
```

Note that dividing by 2 (**cD1/2**) also produces 4 zeros

Continuing on the upper path, we upsample or place zeros between the values and we have at arrow #4

```
dyadup(cD1/2) = 0 0 0 0 0 0 0 0 0
```

At the end of the upper path we have for **D1** (keeping the middle 8 values)

```
D1 = wkeep(conv(dyadup(cD1/2),[1 -1]), 8)
   = 0 0 0 0 0 0 0 0
```

If we add **A1** to **D1** we have

```
 80 80 80 80  0  0  0  0
            +
  0  0  0  0  0  0  0  0
            =
80 80 80 80  0  0  0  0
```

Notice that we have perfectly reconstructed our exams "signal", even with the downsampling and upsampling. This walk-through is not meant to be a proof, but is an example of how the DWT can work OK even with tossing out half the data.

With more complicated data we will need the values of **D1** to cancel out the aliasing in **A1**. However in this simple example **D1** values were all zeros and we ran into no problems. It turns out that values of **D1** that are very small also have small alias cancellation components and that throwing away these small (but non-zero) values will have minimal impact on aliasing.

4.2 Additional Example of Perfect Reconstruction in a Single-Level DWT

Before moving on to the multiple-level DWT, it might be comforting to see if the alias cancellation works with a more arbitrary signal. We use a random number generator to produce the signal

```
S = 0.9794 -0.2656   -0.5484   -0.0963   -1.3807
-0.7284    1.8860 -2.9414
```

We've already walked through the various steps so we'll just show the MATLAB code for the highlights (see Fig. 4.1–1).

```
cD1 = dyaddown(conv(S,[-1  1])) = 1.2451   -0.4521
-0.6523    4.8274
```

```
D1 = conv(dyadup(cD1),[1 -1]) =  0   1.2451   -1.2451  -
0.4521  0.4521  -0.6523  0.6523  4.8274  -4.8274  0
```

```
cA1 = dyaddown(conv(S,[1  1])) = 0.7138   -0.6446
-2.1090   -1.0554
```

```
A1 = conv(dyadup(cA1),[1  1]) = 0  0.7138  0.7138
-0.6446 -0.6446 -2.1090 -2.1090 -1.0554  -1.0554  0

SUM = D1 + A1 =  0    1.9589   -0.5312   -1.0967
-0.1925   -2.7613   -1.4567 3.7720   -5.8828   0
```

If we remove the delay of one by keeping the middle 8 values and then divide by 2 we have the original random-number signal perfectly reconstructed.

Another indication (but not a proof*) of the alias cancellation capabilities of a single-level conventional DWT is to look at the Haar filtering, downsampling, and upsampling as a whole before convolving them with a signal. On the bottom path of the flow diagram at **cA1** (Figure 4.1–1) we have **L** = **[1 1]** downsampled by 2 which gives **cA1** = **[1]**. Upsampling by 2 and filtering by **L'** = **[1 1]** gives **A1** = **[0 1 1]**.

On the top path at **cD1** we have **H** = **[-1 1]** (even) downsampled which gives **cD1** = **[1]**. Upsampling by 2 and filtering by **H'** = **[1 -1]** gives **D1** = **[0 1 -1]**. Combining **A1** and **D1** we have **[0 2 0]**.

Thus the single-level DWT serves to delay the input signal by 1 and multiply it by 2. In other words, it perfectly reconstructs the signal if we remove the delay and divide by 2.†

4.3 Compression and Denoising Example using the Single-Level DWT

Recall that the final result from our simple exam data DWT was **A1** + **D1** and that **A1** was reconstructed from **cA1** while **D1** was reconstructed from **cD1**.

cD1 was [0 0 0 0], and **cA1** (after division by 2) was [**80 80 0 0**]. We could obviously compress the data by saving only **cA1** and later reconstructing the data using zeros. For example, we could transmit only non-zero data and let the DWT on the other end "fill in the blanks" with zeros.

* The conventional DWT is Linear Time Invariant (LTI) which means we can change the order of convolution.

† We could "pre-divide" the filters we would use 1/sqrt(2) instead of the 1.0 values we used for the tutorials. In other words H' would be approx. [0.7071 -0.7071]. Also keeping the middle 8 values (*wkeep* in MATLAB) removes the delay.

Now suppose your *kindly professor* gave at least 1% on exams just for signing your name. The scores would become. [80% 80% 80% 80% 1% 1% 1% 1%]

Using the (conventional) DWT we would have

```
cA1/2 =[80 80 1 1] (averages of the above sets of 2)

cD1/2 =[0 0 0 0] (differences of the above sets of 2)
```

Again, we could compress by transmitting or saving only the 4 **cA1** data points instead of the 8 exam scores. In reconstructing the signal later we simply plug in zeros for the **cD1** coefficients.

But suppose the *Ogre department head* did not allow points for name signing. He/She could "denoise" **cA1** by setting the 1's to 0's (e.g. "If the average score for any 2 exams is less than 2%, disallow it as not earned credit")

Then the *Denoised* **cA1/2** ([**80 80 1 1**]) becomes [**80 80 0 0**]. Reconstructing (with **cD1** = **zeros** = [**0 0 0 0**] through compression) gives us the final result [**80 80 80 80 0 0 0 0**]. In this case we have denoised the signal and used only **cA1** thus achieving a compression ratio of 2:1.

4.4 Multi-Level Conventional (Decimated) DWT of Exam Data using Haar Filters

We can perform further compression with higher level DWTs. Figure 4.4–1 shows a 2-level DWT. Note how the signal is repeatedly shrunk by downsampling and the filters stay the same. In other words, there are only **H**, **H'**, **L**, and **L'**. No $\mathbf{L_{up}}$, $\mathbf{H'_{up}}$, etc. as in the 2-level Undecimated DWT (UDWT).

We can see that the embedded DWT within the oval in Figure 4.4–1 is exactly of the form of the single-level DWT. We completed the walk-through of the single-level in section 4.2 and saw how **S'** was perfectly reconstructed from **S**. In the same manner here **cA1'** is perfectly reconstructed from **cA1** (following the MATLAB software convention of pre-dividing to remove the constant of multiplication and then keeping only the middle points, thus removing the delay).

Further, we can see that **cA1** is still the convolution of **S** and **L** and will be the same as in the single-level DWT. Thus with **cA1'** = **cA1** the diagram simplifies to a single-level DWT and we know we can perfectly reconstruct **S**,

We achieved a compression ratio of 2:1 with the single-level DWT. Let's see if we can do better here. We know

```
cD1/2 =   [0  0  0  0]

cA1/2 = [80   80   0   0]
```

Now we solve for **cD2** and **cA2**:

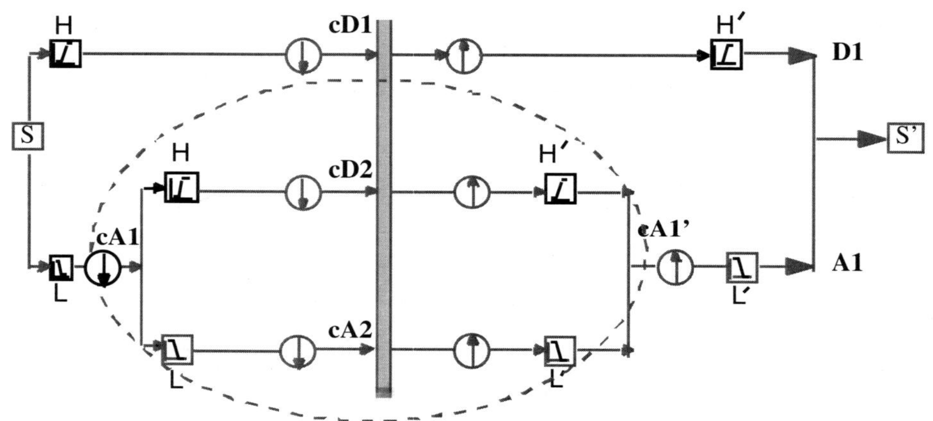

Figure 4.4–1 2-level (conventional, decimated) DWT. Notice the filters, upsampling, and downsampling within the oval are identical to the single-level DWT (Fig. 4.1–1).

```
cD2/4=dyaddown(conv(cA1,[-1 1]))/2 = [0  0]

cA2/4=dyaddown(conv(cA1,[1 1])) = [80 0]
```

Notice that **cA2/4** is the average of first four exam scores and then the average of the last 4. Also notice that **cD2** and **cA2** are only 1/4 the length of the original 8-point signal. With **cD1** being 1/2 the length of the original signal we can see that the total length of all the coefficients in the left or *decomposition* half of the DWT (**cD1**, **cD2**, and **cA2**) is the same as the original signal (4 + 2 + 2 = **8**).

We can indeed do further compression here. We have both **cD2** and **cD1** = **zeros**. Using only 2 data points (**cA2/4** = **[80 0]**) we can reconstruct **[80 80 80 80 0 0 0 0]** using the right or *reconstruction* half* of the DWT as shown in Figure 4.4–2. For example, we could store or transmit only the 2 data points of **cA2** and later assume zeros for the other coefficients. Then after the upsampling and filtering as shown within the dotted lines we can reconstruct our signal.

For our simple example here we have achieved a compression ratio of 4:1. With **cA2/4** = **[80 0]** we could even set things up so that the zero (the 2nd value) was not kept and achieve an 8:1 ratio. In practice, when coefficients are very nearly zero, we can set them to zero with minimal adverse effect as we did in the "Ogre department head" example in the last section.

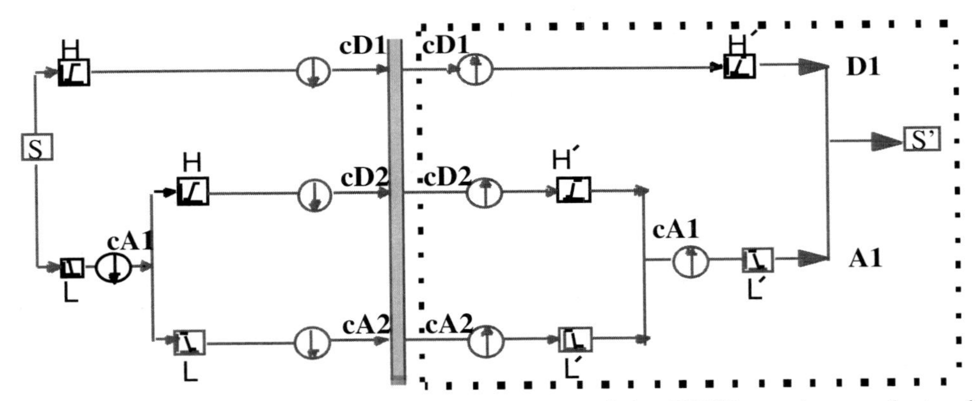

Figure 4.4–2 The right half or *reconstruction* portion of the DWT can be used at a later date (or at a different location) to rebuild the original data starting with the coefficients.

The 3-level DWT shown in Figure 4.4–3 below produces the same values for **cD1**, **cD2**, **cA1**, **cA2** but also produces **cd3/8** = **[40]**, **ca3/8** = **[40]**.

There is no better compression here (we still need 2 values) but notice that **ca3/8** = average of **[80% 80% 80% 80% 0% 0% 0% 0%]** = **40%** or a "D Minus" from the kindly professor.

---

* This right half is referred to in a few texts as an *inverse DWT* with the left half referred to as the *forward* DWT in an attempt to stay with the convention of the popular fast Fourier transform (i.e. IFFT and FFT). The author believes this is a poor analogy in that at the end of the left half of the DWT we have only *coefficients* and unlike an FFT which can use the results directly as the *frequency domain*, we have to work further with the coefficients in a DWT using the right half before we can actually use them in DSP. In other words, these coefficients are not "stand alone" values. They need the so-called "inverse DWT" to provide workable results.

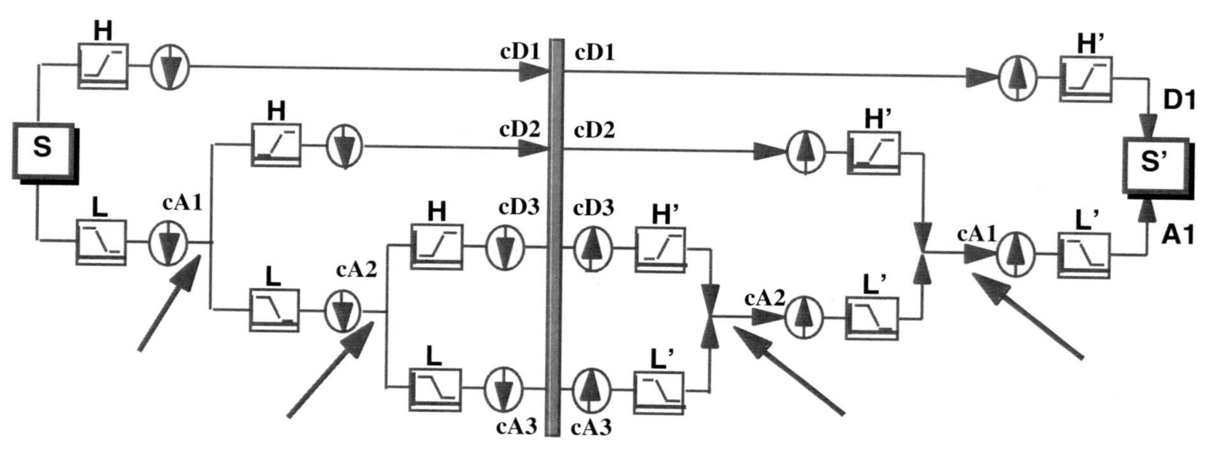

Figure 4.4–3 3-level DWT. Note same 4 filters as in the 2 and single-level DWT. The lower 2 arrows show how **cA2** is reconstructed using an embedded single-level DWT. Upper arrows then show how **cA1** is reconstructed using another embedded single-level DWT (with **cA2** already having been reconstructed). Finally, with **cA1** perfectly reconstructed, we can perfectly reconstruct **S**. Of course we may want to perform denoising or compression using the coefficients **cD1**, **cD2**, **cD3**, and **cA3**, in which case we will not, in general, have perfect reconstruction.

4.5 Frequency Allocation in a (Conventional, Decimated) DWT

The frequency allocation for DWTs is the same as for the UDWTs. For the single-level DWT (Fig. 4.1–1) We have **A1** and **D1** and the frequency allocation is shown in Figure 4.5–1 below.

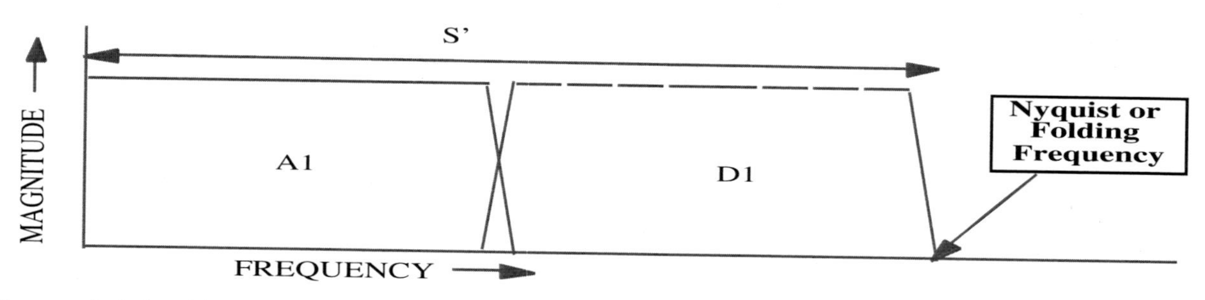

Figure 4.5–1 Frequency allocation after a single-level DWT. Diagram is illustrative only and the actual shape depends on the wavelet filters. Note overlap from non-ideal filtering.

We can re-draw the 2-level DWT as shown in Figure 4.5–2. This allows us flexibility in frequency subdivision. As with the UDWT, we can produce **A2** + **D2** = **A1**, then **A1** + **D1** = **S'**.

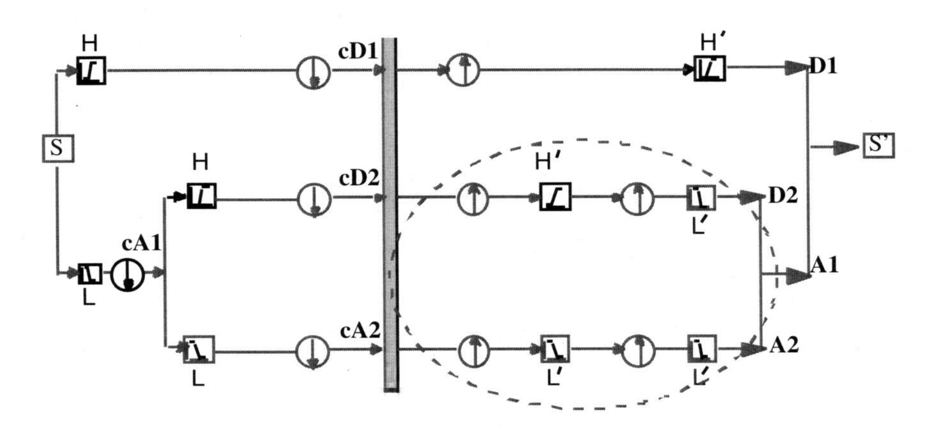

Figure 4.5–2 Equivalent 2-level DWT. Instead of combining the data after **H'** and **L'** (in the oval) and then upsampling and filtering by **L'** as in Fig. 4.4–1, we upsample and filter by **L'** on each path. This give us **D2** and **A2**.

The frequency subbands for the 2-level conventional DWT (or UDWT) are shown in Figure 4.5–3.

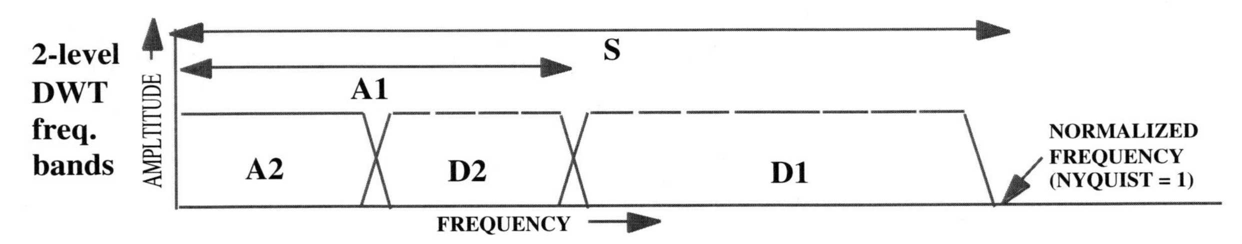

Figure 4.5–3 Frequency allocation for 2-level DWT. **S** = **A1** + **D1** or **S** = **A2** + **D2** + **D1**. This gives us more flexibility

The equivalent diagram for a 3-level DWT (Fig. 4.4–3) can be redrawn as shown here in Figure 4.5–4. The frequency allocation is shown next in Figure 4.5–5.

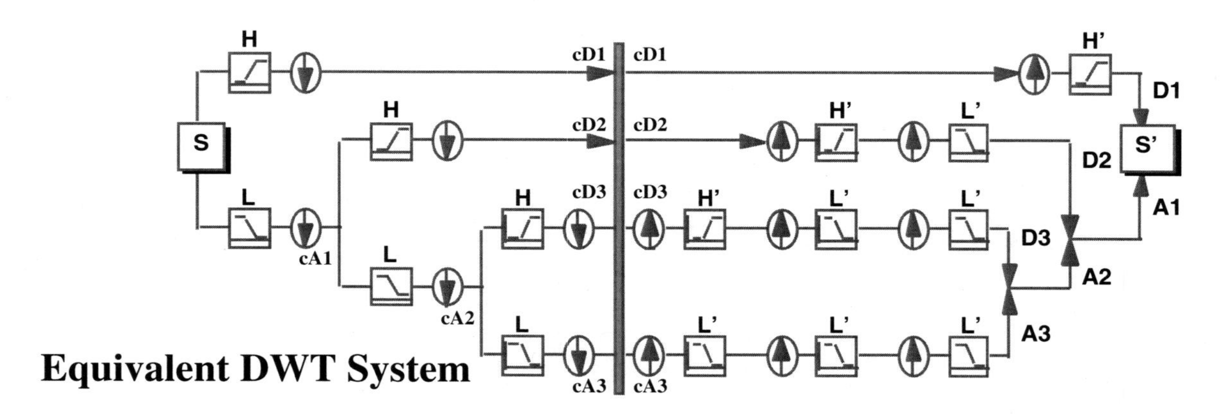

Figure 4.5–4 3-level DWT redrawn to show how the various Approximations and Details can combine. Notice again how **cD1**, **cD2**, **cD3** and **cA3** combined are roughly the size of the original signal due to downsampling.

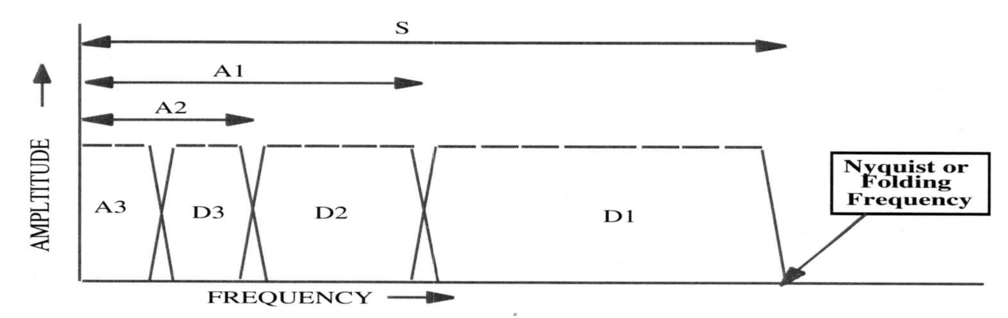

Figure 4.5–5 Frequency allocation for 3-level DWT. **S** = **A1** + **D1**, **A2** + **D2** + **D1** or **A3** + **D3** + **D2** + **D1**.

4.6 Final Approximations and Details and how to read the DWT Display

As with the UDWT, our goal is usually not perfect reconstruction of the original signal, but denoising or compression by removing some unwanted or unneeded components. We showed how to do this with the coefficients (**cD1**, **cA3**, etc.) We now look how to do this with the final Approximations and Details (**D1**, **A3**, etc.) For clarity we can look at a DWT display to see graphically these values. We will stay with our simple example of exam scores for now.

This DWT display (Figure 4.6–1) is patterned after the MATLAB Wavelet Toolbox DWT display but is simpler and better suited to learning here.

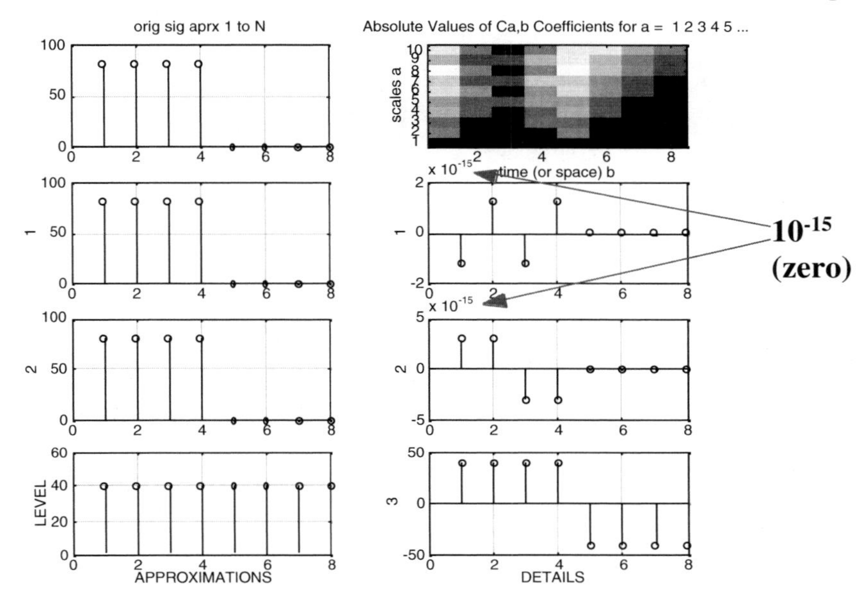

Figure 4.6–1 Display of 3-level DWT of exam scores data. Top row shows original signal and miniature CWT (compare Fig. 2.3–1). 2nd row shows **A1** and **D1**, 3rd row shows **A2** and **D2**, bottom row shows **A3** and **D3**. **Note that pre-divided filters have been used here.**

We will not walk through every path but will instead "spot check" the **D3** details (bottom right graph of Fig. 4.6–1) to see if the display agrees with our numbers. Figure 4.5–4 (next to bottom path) showed that **D3** is built by starting with the coefficients **cD3** then upsampling, filtering (convolving) with **H'**, upsampling, convolving with **L'**, upsampling, then convolving with **L'** again.

From the first part of the walk-through in Section 4.4 we found **cD3** (when divided by 8) was simply = 40. We will use this "pre-divided" result as a starting point and perform the operations depicted in the figure 4.5–4 in 3 steps:

```
step1 = conv(dyadup(40),[1 -1])= 0 40 -40 0

step2 = conv(dyadup(step1),[1  1]) =

0 0 0 40 40 -40 -40 0 0 0
```

```
step3 = conv(dyadup(step2),[1  1]) =
0 0 0 0 0 0 0 40 40 40 40 -40 -40 -40 -40 0 0 0 0 0 0 0
```

Keeping the middle 8 values we have **D3** = **[40 40 40 40 -40 -40 -40 -40]** which agrees with the display for **D3** (bottom right graph).

We can "read" the display to find things about our signal. Looking at the bottom row of Fig. 4.6–1 we see that adding **A3** (bottom left graph) to **D3** will produce **A2**.

Notice that **D1** and **D2** are both zero (to computer precision). This is significant in that it tells us we need only the (non-zero) values in **A3** and **D3** to reconstruct the signal. In other words, **S** = **A3** + **D3** + **D2** + **D1** = **A3** + **D3** + **0** + **0** = **A3** + **D3**. We can see this looks correct on the graph. Adding the values of **D3** = **[40 40 40 40 -40 -40 -40 -40]** to those of **A3** = **[40 40 40 40 40 40 40 40]** gives us **A2** = **[80 80 80 80 0 0 0 0]**. Furthermore, since **D1** and **D2** contain only zeros we have **A2** = **A1** = **S'** which is the same as the original 8 exam scores.

4.7 Denoising using a Multi-Level DWT

Now we know how to read and understand a DWT display, we look at an example of denoising using a conventional DWT.

Assume we have a binary signal that we wish to transmit. For example this could be part of a satellite downlink. We choose an arbitrary 16-bit signal

```
S = [1 0 1 1 0 0 1 0 1 1 1 0 1 0 0 1]
```

Using eight chips per bit* we would have **8 x 16 = 128** data points as shown in Figure 4.7–1. In other words the first bit (1) would be represented by **[1 1 1 1 1 1 1 1]**, the second bit (**0**) would be represented by **[0 0 0 0 0 0 0 0]** and so on.

* The number is usually larger than 8 for better processing gain, but 8 chips/bit will do for our instructional purposes.

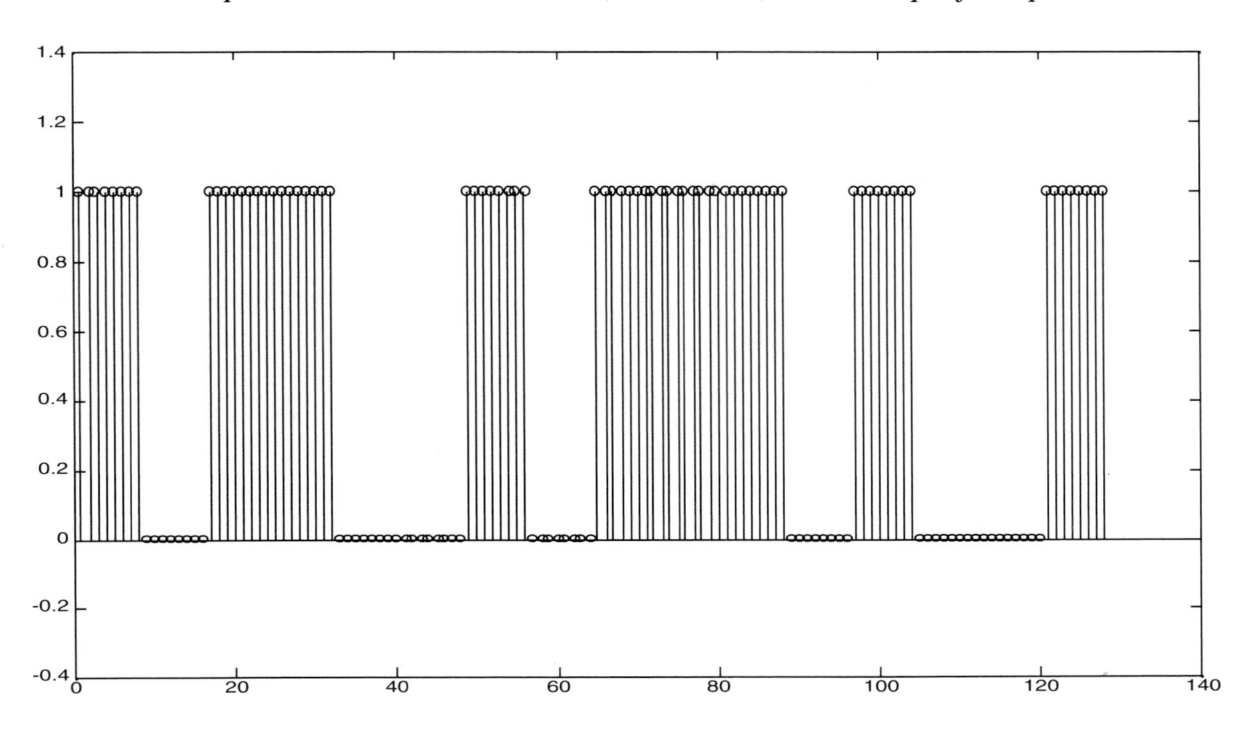

Figure 4.7–1 Representation of a 16-bit binary signal using 8 chips per bit. We use the "stem" utility for better precision rather than "connecting the dots" in a conventional plot. No noise is present at this point.

We proceed to perform a DWT on this noiseless signal. A 4-level DWT is sufficient and the DWT display is shown in Figure 4.7–2.

Notice that **D1**, **D2**, and **D3** are all zero to computer precision. Only **D4** (bottom right) has non-zero values. A look at the frequency allocation diagram for a 4-level DWT (Figure 4.7–3) tells us there are no high frequency values (**D1**, **D2**, **D3 = 0**,) for this noiseless signal.

As with our other examples, we know the answer in advance and use this to check the quality of our denoising. We pretend we don't know the answer and proceed with the denoising. We do know, however, that any similar binary pattern (16 bits with 8 chips/bit), will have zeros for **D1**, **D2**, **D3** (no high frequency components). We can exploit this fact for denoising of a binary pattern that has noise added (e.g. rain fade on the downlink portion of a satellite).

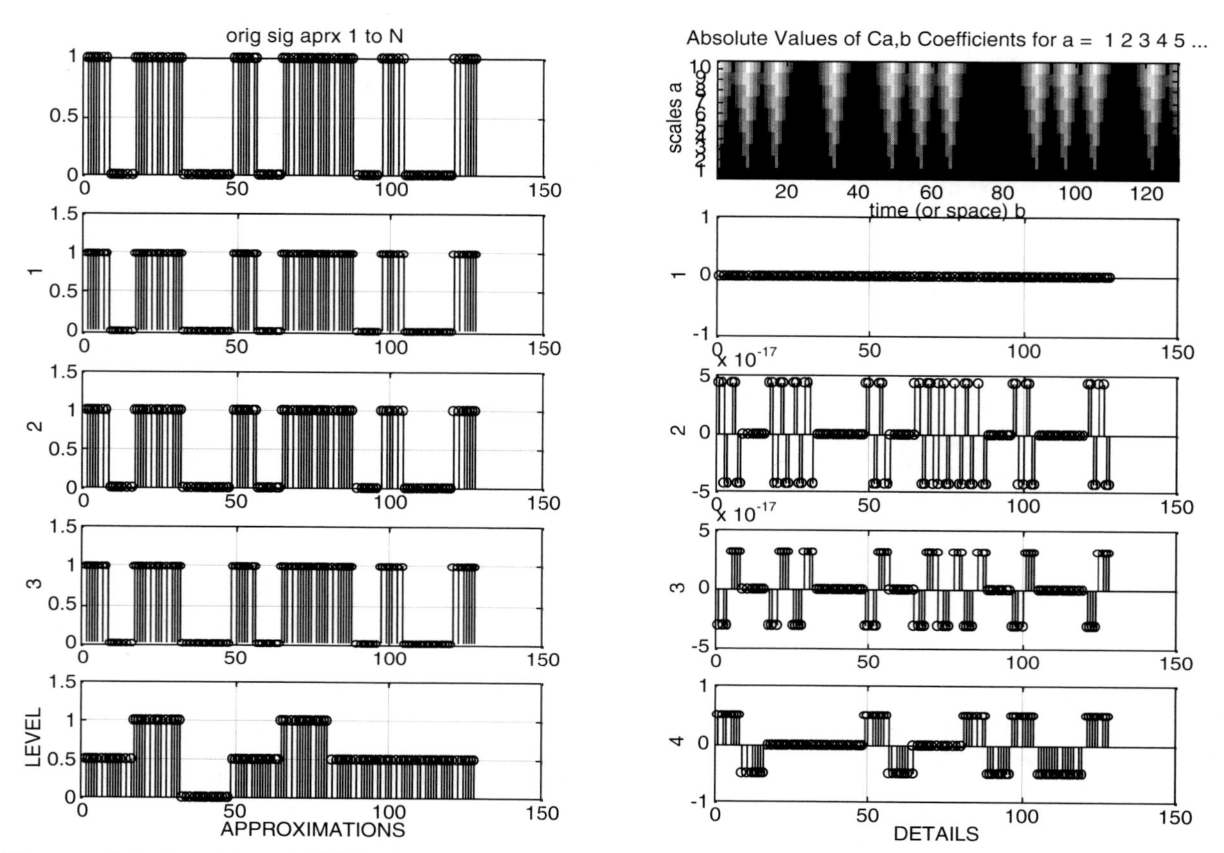

Figure 4.7–2 4-level DWT display of the noiseless signal. The top row shows the signal in miniature and then a miniature display of the CWT. The other rows show the Approximations (left) and Details (right). Note **D1** = **D2** = **D3** = zero to computer precision.

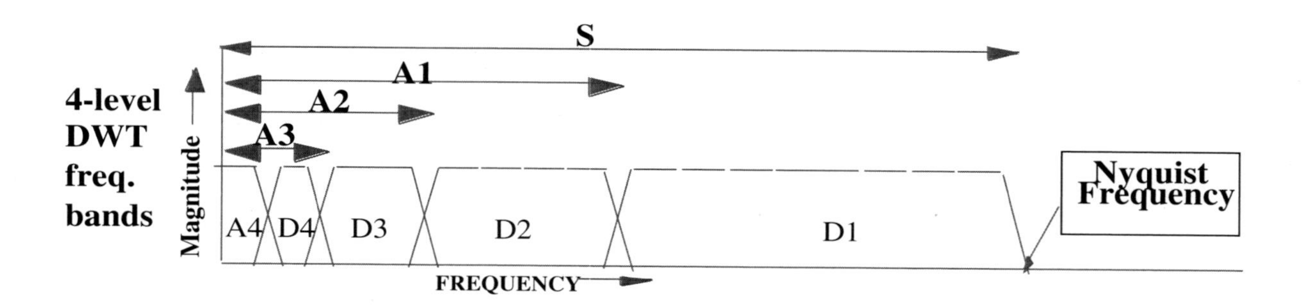

Figure 4.7–3 Frequency allocation of a 4-level DWT. Notice that if **D4** is non-zero, so will **A3**, **A2**, and **A1** be non-zero.

Now we look at same binary signal but with noise added. As shown in Figure 4.74 the noise is heavy enough to make discerning the bits impossible, or at least highly prone to error.

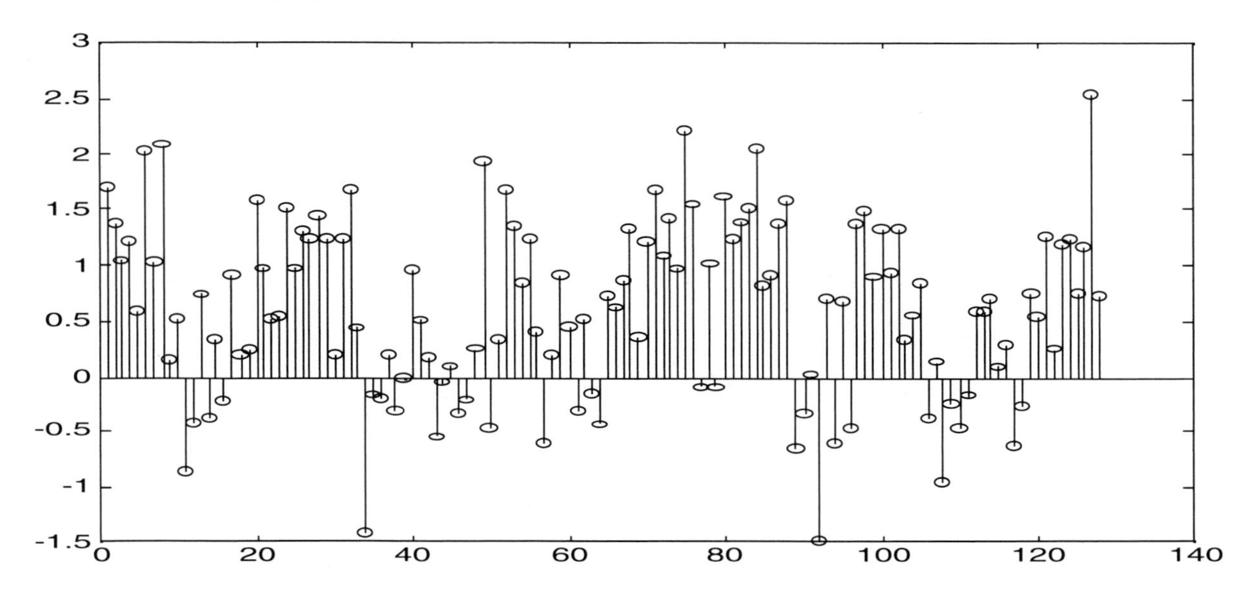

Figure 4.7–4 Original binary signal with 8 chips/bit now is difficult or impossible to decipher.

We next take the DWT of the signal, again using the Haar wavelet filters. The 4-level DWT display is shown in Figure 4.7–5.

Knowing that in a noiseless signal **D1**, **D2**, and **D3** are zero we can use this fact to do some denoising. Specifically, we discovered that in a 4-level DWT

```
S = A4 + D4 + D3 + D2 + D1 = A4 + D4
```

because **D3**, **D2**, and **D1** are zero. We also know that **A4** + **D4** = **A3** so we can use **A3** for the denoised version.

Figure 4.7–6 shows the denoised version of signal by discarding **D1**, **D2**, and **D3** (thus using **A3**). Notice we can now easily discern which bits are one and which are zero (dotted lines at 1.0 and 0.0 are drawn for comparison). We can now discern the signal 1 0 1 1 0 0 1 0 1 1 1 0 1 0 0 1 as shown. This can be done visually or with an algorithm such as comparing with 0.5 to decide if closer to 1.0 or 0.0.

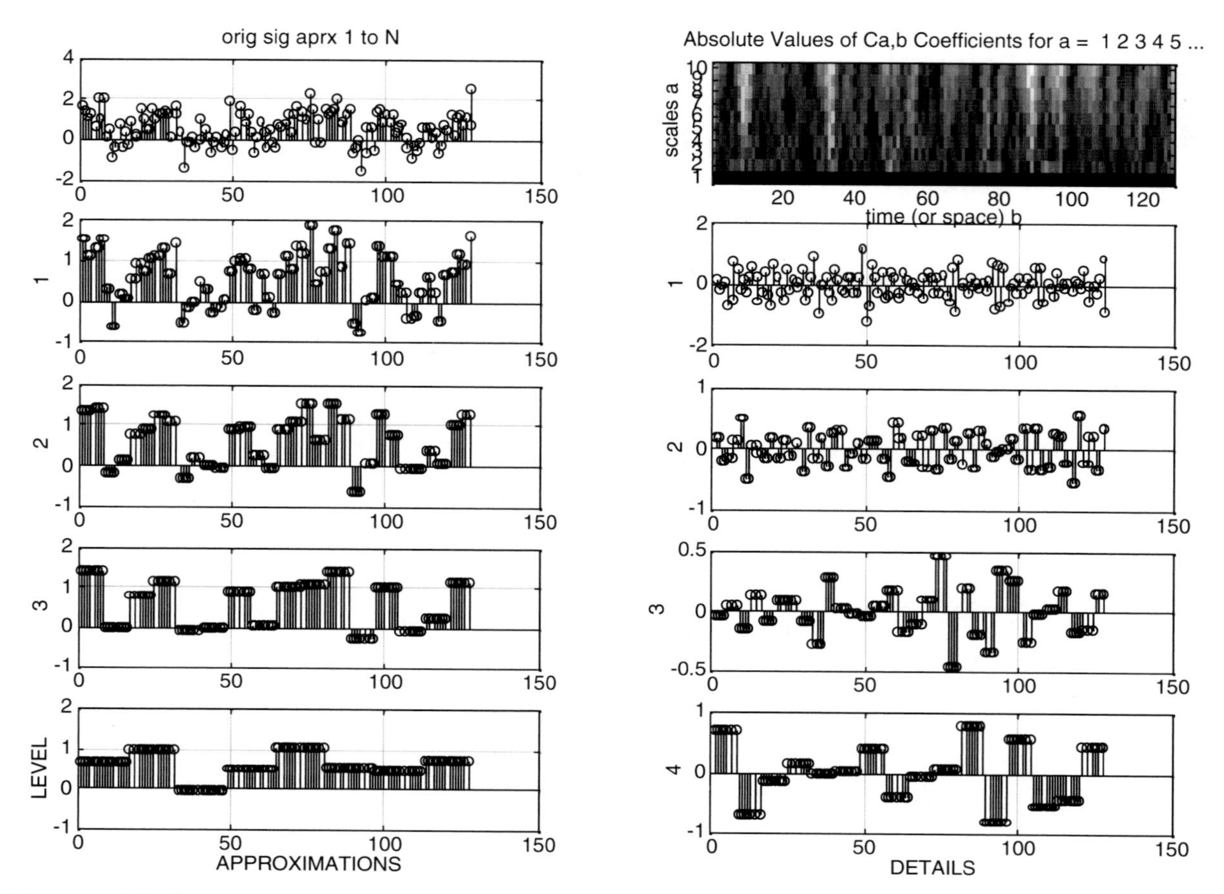

Figure 4.7–5 4-level DWT display of the noisy signal. Notice that the noisy signal now has non-zero components in all of the Details and Approximations.

From looking at **A3** in Figure 4.7–5 we can see that the signal is

```
S = [1 0 1 1 0 0 1 0 1 1 1 0 1 0 0 1]
```

which is indeed the original signal before noise was added.

We will look at a similar real-life example in more detail later. We will also explain how we can also use the CWT (miniature in upper right corner of DWT display) to find bits.

In this simple example we discarded all of **D1**, **D2** and **D3**. Wavelet transforms involve both time and frequency. We will show an example later where we discard **D1** for certain times of the signal, **D2** for other times, etc.

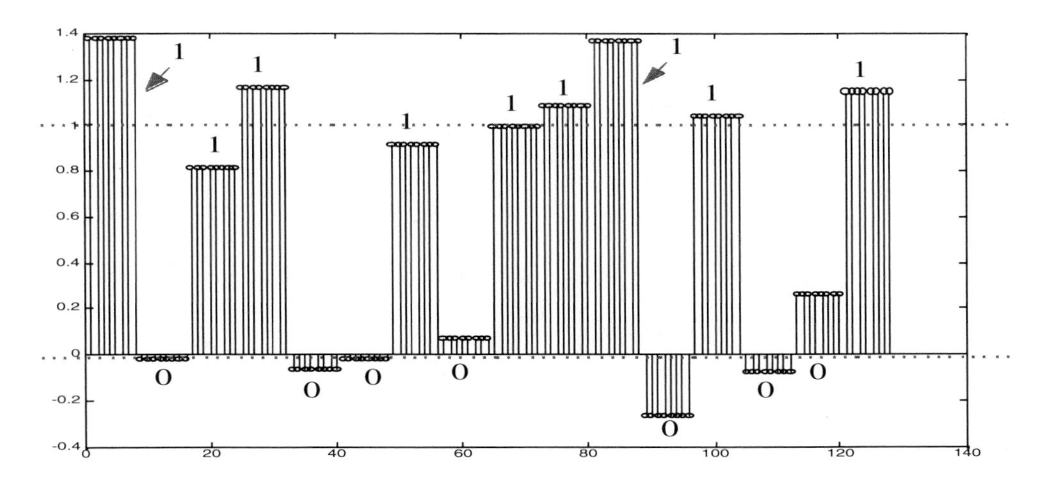

Figure 4.7–6 Denoising using the final values Details and Approximations (specifically $S_{denoised}$ = **A3** + **D3** + **D2** + **D1** = **A3** + **0** + **0** + **0** + **A3**. We can discern the original binary signal.

We can easily write software to discard or limit the Details for a given time interval. The MATLAB Wavelet Toolbox does this with a capability called "Interval Dependent Thresholding"

This is a powerful capability of these wavelet transforms not found in the Fourier transforms. In other words, we could take an FFT of the data and discard certain frequencies, but these frequencies would be discarded for the entire length of the signal.

For noise that is "stationary" (constant frequency for the entire length of the signal), the FFT works fine. However, for real world signals with "events" that start and stop in time (as in most interesting signals) the DWT will often outperform the FFT in compression and denoising.

4.8 Summary

In this chapter we examined the more complicated but better-known *conventional (decimated) discrete wavelet transform*, referred to as simply "the DWT". We saw that for the single-level DWT the signal flow diagram is the same as for the single-level UDWT of the last chapter except the it has downsampling ("decimation") by 2 and upsampling by 2. We learned that we have to deal with the potential adverse effects of aliasing and lack of shift-invariance caused by throwing away every other sample.

We performed a step-by-step walk-through of the single-level conventional DWT and demonstrated that for perfect reconstruction filters, including the simple Haar filters—the same ones as we used in the single-level UDWT—that the aliasing from the top or highpass path cancels that of the bottom or lowpass path. As with the UDWT, we can perfectly reconstruct the original signal to within a delay and a constant of multiplication.

We moved on to multi-level DWTs and could see how, through downsampling, the signal is shrunk rather than the filters stretched. Because of shrinking the signal, the DWT is able to use the same set of filters (**H**, **H'**, **L** and **L'**) throughout. In other words, there are no "A Trous" or upsampled filters (**Hup**, **H'up** etc.) in the conventional DWT and thus no "stretched" filters as in the UDWT.

Using the same series of 8 university exam scores as in previous chapters, we showed how we can achieve substantial compression using a DWT. We didn't have any aliasing problems in this benign example, but in later chapters we will show examples of pathological cases where we will have aliasing problems and how to deal with them.

We learned how to read a DWT display and mentioned that the UDWT display is set up in same way. We showed an example of de-noising a binary signal by eliminating some of the subbands (**D1**, **D2**, and **D3** in our example) and reconstructing a signal that is not perfect, but readable where the noisy signal was not.

We discussed the capability to remove or threshold a specific subband for only a specific length of time. MATLAB Wavelet Toolbox calls this "Interval Dependent Thresholding" and has software to do this. It's not difficult to write software, however, to custom-threshold these specific subbands. This ability to remove unwanted or unneeded data *at a specific time and within a specific frequency subband* is what makes DSP using wavelets so attractive.

We will show in later chapters how to avoid aliasing problems in the conventional DWT by being careful in how we denoise and compress (so as not to throw away the alias cancellation capability). We will learn when to use the UDWT, which has no such problems but requires more data storage (the UDWT has a display identical to the DWT). In addition, we will discuss a hybrid method, using both DWT and UDWT, for speed and minimal aliasing.

CHAPTER 5

Obtaining Discrete Wavelet Filters from "Crude" Wavelet Equations

In Chapter One we showed a preview of wavelets, wavelet filters, and wavelet transforms. In Chapters 2 through 4 we performed a step-by-step walk through of the 3 main wavelet transforms: the Continuous Wavelet Transform or CWT, the Undecimated Discrete Wavelet Transform or UDWT, and the conventional (decimated or downsampled) Discrete Wavelet Transform or DWT.

We showed in these chapters how these transforms are based on correlations of the data with the wavelet filters. In this chapter we will discuss one method of generating wavelet filters using equations. Note: This method generates filters that can be used in the CWT but not in the DWTs. We will show how to generate "DWT-worthy" filters in the next chapter.

5.1 Review of Familiar DSP Truncated Sinc Function

We recall from digital signal processing (DSP) that an ideal Lowpass Filter (LPF) in the frequency domain is an infinitely long Sinc function (sin(t)/t)* in the time domain.

In the real world of finite data, we must of course truncate the length. Simply discarding points at both ends of the Sinc function can be thought of as multiplying by a rectangular or "boxcar" window (We often use other windows such as Hamming, Blackman, Hanning, etc. to reduce Gibb's effect ripples). Thus one way to add extra points to a truncated Sinc function is the familiar method from DSP of extending the rectangular "window" that truncates the infinitely long Sinc function. We first look at a Sinc function that is truncated (windowed) to 81-points as shown in Figure 5.1–1:

---

* This is the classical expression. With computers we use sin(πt)/(πt). When t is zero we have sin(0)/(0) so we use L'Hospitals rule (derivatives of numerator and denominator) and we have πcos(0)/π = π/π = 1. Also, results are often normalized so at t=0 the peak value may be 1, .1/sqrt(2), 1/2 or some other value but the ratio of the sidelobes to the main lobe is the same.

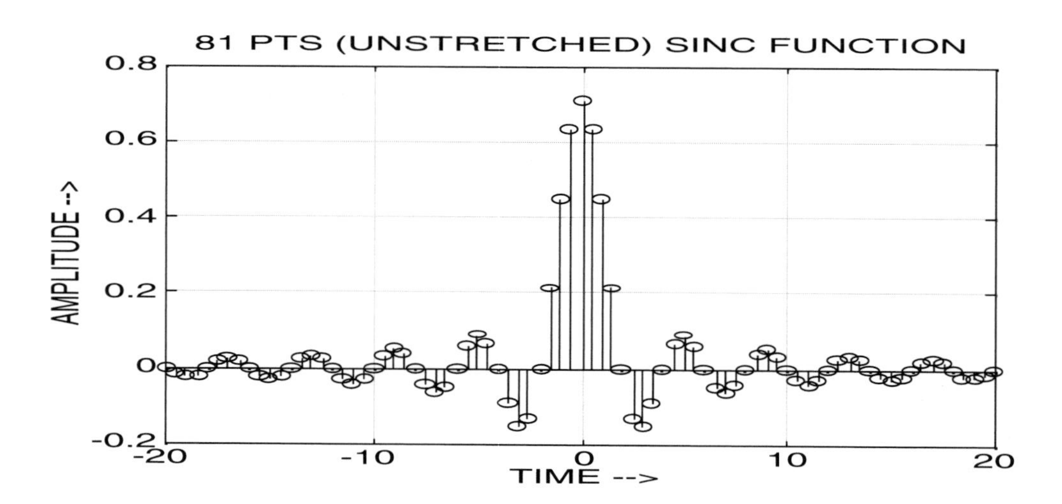

Figure 5.1–1 81 point Sinc Function. Note the abrupt truncation at -20 and +20. This is referred to in DSP as a "boxcar" window. Points are equispaced at 1/2 integers from –20 to +20 (including the peak at time = zero).

The Frequency Response is shown below in Figure 5.1–2. A Discrete Fourier Transform (also the functionally equivalent but faster FFT) is discrete and periodic in both the time and frequency domains. We show here the frequency response from 0 to twice the folding or Nyquist frequency (or equivalently 0 to 2π on the unit circle). Note the ripples (Gibbs effect) and that the cutoff frequency (0.25 Nyquist) is not precise.

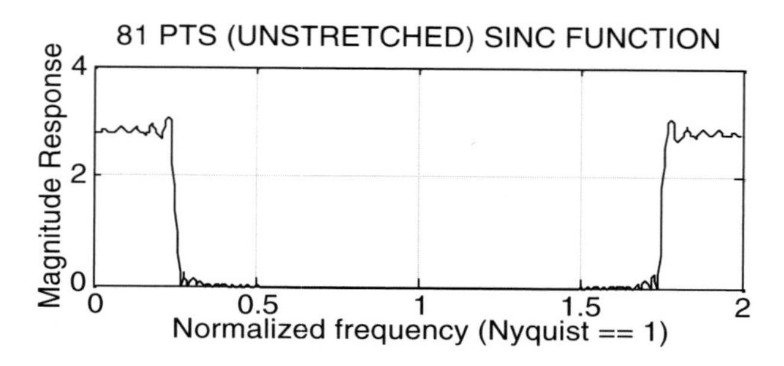

Figure 5.1–2 Frequency response of the 81 point Sinc Function truncated to 81 points.

5.2 Adding More Points at the Ends for Better Filter Performance

Continuing with our review of conventional DSP, we look at what happens when we "open the window" by making the truncation of the infinitely long Sinc function less severe. We add 40 points on each end extending the window from 81 points to 161 points as shown in Figure 5.2–1.

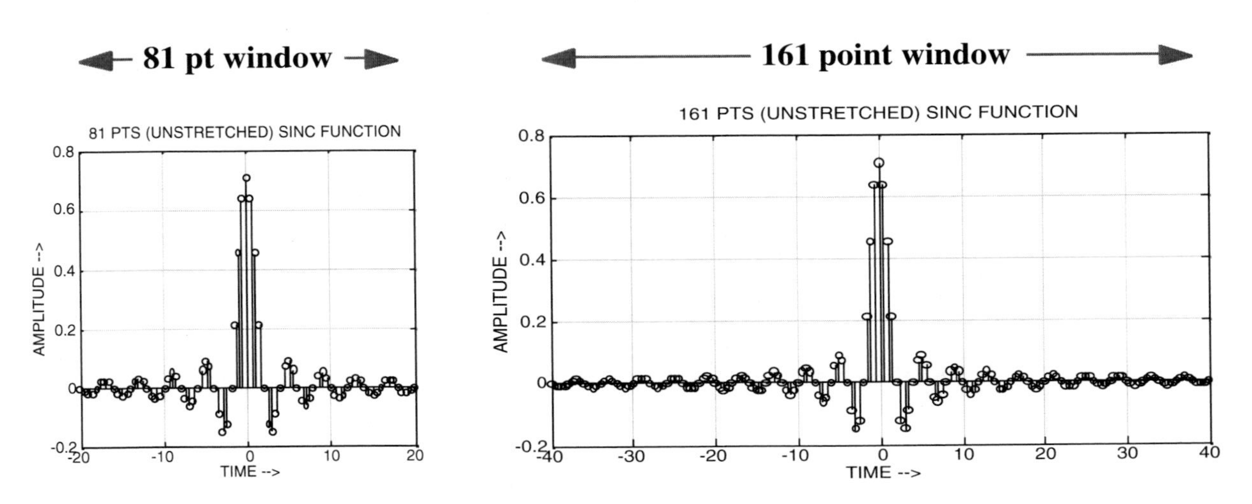

Figure 5.2–1 "Boxcar" window is opened up from 81 points to 161 points. Points are still 1/2 integer apart but now begin at –40 and end at +40. Note that the number of points in the main lobe (9 pts. from –2 to +2) remains the same for the larger window.

The Frequency Response for the less-severely truncated Sinc function is shown in Figure 5.2–2. Note we still have the Gibbs effect—we have not changed the form of the window, only its size—but the frequency cutoff (*still at 0.25 Nyquist*) is now more precise (i.e. the *transition band* is shorter).

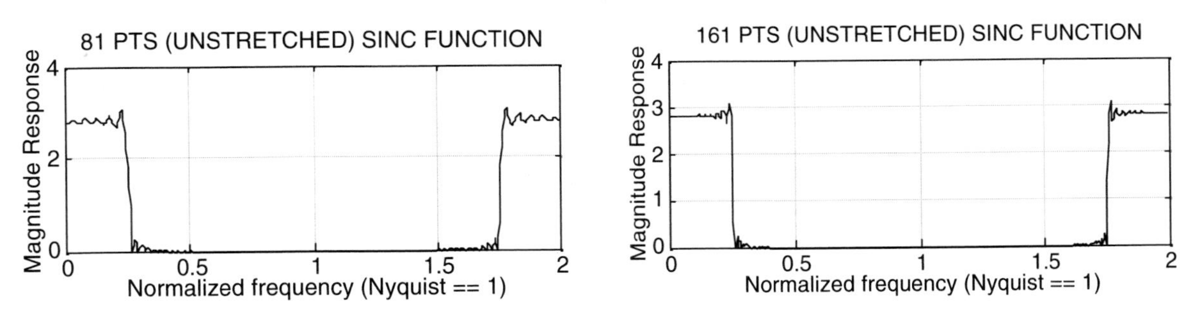

Figure 5.2–2 Comparison of the Frequency response of the Sinc function truncated to 81 points and then to 161 points Note that the bigger window allows for sharper cutoff frequency (still at 0.25 Nyquist) as seen at right.

5.3 Adding More Points by Interpolation for Lower Cutoff Frequency

We just reviewed the conventional DSP method of adding more points at the ends of the Sinc function by "opening the (boxcar) window". In *wavelet* processing we add more points but do so by stretching ("dilating" in wavelet terminology) the Sinc function.

Figure 5.3–1 shows the original 81 point Sinc function stretched to 161 points by *interpolating* a point between each of the 80 original points. Note that the number of points in the main lobe has increased from 9 points (1/2 integers from –2 to +2) to 17 points (1/2 integers from –4 to +4). Note also that there are the same number of lobes, but with more points per lobe. In other words, the points are still equispaced and still 1/2 integer apart but now serve to *stretch* the Sinc function rather than *extend* the boxcar window as is often done in conventional DSP.

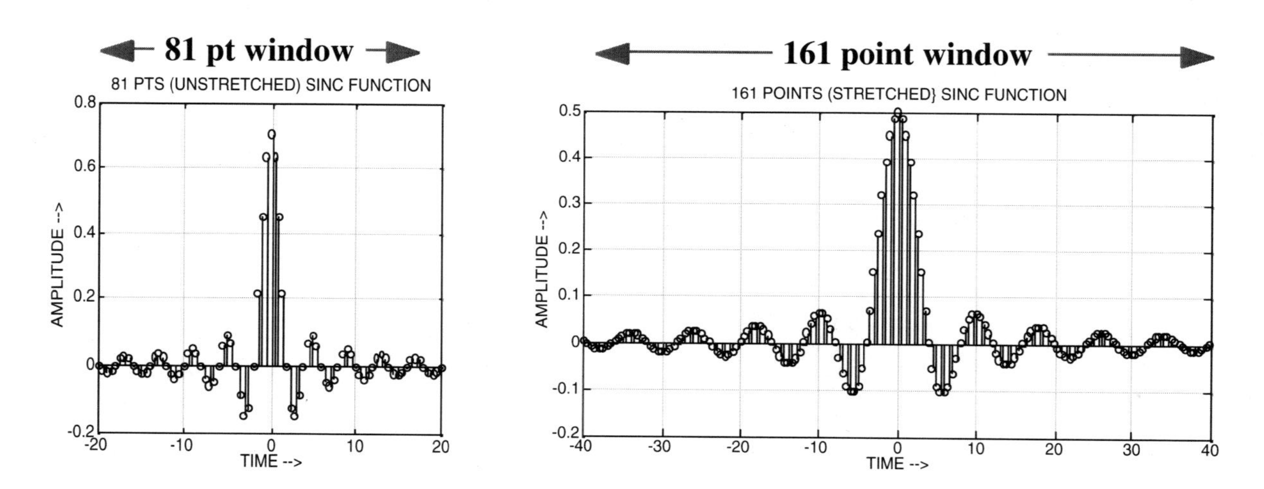

Figure 5.3–1 By stretching the truncated Sinc function (rather than making the "boxcar" truncation less severe) we have the same waveform but with more points in each lobe.

We now look at the Frequency Response of this stretched waveform. Figure 5.3–2 shows the cutoff frequency has changed from 0.25 Nyquist to 0.125 Nyquist. Note also that the height of the passband has increased.

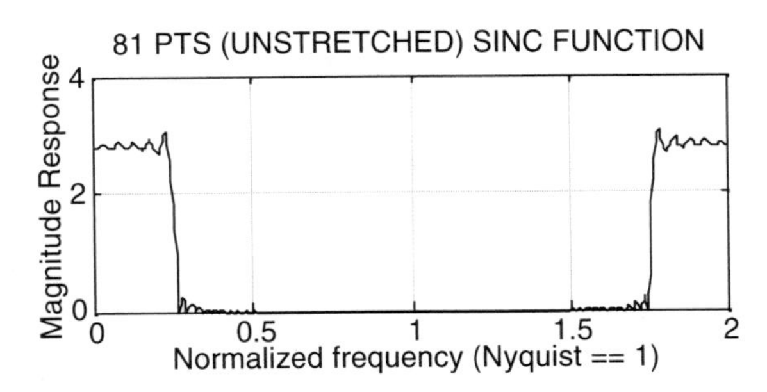

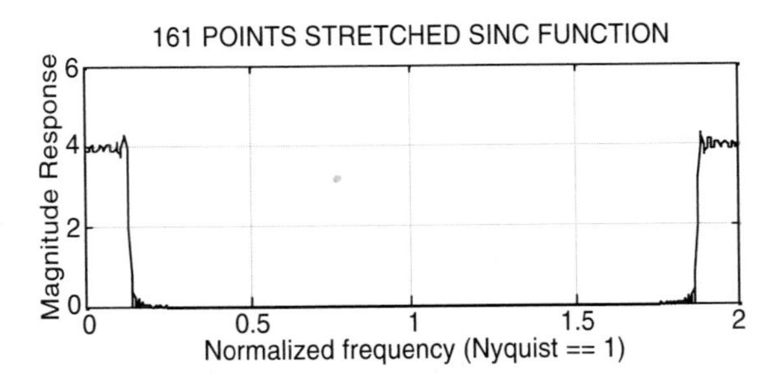

Figure 5.3–2 Comparison of the frequency response of the original 81-point Sinc function and a stretched ("dilated") 161 point Sinc function. Note the cutoff frequency has changed from 0.25 Nyquist to 0.125 Nyquist. In other words, stretching the Sinc function by 2 divides the cut-off frequency by 2.

5.4 Multi-Point Stretched Filters ("Crude Wavelets") from Explicit Equations

Using wavelets in the real world of digital computers requires us to obtain finite, discrete *wavelet filters* for correlation with our finite, discrete data. These filters start out as a few points and then are stretched by adding more points by interpolation as we just did with the familiar Sinc function. This was very simple for the Sinc function because it had an explicit mathematical equation: **Sinc(t) = sin(t)/t**. (Not all wavelets have an explicit equation and in the next chapter we will discuss those).

As introduced in Chapter One, this type of equation-generated wavelet is referred to in wavelet literature as a *crude* wavelet. As we just demonstrated, we can stretch the Sinc function by interpolating (placing extra points between the existing ones). This type of interpolation is very simple—we input desired values of t into the equation and we obtain the $sin(t)/t$ value directly.

The Sinc function (with some modifications) can be a wavelet. Obtaining Sinc values at any desired time is no problem. However, we saw one limitation in that it is theoretically infinite in length and therefore must somehow be truncated. In other words it does not "die out" in time without windowing. We will return to the Sinc function as a possible wavelet later, but for now we will look at some Crude Wavelets which, although theoretically infinite in length (or time), soon die out (go toward zero as time increases) and thus become useable to produce finite wavelet filters.

5.5 Mexican Hat Wavelet Filter as an Example of a Stretched Crude Filter

Like the Sinc function, the Mexican Hat wavelet (Fig. 5.5–1) is defined by an explicit mathematical equation:

```
mexh(t) = 2/(sqrt(3)*pi^0.25) x exp(-t^2/2) x (1-t^2)
```

Although theoretically infinite in length, values outside the "effective length" of –5 to +5 are essentially zero. For example, the value at 5.1 = 3.6939e–06.

Jargon Alert: *Effective Length* is sometimes referred to as *"Effective Support"*

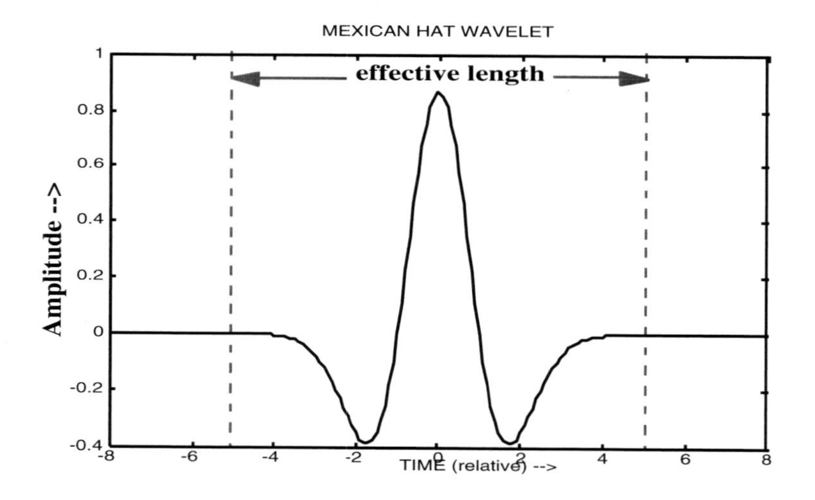

Figure 5.5–1 Mexican Hat Wavelet. Note that outside the range of –5 to +5 that the amplitude appears to be zero.

The first thing we notice about this waveform (other than its similarity to a Mexican Sombrero) is that it looks a little like a severely windowed Sinc function. If we look at the above equation we see three terms. The first term is just a constant, the 2nd term is an exponential decay, and the 3rd term is an inverted parabola offset by +1. Figure 5.5–2 shows these last 2 terms.

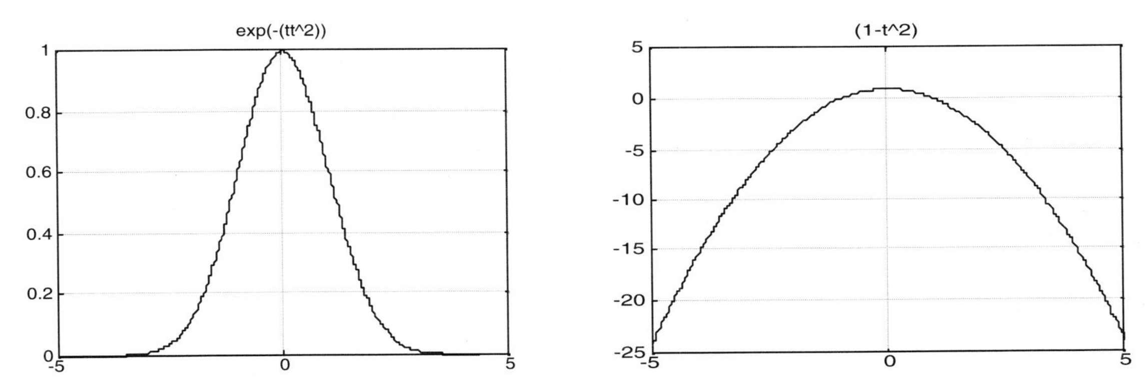

Figure 5.5–2 The non-constant components of the Mexican Hat wavelet. Notice that the rapidly decaying exponential at left allows only the middle part of the term at right to remain.

Even with explicit, continuous, theoretically infinite mathematical equations for these crude wavelets, we still must produce discrete, finite filters for convolving with our signal in the time domain. We have learned that the CWT begins with a short filter and then stretches it. For the Mexican Hat or *mexh** we will follow the example in the MATLAB Wavelet Toolbox Software and start with 17 points at the integer values (Fig. 5.5–3).

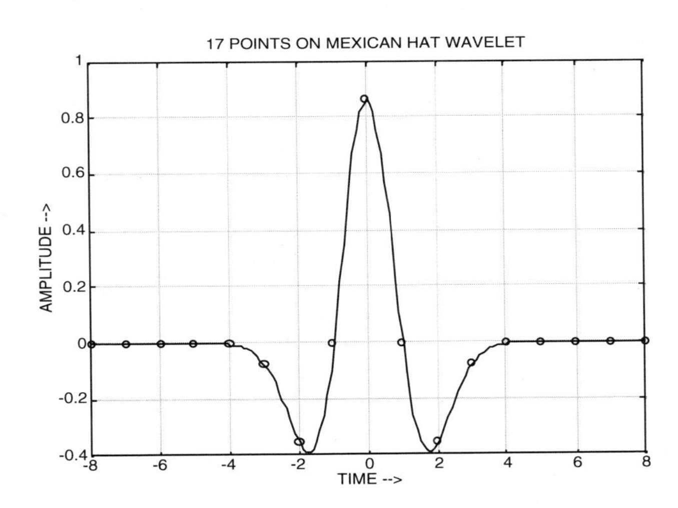

Figure 5.5–3 Mexican Hat Wavelet with 17 equispaced points on the interval from –8 to +8. The points were derived by evaluating the explicit equation at integer points (including zero).

* MATLAB refers to this filter as *mexh* There is other wavelet software available but the Mathworks MATLAB Wavelet Toolbox is the most familiar

After comparing this 17-point "filter" (**scale** = **a** = **1**) with the signal, the MATLAB CWT software "stretches" it to 33 points corresponding to values of *mexh* at the 1/2 integer points from −8 to +8 (−8, −7.5, −7 . . . +7.5, +8) as shown at the left of Figure 5.5–4. This is **scale** = **2**. The next stretching (**scale** = **3**) is the 49 points corresponding to 1/3 integer values in the same interval as shown at the right of Figure 5.5–4.

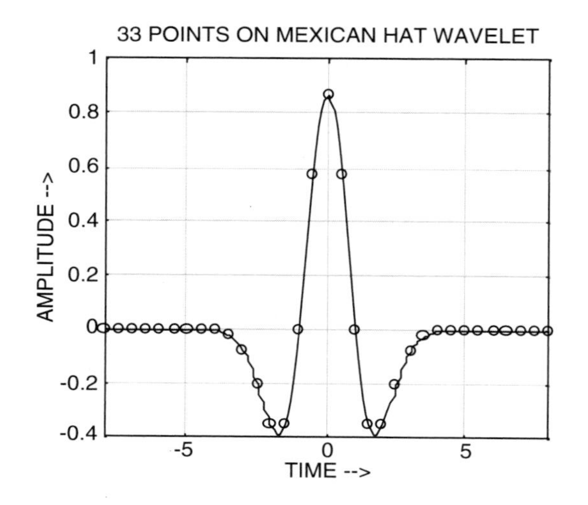

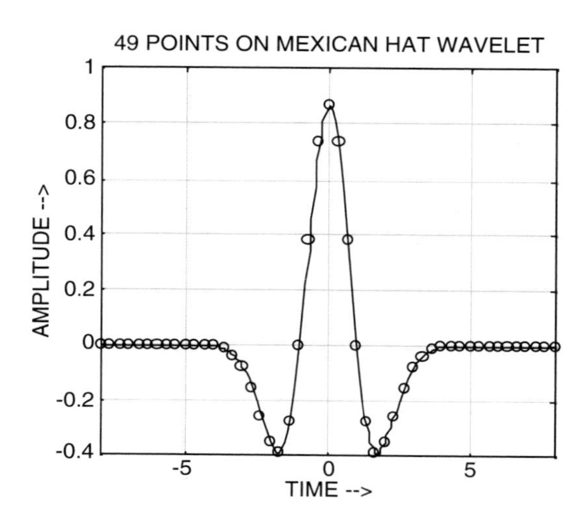

Figure 5.5–4 Mexican Hat Wavelet with 33 equispaced points (1/2 integer) and then 49 points (1/3 integer) on the interval from −8 to +8.

We continue to stretch the wavelet by generating 65 points at 1/4 integer intervals as shown in Figure 5.5–5. This corresponds to **scale** = **a** = **4** in our CWT processing. A reminder is in order that even though there may exist an explicit mathematical equation as we have here, in the real world of digital signal processing with discrete data we never actually use a continuous wavelet. We can see from the above figures that as we "dilate" (stretch) the wavelet further and obtain more points that these points create an estimation (approximation) of the so-called "continuous wavelet". We will see more of this technique in the next chapter.

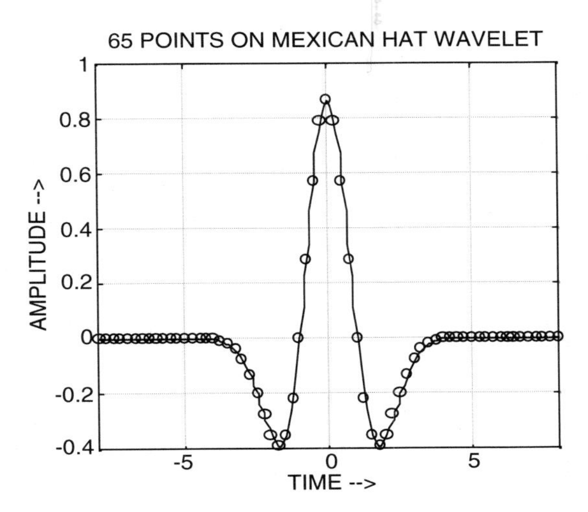

Figure 5.5–5 Mexican Hat Wavelet with 65 equispaced points (1/4 integer apart) superimposed on the explicit "point generator" equation

Note that the wavelet is becoming longer (more points) even though the shape is the same. As the stretching continues, the correlation of these "crude filters" with the signal in a CWT will produce a result longer than the original signal. You may recall from DSP that

Total Length = Signal Length + Filter Length -1

or

Total Length = L+M-1

The result is usually truncated to the signal length for the Continuous Wavelet Transform (CWT).

To show the Mexican Hat Wavelet "in action" we generate a test signal. Figure 5.5–6 shows a 1024 point real split sine signal* in the time and frequency domains. We will use this test signal to evaluate the performance of some wavelet filters. The first 512 points of the signal are at 0.125 Nyquist and the second half is at 0.25 Nyquist as seen in the FFT of the data at right.

A closeup of the signal is shown at the left of Figure 5.5–7. We can see at left 16 points per cycle in the first half then changing to 8 points per cycle. At the right we show again the Mexican Hat 33 point wavelet filter. We can see that

* We use a Cosine as our sinusoidal signal to obtain an FFT with real values.

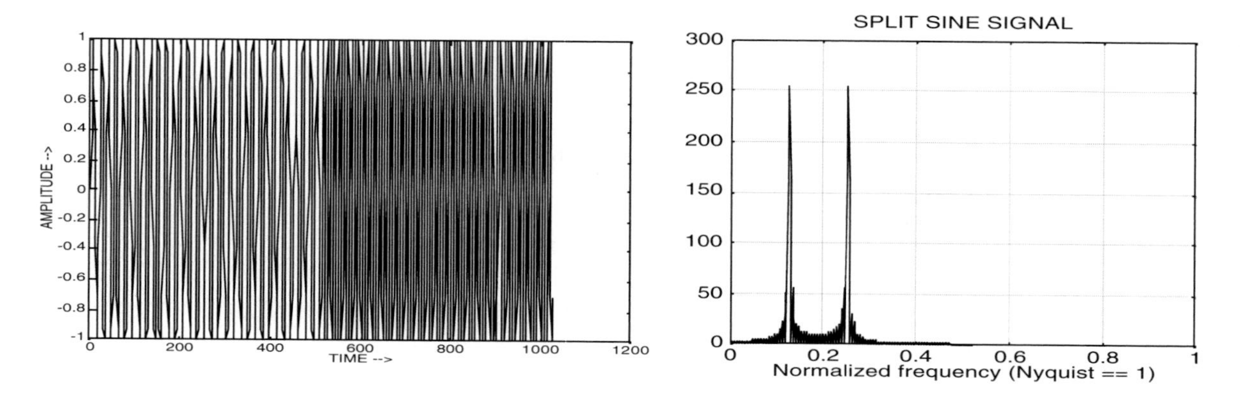

Figure 5.5-6 1024 point "split sine" test signal. The first 512 points are at 1/8 Nyquist and the last 512 points are at 1/4 Nyquist. The FFT of this signal is shown at right.

the filter has a "cycle" from negative to positive and back to negative again and in that cycle there are about 8 points. We can thus predict that when the filter has about 8 points in its cycle (scale = 2 for the *mexh*) that we should obtain an optimal correlation. We will see shortly that this is the case.

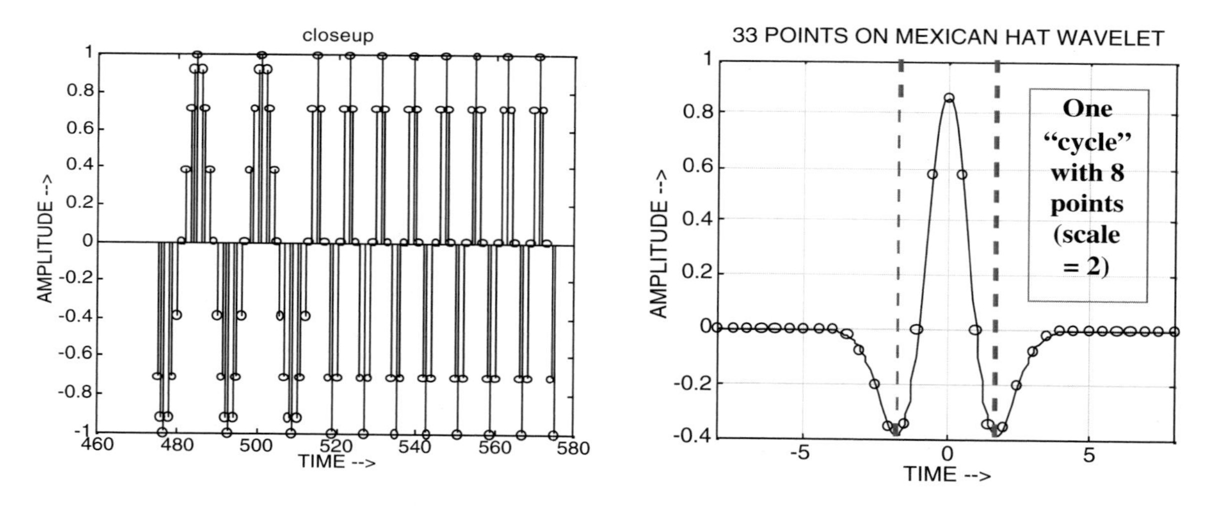

Figure 5.5-7 Closeup of the spilt sine test signal in the area where the frequency changes from 1/8 to 1/4 Nyquist. Note that in the signal there are 16 points per cycle in the left half and 8 points in the right half. The Mexican Hat 33-point filter reproduced at right has roughly 8 points in it's "cycle" (between 7 and 9 points depending on where you start and stop the cycle).

Stretched to 65 points at **scale** = **4**, the mexh wavelet has about 16 points per "cycle" as shown in figure 5.5–8 and correlates fairly well the left half of the split sine signal (16 pts/cycle in the left half).shown again here for reference.

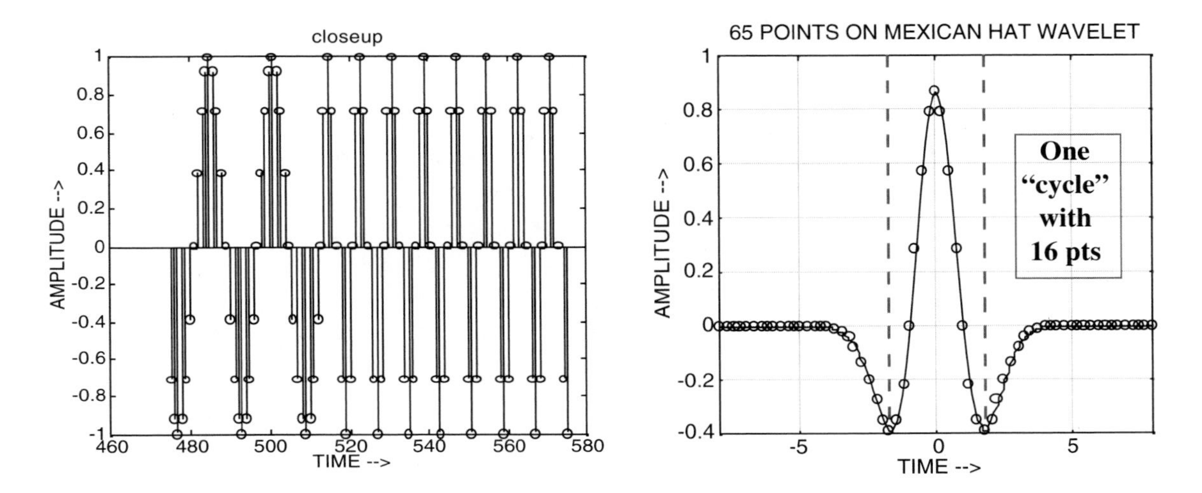

Figure 5.5–8 The Mexican Hat 65-point filter reproduced at right now has roughly 16 points in it's "cycle" and should correlate well with the left half of the split sine test signal redrawn at left.

The CWT of the spilt sine test signal using the Mexican Hat wavelet is shown below in figure 5.5–9. This shows the excellent correlation of the right half (higher frequency) at **scale** = **2** and the left half (lower frequency) at **scale** = **4**.

Looking at **scale** = **2** (the lower dotted line) we notice that although the right half shows much better correlation of the wavelet filter with the test signal (at this particular stretching) as indicated by the brightness, that there is still some correlation with the left half as indicated by a darker shade of gray, but not black. This has to do with the bandpass characteristics of this filter and the fact that the shape of the Mexican Hat filter (at any stretching) does not match that of our sinusoid test signal very well.

In the next 2 sections we will look at a filter that is a better match for sinusoids and then look at the bandpass characteristics of both of these filters in detail.

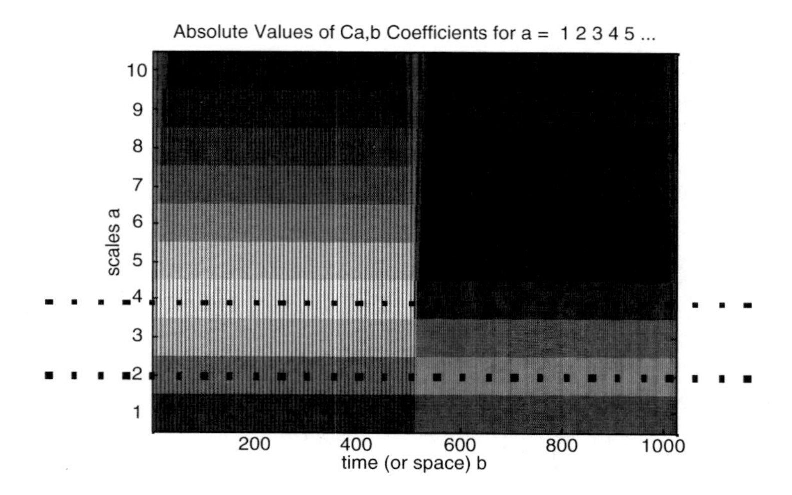

Figure 5.5–9 Continuous *mexh* Wavelet Transform of the spilt sine signal. For the higher frequency part, the best correlation is found when the Mexican Hat wavelet filter is at 33 total points (**scale** = 2) as shown by the lower horizontal dotted line. For the lower frequency portion (the first 512 points) the best correlation with the 65 point mexh is at **scale** = **4** (upper dotted line).

5.6 Morlet Wavelet as another example of Stretched Crude Filters

Let us look at one more example of Crude Wavelet Filters generated from an explicit mathematical equation. Consider the real Morlet wavelet or *morl* (there is a complex version we will discuss later). Like the *mexh* it is simply derived from a mathematical equation.

morl(t) = exp(-t^2/2) x cos(5t)

Figure 5.6–1 shows the "continuous" wavelet produced by this equation. We notice that this simple equation is similar to the equation for the *mexh* wavelet function we saw in the last section. It is simply a cosine modulated by an exponential.*

* Some authors refer to the Morlet as the "original wavelet". Even though Alfred Haar's wavelet (Haar) predates the Morlet by decades, this wavelet formulated by Jean Morlet et. al. in the 1980's helped lay the foundation of this powerful method of time-frequency analysis. Also the Haar is defined, but does not come from an explicit mathematical equation.

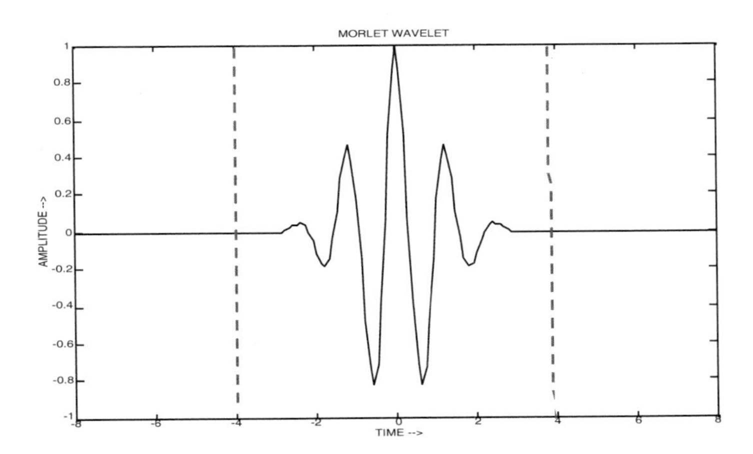

Figure 5.6–1 Representation of the Morlet Wavelet. Note that because of the exponential term attenuating the cosine term that values outside the range –4 to +4 are essentially zero (effective support = –4 to +4).

Figure 5.6–2 shows the 2 components. The *effective length* (or "*effective support*") is from –4 to +4. Outside this range, the values become zero to computer precision. For example at –4.1, the value is –1.7802e-05.

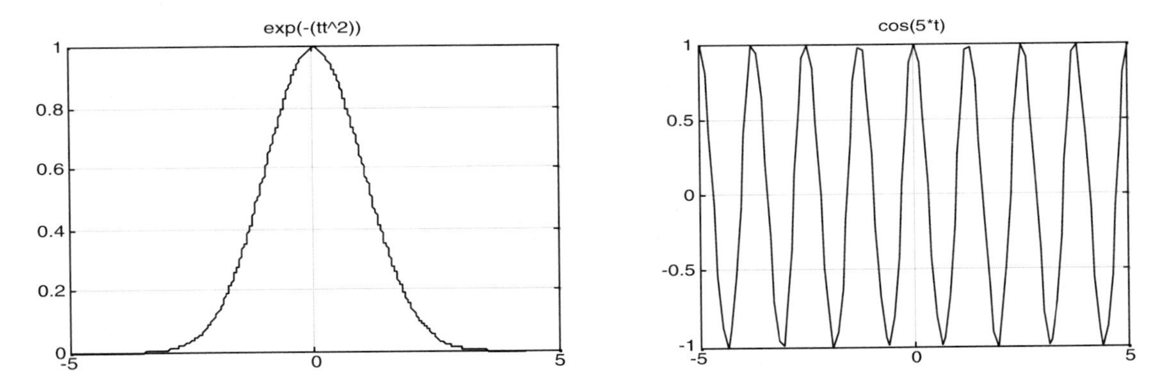

Figure 5.6–2 The components of the Morlet Wavelet. Notice that the rapidly decaying exponential at left rapidly attenuates the cosine function shown at right.

The MATLAB software to generate the Morlet or *morl* wavelet filter begins exactly the same as with the Mexican Hat wavelet with 17 points placed at the integers from –8 to +8 as shown in Figure 5.6–3 for **scale** = **a** = **1**. Although this allows for consistency with the *mexh*, we can see that these points poorly define the shape. 33 points at 1/2 integer spacing (**scale** = 2) at the right of the figure do a slightly better job of defining the shape.

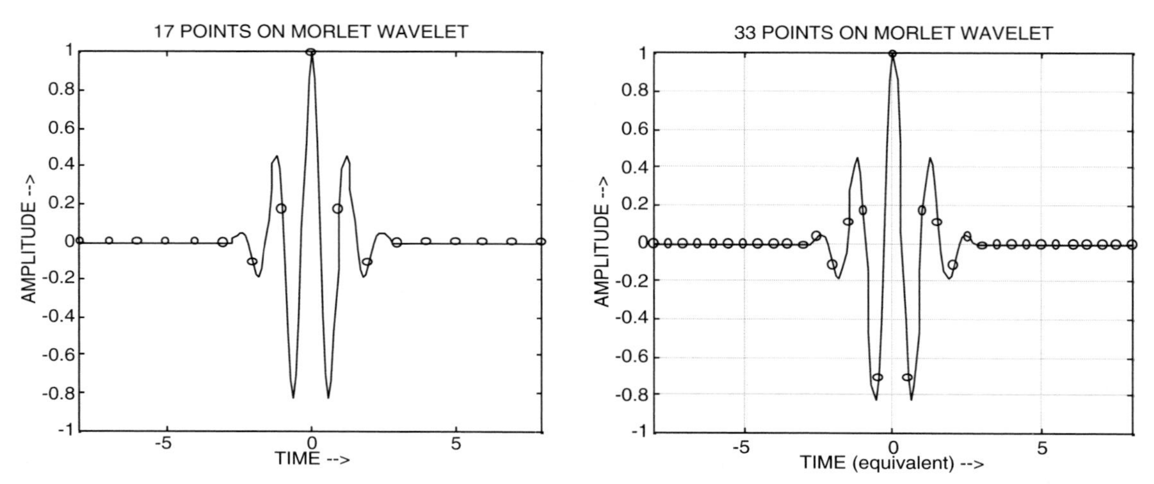

Figure 5.6–3 Morlet Wavelet with 17 equispaced points (integer apart) superimposed on the explicit "point generator" equation. Stretched to 33 points (1/2 integer apart, at right) the filter better fits the shape of the wavelet (but we still only have about 3 points per cycle).

Placing points at 1/3, 1/4, 1/5 and 1/6 integers (**scale** = **a** = **3**, **4**, **5**, and **6**) produces further stretching. Of particular interest are 97 points at **scale** = **6** and 193 points at **scale** = **a** = **12** as shown in Figure 5.6–4.

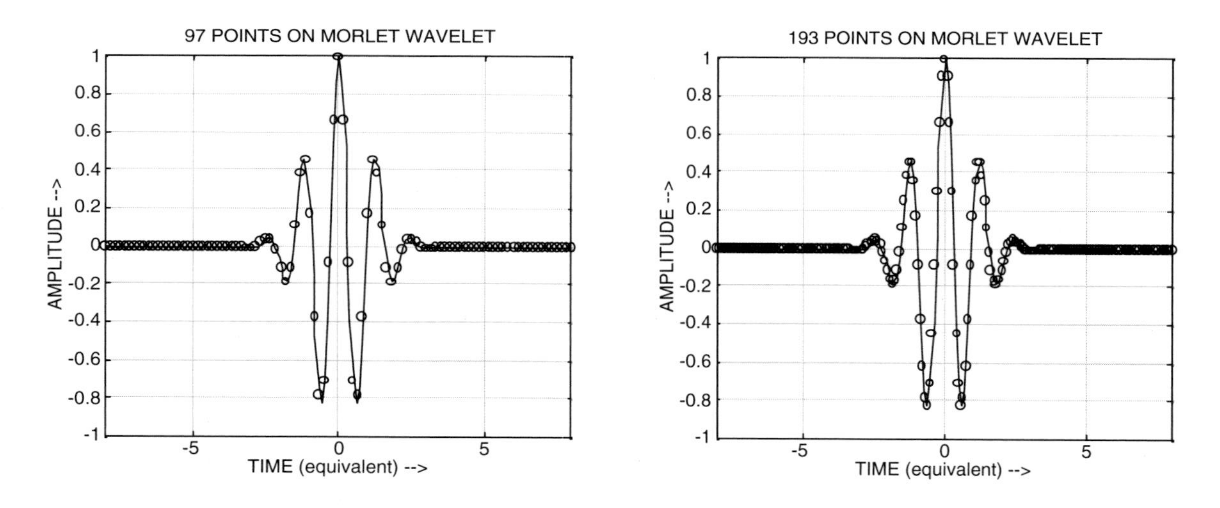

Figure 5.6–4 Morlet Wavelet with 97 equispaced points (1/6 integer apart) superimposed on the explicit "point generator" equation. Note there are about 8 points per cycle at left. At right is shown 193 points at 1/12 integer apart (**scale** = **12**). Not there are now about 16 points in a cycle.

This Morlet wavelet looks more like our split sine (cosine) test signal than the earlier Mexican Hat representation. We could thus expect an even better correlation when the number of points in a wavelet cycle matches those in a cycles of the test signal. We just saw (Fig. 5.6–4) that at **a** = **6** we have roughly 8 points in a wavelet cycle and that at **a** = **12** we have about 16 points. As expected, we see in Figure 5.6–5 a bright band in the right half at **a** = **6** and another bright band in the left half at **a** = **12**.

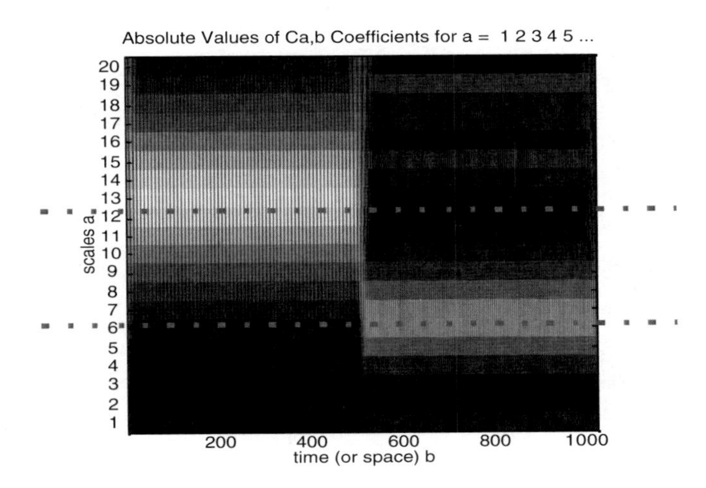

Figure 5.6–5 Continuous *morl* wavelet transform display of the spilt sine signal. For the higher frequency part, the best correlation is found when the Morlet wavelet filter is at 97 total points (**scale** = **6**) as shown by the lower horizontal dotted line. This allows roughly 8 points in each "cycle" of the Morlet wavelet. For the higher frequency portion (the first 512 points) the best correlation with the 193 point *morl* is at **scale** = **12** (upper dotted line).

In comparing the CWT display for the Mexican Hat filters (Fig. 5.5–9) to that of the Morlet Wavelet filters (Fig. 5.6–5) we also notice that the Morlet seems to do a better job of discrimination of the 2 parts of the signal. This is also because the Morlet wavelet filters "match" the sinusoidal test signal better than the Mexican Hat filters.

Figure 5.6–6 below shows the correlation of the Mexican Hat wavelet with the split sine test signal at levels 2 and 4. Note: This is the same data as can be seen in the 3-D (Time, Scale, Magnitude) CWT plots above but plotted in a more conventional 2-D plot (Time and Magnitude for a given scale).

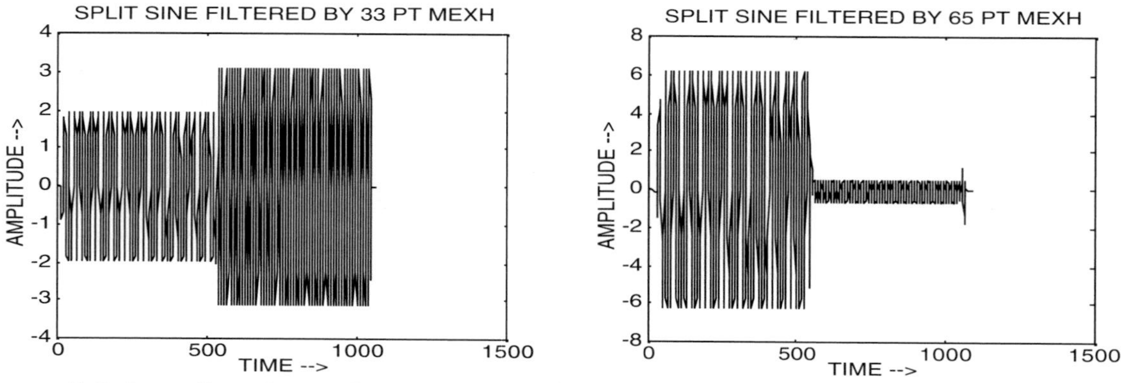

Figure 5.6–6 Correlation Strength of the Mexican Hat Wavelet Transform of the split sine test signal for **scale** = **a** = **2** (left) and **scale** = **a** = **4** (right). We see the same bandpass filtering as in the functionally equivalent CWT display (see Fig. 5.5–9). We notice, however, that the frequency discrimination could be improved, especially for the **scale** = **2** graph at left.

The Morlet wavelet filters show better bandpass characteristics for our test signal than the Mexican Hat. We saw this in the CWT display (Fig. 5.6–4) and can verify this in the conventional 2-D plots of Time vs. Amplitude shown below in Figure 5.6–7.

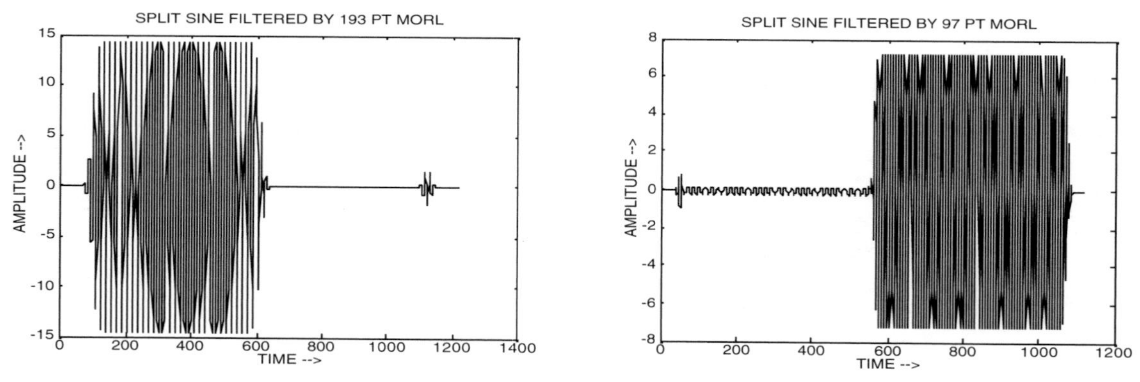

Figure 5.6–7 Correlation Strength of the Morlet wavelet transform of the split sine test signal for **scale** = **a** = **6** (left) and **scale** = **a** = **12** (right). As with the functionally equivalent CWT display (Fig. 5.6–5), we can now see a much better job of bandpass filtering for this test signal than we saw with the Mexican Hat wavelet (compare with Figure 5.5–9 or Figure 5.6–6).

5.7 Bandpass Characteristics of the Mexican Hat and Morlet Wavelet Filters

The power of wavelets in DSP is the ability to simultaneously discriminate in both *time* and *frequency*. We have demonstrated how their finite length allows them to discriminate in *time* and thus identify *when* an event occurred. To also discriminate in *frequency* these filters must be *bandpass*.* We have discussed earlier how stretching or "scaling" the wavelet filters lowers their (pseudo) frequency. We also demonstrated in the previous sections of this chapter how the Mexican Hat and Morlet wavelets can function as bandpass filters. For our test signal the Morlet wavelet filters did an excellent job of both *time* and *frequency* discrimination (reference Fig. 5.6–5).

We now look at the frequency characteristics of these 2 filters in general. We look first at the Mexican Hat filter at **scale** = **a** = 2. The original 17 point filter is stretched to 33 equation-generated points (see Fig. 5.5–7). Figure 5.7–1 below shows the bandpass nature. With a center frequency of about 0.27 Nyquist we can now see how the right half or last 512 points of the split sine test signal (frequency = 0,25 Nyquist) was allowed to pass. We can also see from this frequency response why the left half of the test signal (frequency 0.125 Nyquist) is attenuated, but not completely removed.

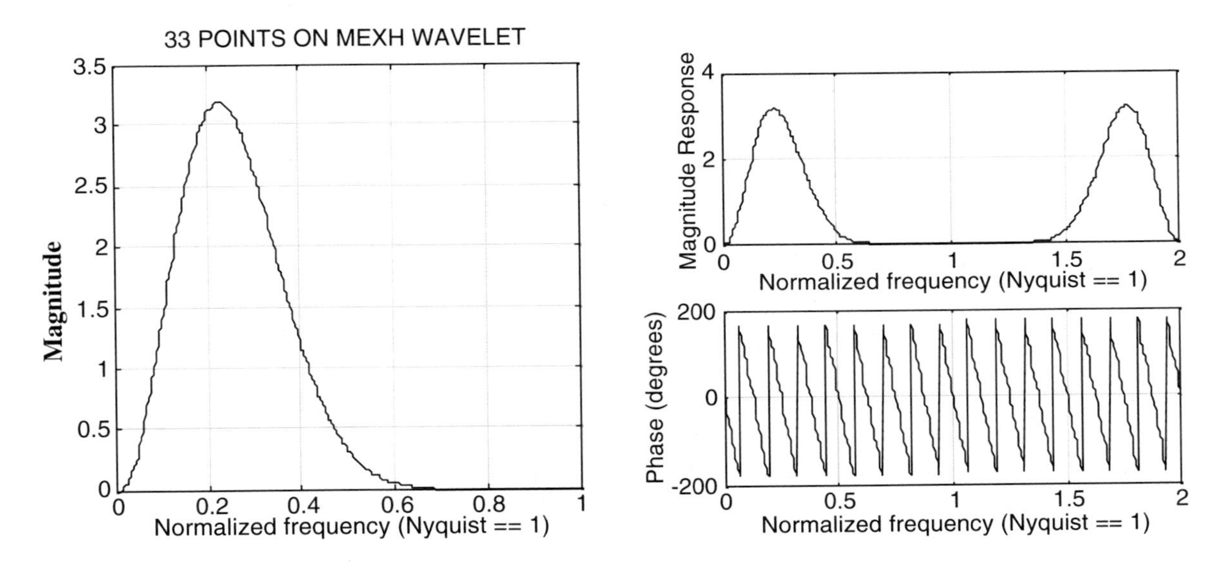

Figure 5.7–1 Frequency response of the 33 point Mexican Hat wavelet filter shown from zero to Nyquist (π radians) at left and from zero to 2π at right. Phase is also shown.

* We saw earlier in this chapter that the Sinc function is lowpass, not bandpass. Later in the book we will show how to make a *complex Sinc function wavelet* (filter) and *band-shift* it so it can be used as a *bandpass* filter.

We next look at the frequency response of the Mexican Hat wavelet filter stretched to 65 equation-generated points at **scale** = **a** = **4** (see Figure 5.5–8). Figure 5.7–2 shows that the bandpass nature has changed and that it now has a center frequency of about 0.12 Nyquist. We now see why the lower frequency (0.125 Nyquist) left half of the test signal was allowed to pass while the right half was attenuated.

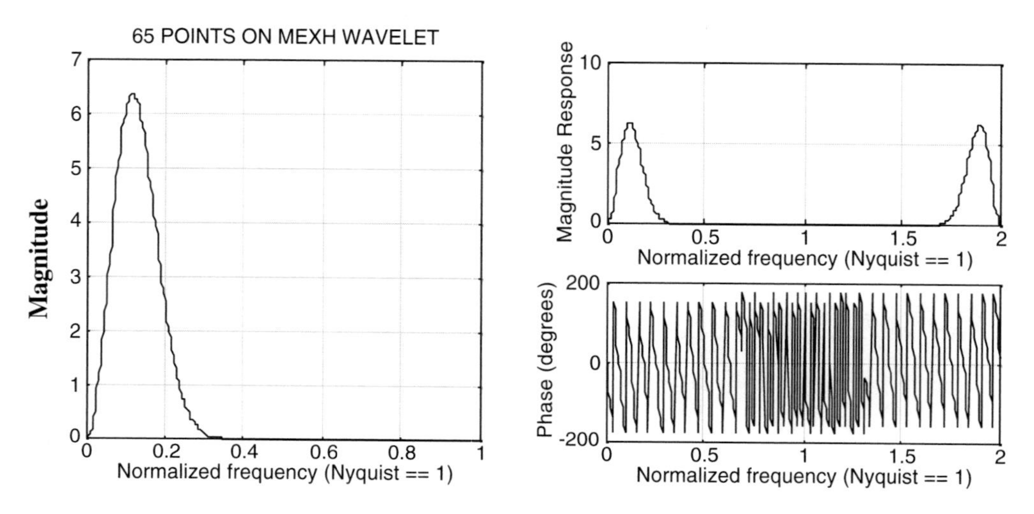

Figure 5.7–2 Frequency response of the 65 point Mexican Hat wavelet filter shown from zero to Nyquist (π radians) at left and from zero to 2π at right. We recall that when magnitude is near zero, as is the case here for high frequencies, that the phase becomes incoherent.

Comparing with Figure 5.7–1 we see the height has doubled while both the center frequency and the width of the passband are halved. This property is known as *Constant Q*. Because of the narrower passband we have the better discrimination we saw earlier in the right graph of Figure 5.6–5.

The Morlet wavelet filter is now examined. Figure 5.7–3 shows the frequency response of the (equation-generated) 97 point Morlet wavelet filter that corresponds to **scale** = **6**. As with the Mexican Hat filter when it was stretched to correspond with the higher frequency half of the test signal, we see the center of the passband to be about 0.27 Nyquist. Thus the higher frequency half of the test signal (0.25 Nyquist) is passed. The lower frequency half of the test signal (0.125) Nyquist will be fairly severely attenuated as was demonstrated in the previous section.

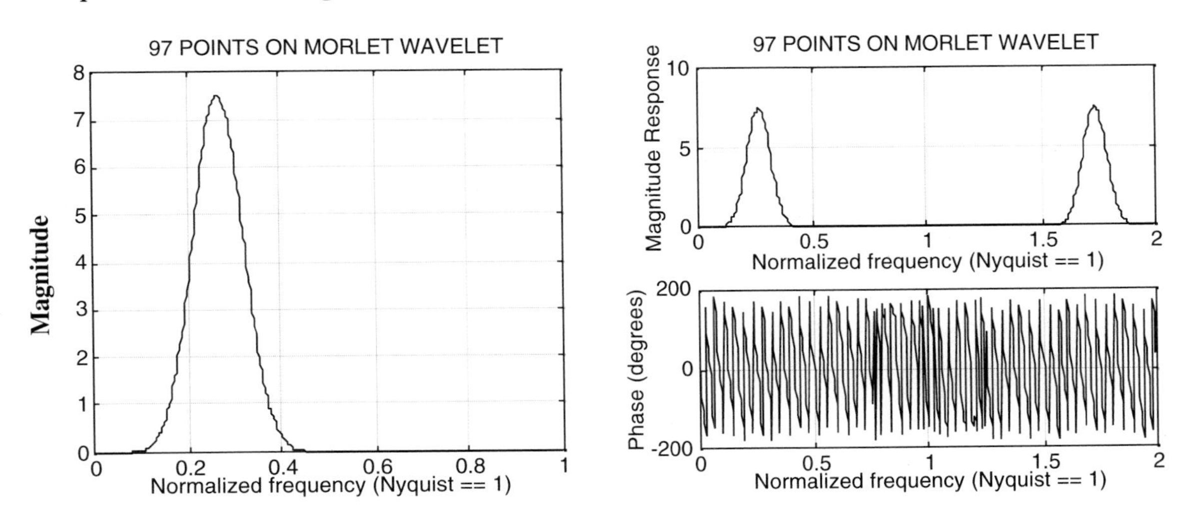

Figure 5.7–3 Frequency response of the 97 point Morlet wavelet filter (**scale** = **a** = **6**) shown from zero to Nyquist (π radians) at left and from zero to 2π at right.

As we stretch the Morlet filter by a factor of 2 to 193 points we see the same *constant Q* behavior with a higher peak as both the center frequency and the passband are reduced by a factor of 2 as shown in Figure 5.7–4. Notice that the center frequency of this bandpass filter is about 0,12 and that it will allow the lower frequency left half of the test signal to pass while severely attenuating the right half with its 0.25 Nyquist frequency (see Figure 5.6–5).

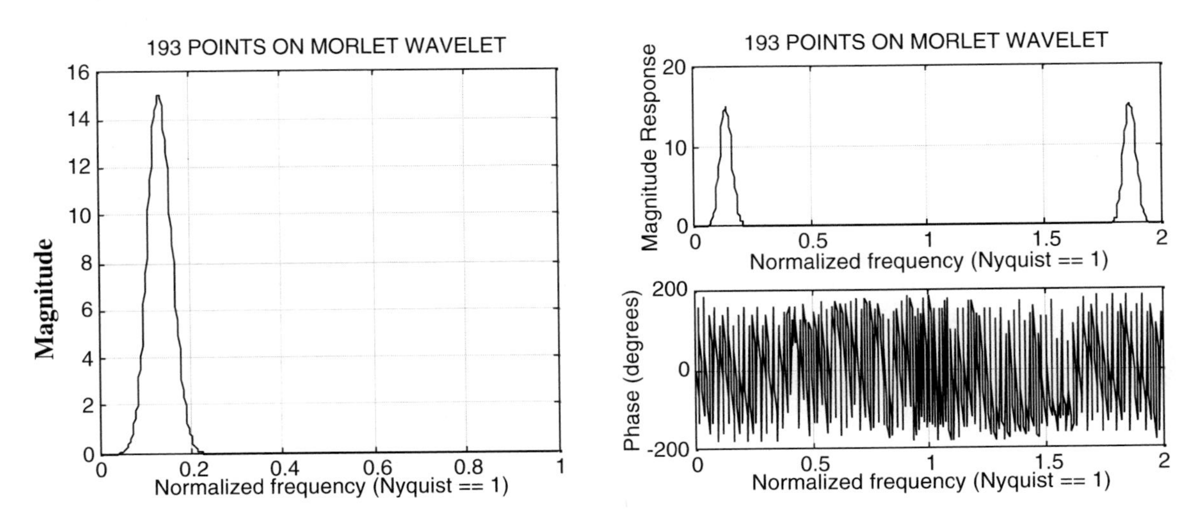

Figure 5.7–4 Frequency response of the 193 point Morlet wavelet filter (**scale** = **a** = **12**).

A reminder is in order here: We can design bandpass filters using conventional DSP methods with sharper cutoffs, better transition bands, etc. than either the Mexican Hat or the Morlet filters. However, the strength of these *wavelet* filters can be seen in their ability to provide simultaneous time, frequency, and waveform shape information about an unknown signal. For example, the CWT displays in this chapter have shown us information about the *frequency* of our test signal *at any given time*. With the better discrimination of the Morlet wavelet than the Mexican Hat we can even tell the *shape(s)* of the test signal (2 time-sequential sinusoids). As we proceed we will discover more of the capabilities of these powerful tools.

5.8 Summary

In this chapter we got our bearings by reviewing the traditional DSP methods of working with a Sinc function. Specifically, we added extra points on the "ends" which is, in effect, using a larger "boxcar" window on the (infinitely long) Sinc function filter. The Sinc function looked the same but had more "lobes" on the ends. Looking in the frequency domain, we saw the familiar result of a shorter transition band at the same cutoff frequency.

We than contrasted this familiar method with that used in wavelet processing to stretch (*dilate*) the filter by interpolating. We showed an interpolated Sinc filter with the additional points placed *between* the existing points rather than at the *ends* as before. The Sinc function now has the same number of lobes as the original, but there were twice as many points in each lobe. Looking again in the frequency domain, we saw that these additional points did not affect the transition band, but changed the cutoff frequency by a factor of 2 (from 0.25 to 0.125 Nyquist).

We discussed *crude* wavelets where the filter points can be generated directly from an explicit mathematical equation. We learned that these wavelets, although theoretically infinite in length, have an *effective support* or very limited range where they are non-zero to computer precision.

We looked first at the whimsically named *Mexican Hat* wavelet filter. Following the example of the MATLAB software, we generated 17 integer-spaced points on the interval from –8 to +8. This is **scale** = **a** = **1**. By using the same equation at 1/2 integer intervals, we generated 33 points to produce a "stretched" filter at **scale** = **2**. We continued this process to produce a 49 point and then a 65 point filter at scales 3 and 4.

We generated a 1024 point spilt-sine (cosine) test signal with 16 points per cycle for the first 512 points and 8 points per cycle for the last 512. When the Mexican Hat filter is stretched to 33 total points (in the interval from –8 to +8) it has about 8 points in its :"cycle". When it is stretched to 65 points (**scale** = **4**) it has about 16 points in its "cycle". The CWT display for the Mexican Hat wavelet filter showed a strong correlation with the right half of the test signal at **scale** = **2** when the 8 points in the wavelet "cycle" matched the 8 points in a cycle of the test signal. It also showed a strong correlation with the left half of the signal at **scale** = **4** when both the wavelet and the left half of the signal had 16 points.

We performed a similar analysis using the sinusoidal Morlet wavelet filter on the same split sine test signal. This is another *crude wavelet* that uses an equation to generate points. As the original or "mother" wavelet filter of 17 points in the interval from –8 to +8 is stretched to 97 points (**scale** = **6**) we saw that we had about 8 points in a cycle. As we stretched further to 193 points (**scale** = **12**) we saw we had about 16 points in a cycle.

A Continuous Wavelet Transform (CWT) using the Morlet wavelet filter with the sinusoidal test signal showed an excellent correlation with the left and right halves at scales 12 and 6 respectively. Noting that the correlation was stronger and the discrimination was better when substituting the Morlet for the Mexican Hat, we pointed out that from the CWT and its display we could determine the *specific frequency* at *specific times* and even know what the signal *looked like* (sinusoidal). In other words, the CWT did a good job of describing our pre-known test signal and thus should be able to be used with real-life unknown, time-varying signals.

Finally, we looked at the frequency characteristics of the Mexican Hat and Morlet filters. We saw they are *bandpass filters* and that by stretching them we decrease the center frequency (similar to the analysis we performed on the lowpass Sinc Function at the start of this chapter). We also saw the passband become narrower as the filters were stretched.

We now move on to wavelet filters of any desired length that are *not* generated from equations, but are built from a very few points in the "mother" wavelet filter.

"Tell me and I forget. Teach me and I remember. Involve me and I learn."

—Benjamin Franklin

CHAPTER 6

Obtaining Variable Length Filters from Basic Fixed Length Filters

In Chapter 5 we discussed crude wavelets with explicit equations which allowed us to stretch a wavelet filter to any desired length by simply "plugging in" equispaced values into the equation. For example, if we had 17 equispaced values at integers from –8 to +8 (including zero) we could easily interpolate to 33 values by inputting half-integer values (–8.0, –7.5, –7.0 . . .). This process is of course automated as the filters are stretched and then correlated with the data by the CWT display software.

In this chapter we will look at wavelet filters that are not defined by an explicit mathematical equation. These mother wavelet filters have very few points. We will study a method of interpolation that allows us to first stretch these filters to hundreds of points and thus create an estimation of a "continuous" wavelet.

Next, we will show how to create a wavelet filter of any desired length from these hundreds of points using numerical analysis techniques—a reminder that we neither need nor use a "continuous" wavelet in the real world of digital computers.

This process of interpolation to hundreds of points and then creating filters of any desired length is also usually automated in the existing software.

6.1 Review of Conventional Interpolation Techniques from DSP

A familiar method of interpolation from DSP is to upsample by 2 and then lowpass filter the result.

Jargon Alert: ***To upsample by 2 means to place zeros between the existing values. For example, [1 2 3 4 5] upsampled by 2 becomes [1 0 2 0 3 0 4 0 5]* This process is often referred to as "dyadic upsampling"***

As an example, we use a familiar Gaussian function with 17 points as shown at the left of Figure 6.1–1. If we "connect the dots" or plot the data as shown at right, we see a need for more points to provide a smoother representation.

* In some instances a zero is also placed on one or both ends. For example, **[0 1 0 2 0 3 0 4 0 5 0]**

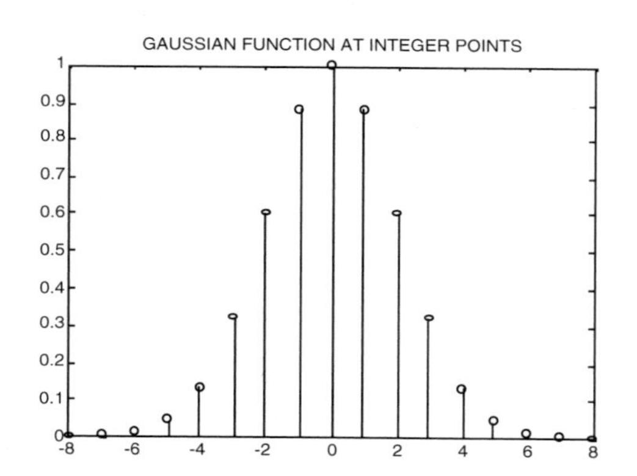

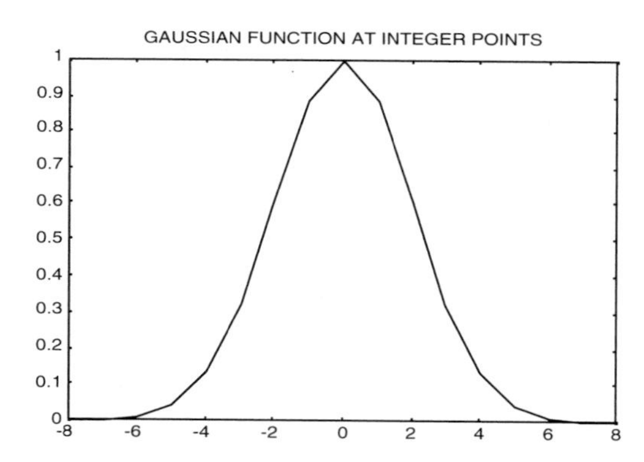

Figure 6.1–1 17 points of a Gaussian function at left. Connecting the points shows abrupt changes in the slope (discontinuities in the 1st derivative) and the need for further smoothing.

To obtain more points we could "cheat" by using the same equation* that generated the 17 points to produce an additional 16 values in between the existing points. Instead, however, let us follow the example of the many wavelet filters that are *not* defined by an equation and interpolate directly. We begin by upsampling by 2 as shown in Figure 6.1–2.

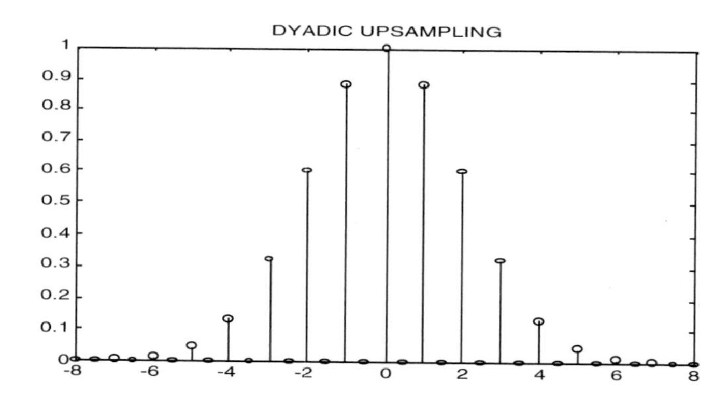

Figure 6.1–2 16 zeros placed between the original 17 points (dyadic upsampling)

The next step is to lowpass filter. We can see why this step is essential by looking at the FFT of the upsampled signal as shown in Figure 6.1–3. Placing zeros between the points has resulted in a high-frequency image centered around Nyquist (we will provided insights as to why this is so in a moment).

---

* The equation used here is **y(n) = exp(–(n^2)/8)**

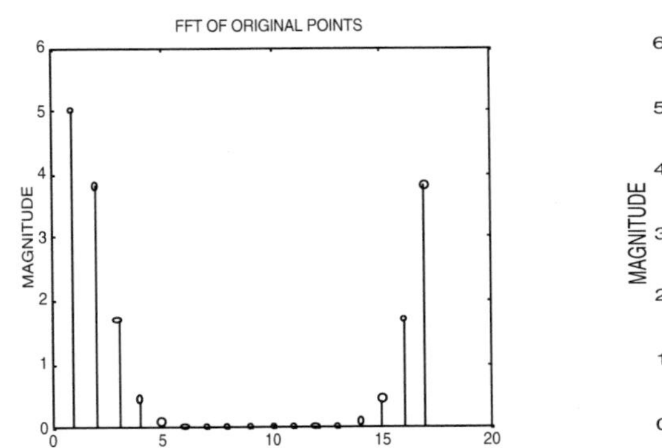

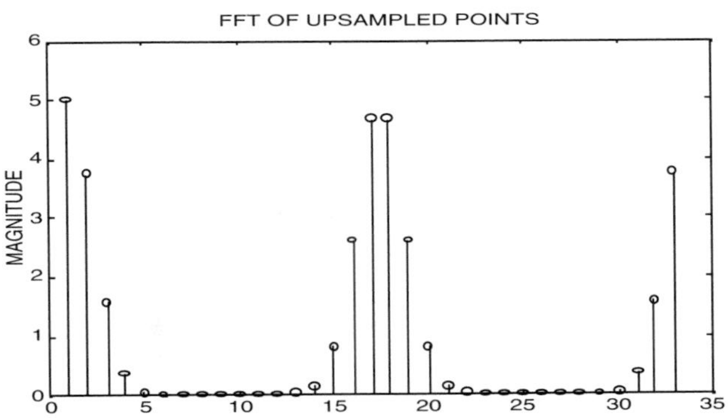

Figure 6.1–3 FFT of original 17 points (left) and FFT of upsampled 33 points (right). Note the high-frequency image around Nyquist.

We proceed to lowpass filter. As we keep only the low frequency components we have in the frequency domain the data as shown in Figure 6.1–4.*

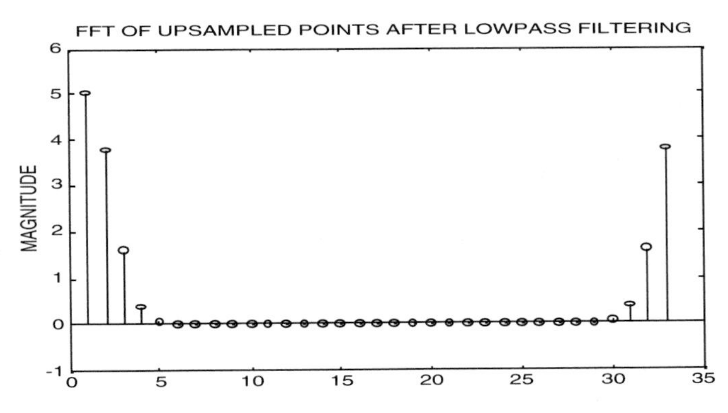

Figure 6.1–4 FFT of upsampled points after lowpass filtering. Note the high-frequency image is gone (replaced by zeros).

After lowpass filtering we obtain the set of 33 points as shown at the left of Figure 6.1–5. As we plot these points and compare with the original plot (Fig. 6.1–1) we can see that we have successfully interpolated without using the original equation.

* Those familiar with interpolation will recall that an alternative to upsampling and lowpass filtering is to take the FFT of the original points, zero-pad in the middle, and take the Inverse FFT. This gives us the same result as shown above.

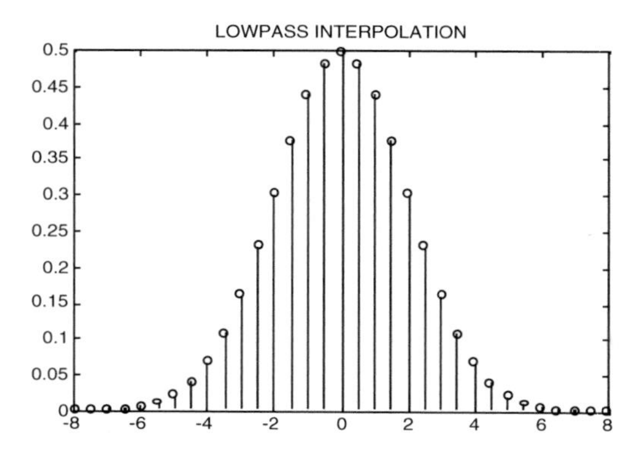

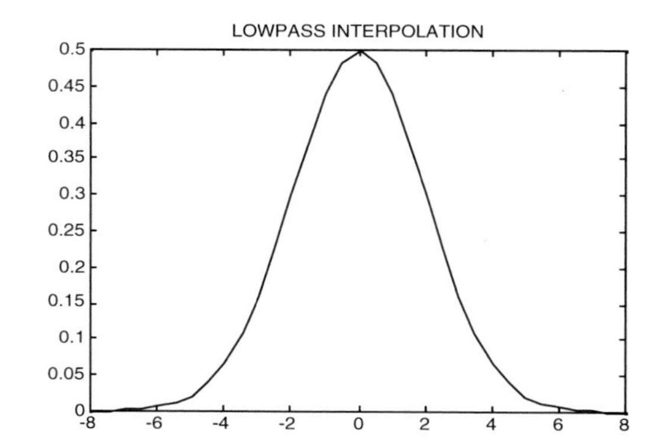

Figure 6.1–5 Interpolation results. We have the original 17 points at the integer values and 16 interpolated points at the half integer values (left). Note the smoother estimation (approximation) of a Gaussian function we obtain by plotting these 33 values (right).

Note that this interpolation process can be done entirely in the time domain. We start with the original points, upsample by a factor of 2, convolve the data with a lowpass filter, and use a constant of multiplication* later to normalize (adjust for the change of magnitude due to the extra points). We will use this same process in of upsampling, lowpass filtering, and multiplication by a constant to interpolate wavelet filters that are not defined by an equation.

We did use the FFT in the above example for purposes of illustration. Figure 6.1–3 showed a high-frequency component introduced by the upsampling. We can also demonstrate this concept in the time domain. The dyadically upsampled data can be thought of as the sum of an interpolated low-frequency signal (our final result as shown in Fig. 6.1–5) and an interpolated high-frequency signal as shown below in Figure 6.1–6. Notice that this high-frequency signal is the same as our low frequency signal except that the signs alternate.†

* In conventional DSP we would refer to this as *scaling* the data. However the wavelet literature uses the terms *scale* and *scaling* to describe stretching or changing the frequency rather than the magnitude. Thus we reserve these words and use *constant of multiplication* instead.

† This process is familiar in DSP. To transform a lowpass filter to a highpass filter we use the equation

$$\mathbf{h_{HP}(n) = (-1)^n h_{LP}(n).}$$

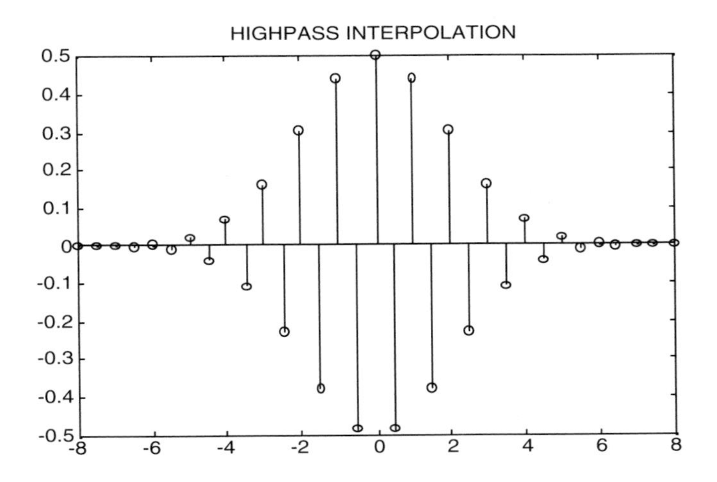

Figure 6.1–6 Interpolation results we would obtain using a highpass filter on the upsampled data. The original 17 points are positive and the 16 interpolated points are negative.

We can see that as we add the high and low frequency signals together in the time domain that every other term will cancel and we will have the earlier *dyadically upsampled* signal with zeros in the odd terms. This process is illustrated in Figure 6.1–7.

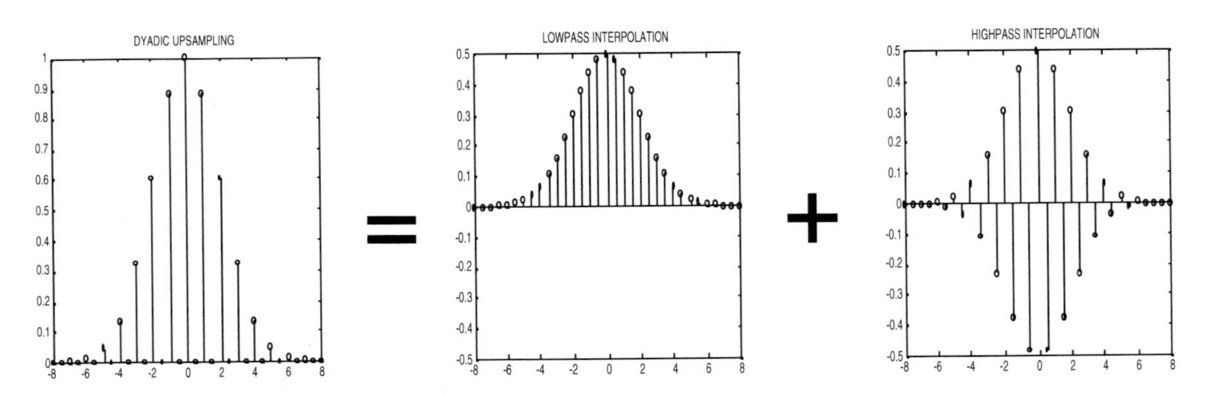

Figure 6.1–7 Demonstration in the time domain of how an upsampled signal is the sum of the low-frequency and the high-frequency interpolation. As we remove the high-frequency part from the upsampled signal, we have the desired smoothing.

6.2 Interpolating the Basic ("Mother") Wavelet by Upsampling and Lowpass Filtering

The *Discrete Wavelet Transform* (DWT), used for compression and denoising (among other things), requires additional constraints on the filters such as Perfect Reconstruction and Alias Cancellation capabilities. We will begin with the simplest and shortest of these wavelet filters: The Haar, named for Alfred Haar. We used these filters in earlier chapters and they may be familiar to some students as *block averagers* or *block differentiators.* The basic wavelet filter or *mother* wavelet filter is the 2-point vector **[1 –1]**. It has a lowpass counterpart or *scaling function* filter **[1 1]**. These filters were introduced in Chapter 1 as the *highpass reconstruction filter* and the *lowpass reconstruction filter*, respectively*.

It will be recalled that the Continuous Wavelet Transform (CWT) used only one filter and stretched it. For the Haar the basic filter was **[1 –1]**. Stretching a *crude wavelet filter* was easy—we simply used the defining equation. To stretch a Haar wavelet filter to any desired length we must first construct an estimation of the theoretically continuous wavelet using the interpolation techniques described in the last section. (This process will be seen later in the functional diagram of a DWT and also in the "Dilation Equation").

For now we proceed to simply build an estimation† of a continuous wavelet by upsampling and lowpass filtering. We begin by upsampling the Haar wavelet filter. This becomes **[1 0 –1]**. We next lowpass filter by convolving with the scaling function filter **[1 1]**. This gives us a stretched or dilated filter **[1 1 –1 –1]**. The basic filter has been stretched by a factor of 2.

Figure 6.2–1. shows the basic 2-point Haar wavelet filter followed by the results of upsampling and convolving with the basic Haar scaling function lowpass filter ([1 1])

* There are 2 other filters used in the DWT—the *highpass decomposition filter* given by **[–1 1]** (for the Haar) and the *lowpass decomposition filter* given by **[1 1]**.
† The word *approximation* is more descriptive than *estimation* but "Approximation" is reserved in wavelet literature to describe the ends of lowpass paths in wavelet transforms (e.g. A1, A2, A3, etc.).

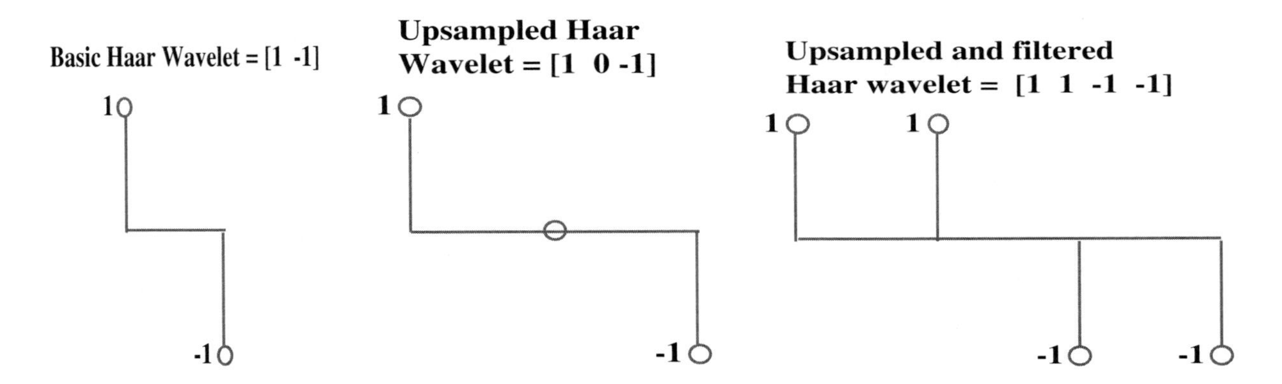

Figure 6.2–1 Stretching the basic 2-point Haar wavelet filter by a factor of 2 to become 4 points long by upsampling and then convolving with the Haar scaling function (lowpass) filter ([1 1]). Note the length will be **L** + **M** – **1** = **4** with **L** = **3** and **M** = **2**.

If we upsample this latest result we have **[1 0 1 0 –1 0 –1]**. Convolving with the lowpass filter (**[1 1]**) gives us **[1 1 1 1 –1 –1 –1 –1]**. We have now stretched the Haar wavelet filter from its original 2 points to 4 and then 8 points as depicted in Figure 6.2–2.

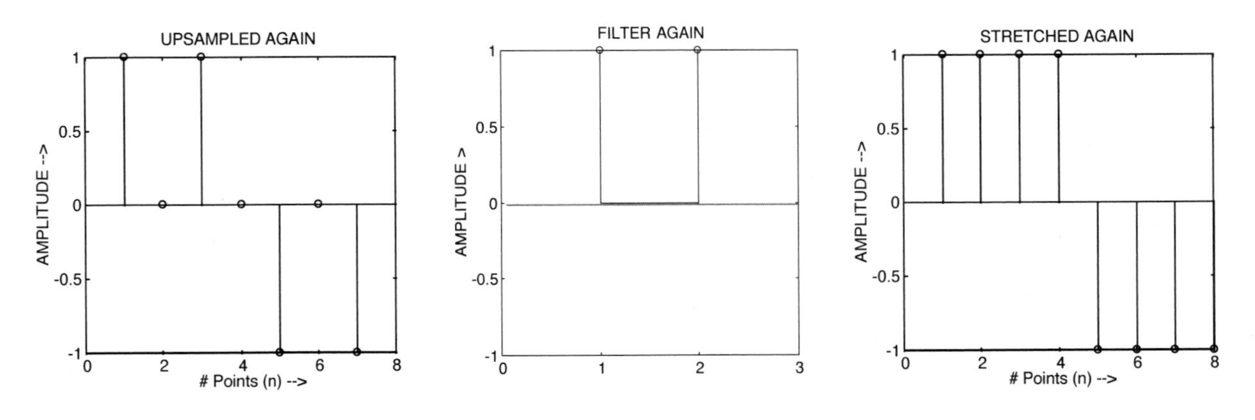

Figure 6.2–2 Stretching the 4-point Haar wavelet filter by a factor of 2 to 8 points by upsampling and then convolving with the Haar scaling function (lowpass) filter (**[1 1]**). Note the length will be **L + M – 1 = 8** with **L = 7** and **M = 2** again.

We upsample and filter (by [1 1]) 4 more times. The wavelet filter is stretched to 16, 32, 64, then 128 points as shown in Figure 6.2–3.

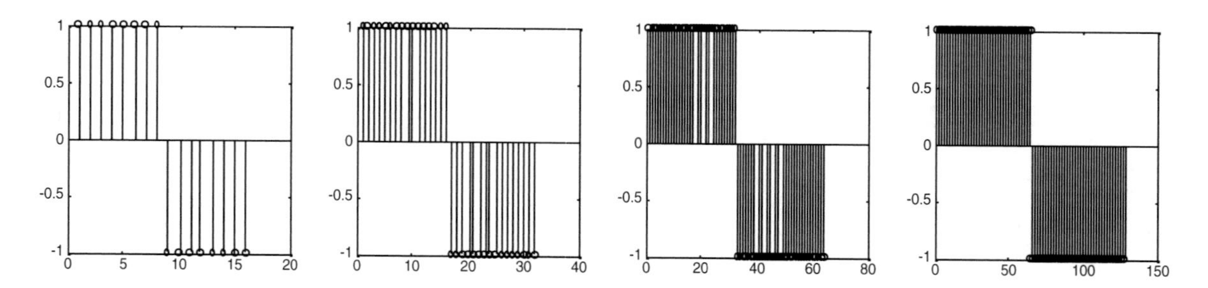

Figure 6.2–3 Further upsampling and convolving with the lowpass filter produces 16, 32, 64 and 128 points.

With one more upsampling and convolving we have 256 points. If we now plot the points we have a good estimation of what a continuous wavelet would look like. Ending the stretching at 256 points is an arbitrary, but conservative choice. MATLAB chooses this number as sufficient to estimate (approximate) a "continuous" wavelet function. MATLAB also adds a zero point to each end for reasons we will soon explain so we actually have 258 points to approximate the Haar wavelet function.

In the real world of digital computers we work with filters of various lengths and not with the mathematical representation. We saw this in the last chapter as we constructed "crude" filters of various lengths from the explicit equations. The literature, however, often shows the "Haar Wavelet Function" or *psi* as having a "length" of 1. The value of the psi is often given as

```
psi(t) = 1, 0 < t < 0.5;  -1, 0.5 < t <1; 0 otherwise
```

We now "connect the dots" and plot the 258 points on a time scale from 0 to 1 (with zero elsewhere). This is shown in Figure 6.2–4 below. Note the discontinuities at times 0, 1/2 and 1.

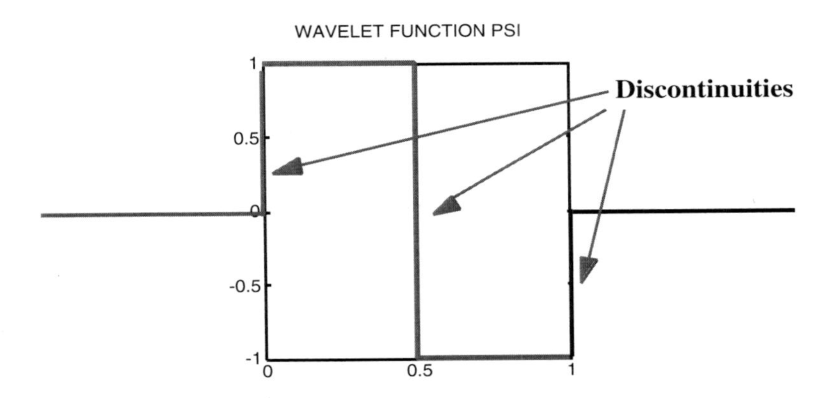

Figure 6.2–4 258 points mapped onto the interval from 0 to 1 provide an estimation of a "continuous" wavelet function. Note the values outside this interval are zero and that there are 3 discontinuities.

Now we have a detailed estimation of the wavelet function we can proceed to produce a filter of *any desired length* as is required for a CWT.

One method would be to simply interpolate the desired filter points from the existing points. The problem is with the discontinuities at **t** = **0**, **t** = **0.5**, and **t** = **1**. Discontinuities in the function or its derivatives is common for many of these wavelet functions built from wavelet filters.

A way around this is to first do a numerical integration of the wavelet function and then differentiate (look at the slopes) to find the filter points. For example, at **scale** = **5** the Haar CWT requires a 5 point equispaced filter. If we were to interpolate directly we would choose the values at **t** = **0**, **1/4**, **1/2**, **3/4**, and **1**. The values at **t** = **1/4** and **3/4** are clearly one and minus one, respectively, but those at **t** = **0**, **1/2**, and **1** are at discontinuities.

If we perform a numerical integration of these 258 points we accumulate a pyramid that increases till **t** = **0.5** and then decreases back to zero as shown in Figure 6.2–5. We then chose 6 points over the interval from 0 to 1 and look at the slopes (differentiate). The slope between the first 2 points is 1 as is the slope between the second 2 points. The slope between the 3rd set of 2 points, however, is zero. The last two slopes are –1 and –1. Thus for **scale** = **5** for the Haar CWT we have the wavelet filter (**[1 –1]**) stretched to **[1 1 0 –1 –1]**.* We can now calculate a Haar filter for any number of points.

* This is intuitively correct because one requirement is that the coefficients of wavelet filters should sum to zero. The first 2 and last 2 values cancel thus the center value should be zero.

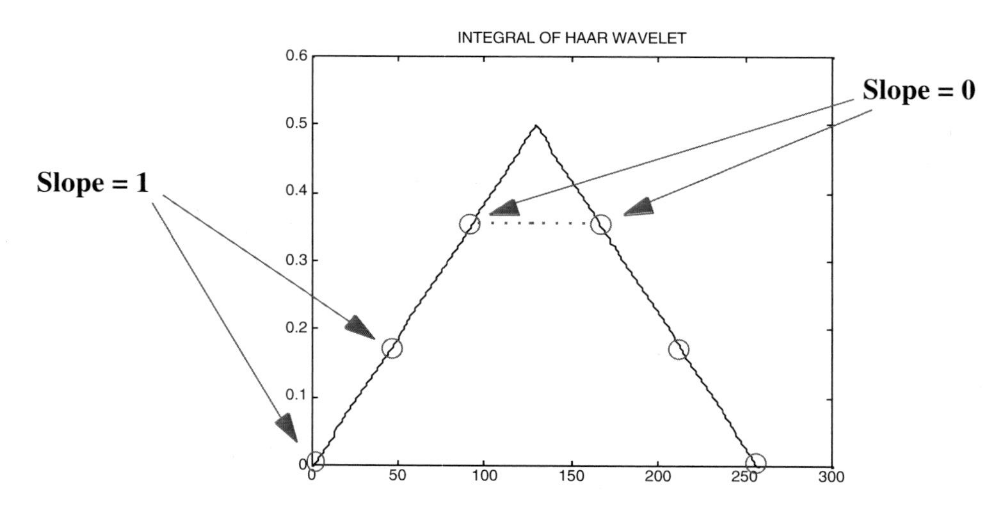

Figure 6.2–5 Numerical integration of the 258 point estimation of the Haar wavelet function builds the pyramid shape. We then differentiate using equispaced points. If we desire a 5 point Haar filter we use 6 equispaced points and calculate the slope as shown here.

6.3 Frequency Characteristics of the Basic and Stretched Haar Filters

We begin with the 2 point basic Haar wavelet filter and the 4 point stretched Haar wavelet filter. Note that the bandpass nature begins to be seen in the 4 point filters shown below in Figure 6.3–1.

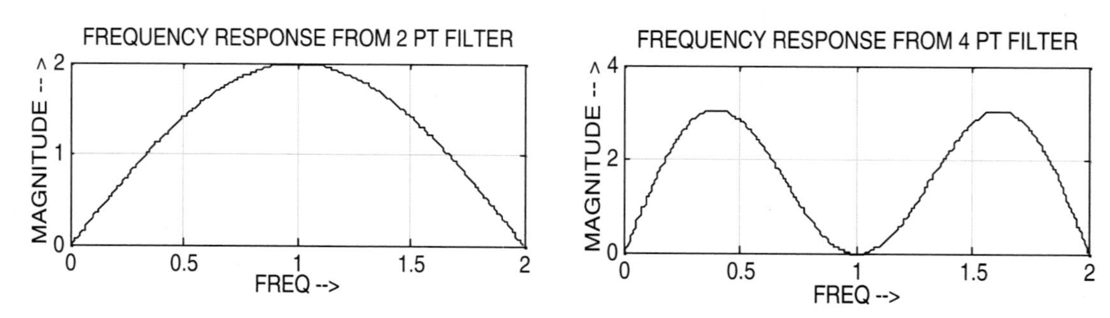

Figure 6.3–1 Frequency Response of 2-point Haar filter (**[1 –1]**) and 4-point stretched version (**[1 1 –1 –1]**). 2-point is “bandpass” only in the interval 0 to 2π.

The frequency response for the 8, and 16-point filter is shown in Figure 6.3–2.

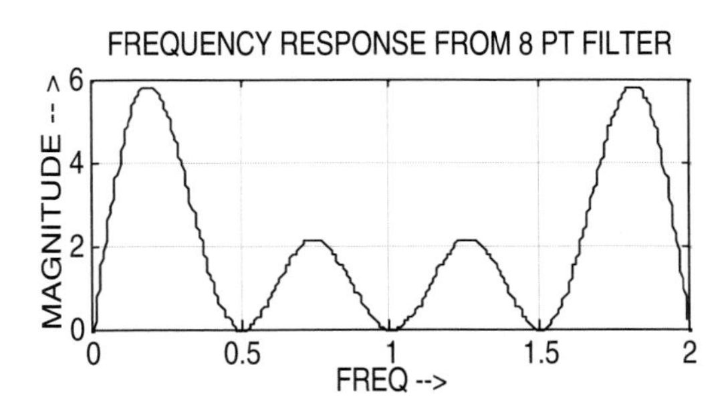

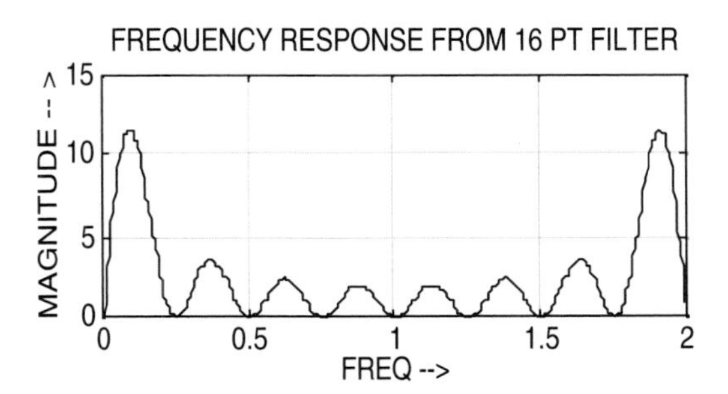

Figure 6.3–2 Frequency Response of 8, and 16-point stretched Haar wavelet filter. Note that as the height of the main lobe doubles, the passband center frequency and the width of the passband are cut in half.

The 2-point Haar wavelet filter is the simplest of the Daubechies (Db) family of wavelet filters. The next "filters from wavelets from filters" we will look at is the Db4, having 4 filter points. This is sometimes referred to as a Db2, with the "2" signifying an exponent in a process called *spectral factorization* or as the number of *vanishing moments* (we will discuss these topics later in the book).

The 4 values for the Db4 wavelet filter will be derived later as we discuss the DWT. For now they will remain "Magic Numbers"* as sometimes referred to in the literature. An analogy is in order here: Finding the square root of 103,041 is difficult without a computer or hand calculator. Verifying that 321 is the correct answer, however, is very easy. Similarly, finding the "magic numbers" is hard but verifying them is easy as we will soon show.

Instead of **[1 –1]** for the Haar, the Db4 wavelet function filter is given by

```
[-0.1294   -0.2241    0.8365   -0.4830]
```

Instead of **[1 1]** for the Haar, the LPF or basic Db4 scaling function filter is closely related to the above 4 numbers and is given by

```
[0.4830    0.8365    0.2241    -0.1294]
```

Note that the 4 numbers do not change, only the position and the signs. We will explore this relationship later. Figure 6.3–3 shows these 2 filters.

* Ingrid Daubechies is credited for discovering these numbers through the Spectral Factorization of a halfband filter.

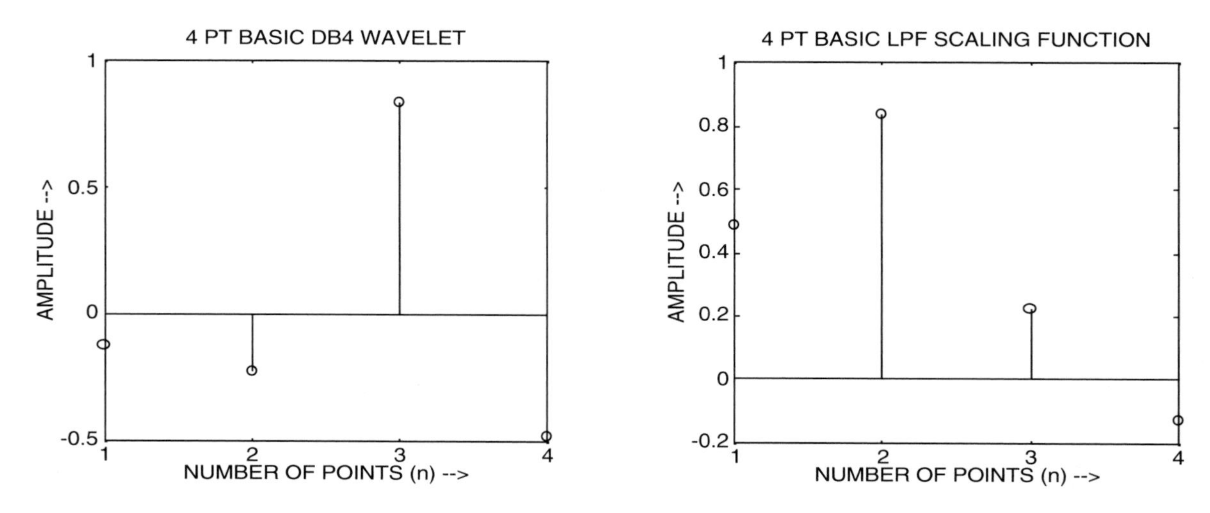

Figure 6.3–3 Daubechies 4 basic or *mother* wavelet filter and scaling function filter.

We perform the same series of upsampling and lowpass filtering as we did for the Db2 or Haar wavelet function filter (**[1 –1]**).* Figure 6.3–4 shows the basic wavelet filter upsampled with zeros between the existing points and the result of lowpass filtering by the 4-point Db4 scaling function.

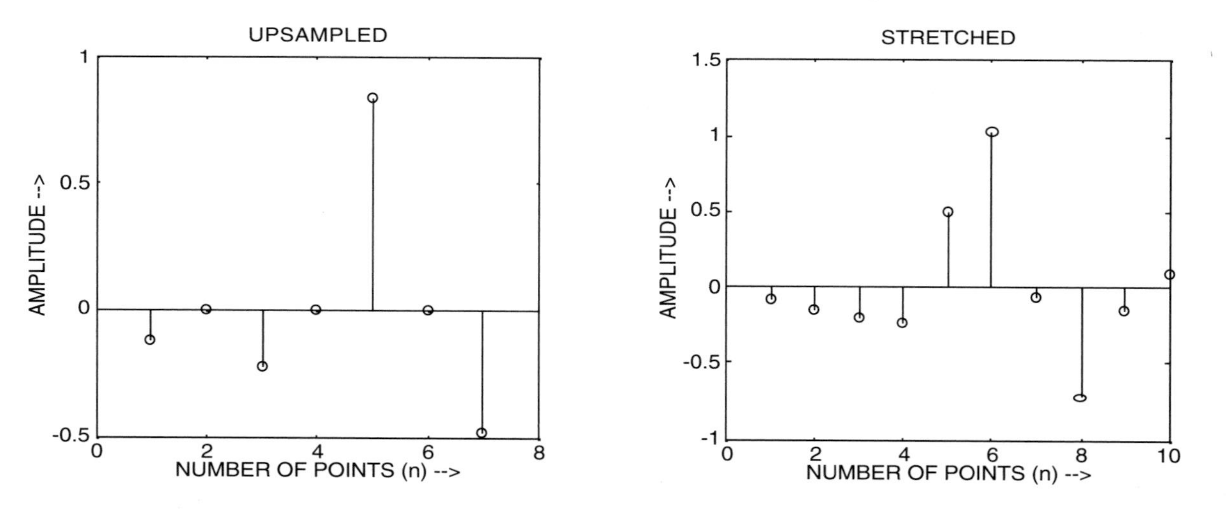

Figure 6.3–4 Four-point Db4 wavelet filter first upsampled (7 points) and then lowpass filtered by the Db4 scaling function filter to produce the 10-point stretched Db4 filter (**7+4–1=10**).

* We multiply by sqrt(2) at each step to account for the increased energy.

We continue the process of upsampling and lowpass filtering to produce increasingly stretched wavelet filters with 22, 46 94, 190, 382 and finally 766 points. At this stage software such as MATLAB deems this a sufficient estimation of a "continuous" function. The stretching to 46, 190, and 766 points is shown below in Figure 6.3–5

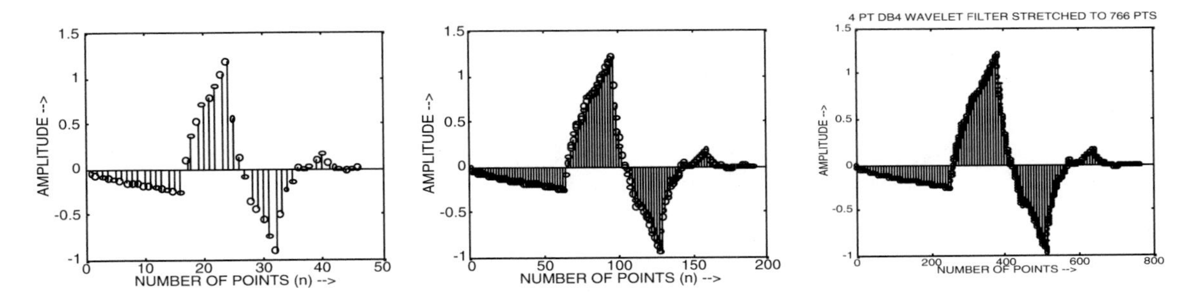

Figure 6.3–5 Four-point Db4 wavelet filter when stretched to 46 points, 190 points, and 766 points.

If we "connect the dots" by plotting the 766 points we now have an approximation or estimation of a Db4 "continuous" wavelet function built from the original 4 points. Similar to the way we mapped the 258-point Haar wavelet filter to an interval from 0 to 1, we now map the 766 point Db4 wavelet filter onto an interval from 0 to 3. This is shown below in Figure 6.3–6.

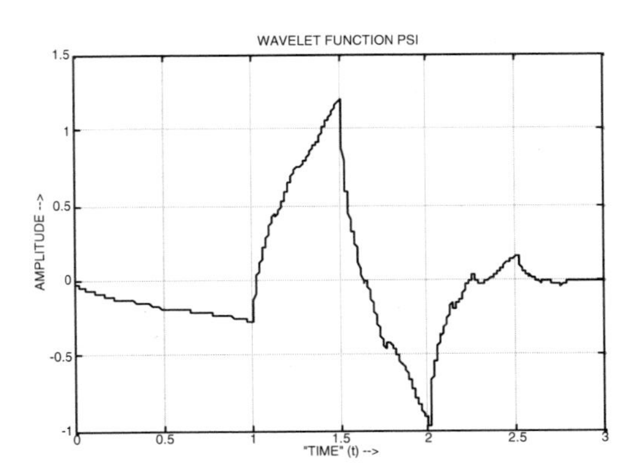

Figure 6.3–6 766-point estimation of a continuous wavelet function mapped onto an interval from 0 to 3. The value outside this interval is zero.

The literature usually depicts the Db4 wavelet function or *psi* (ψ) as having a "length" of 3 and starting at **t** = **0**. One reason for this choice of length can be seen from having discontinuities in the slope appear at 1.0, 1.5, 2.0 and 2.5. Another reminder is in order here that we never actually use the wavelet function. The 766 points connected here are used to produce filters with any desired number of points. In other words, it is when you convolve your data with a filter having a certain number of points that you determine the real-world time axis.

As with the Haar, we have the Wavelet Function (estimation) and can proceed to produce a filter of any desired length as is required for a CWT. Also as with the Haar, we use an integral of the wavelet function* and then use slopes (differentiate) to handle any discontinuities (Fig. 6.3–7).

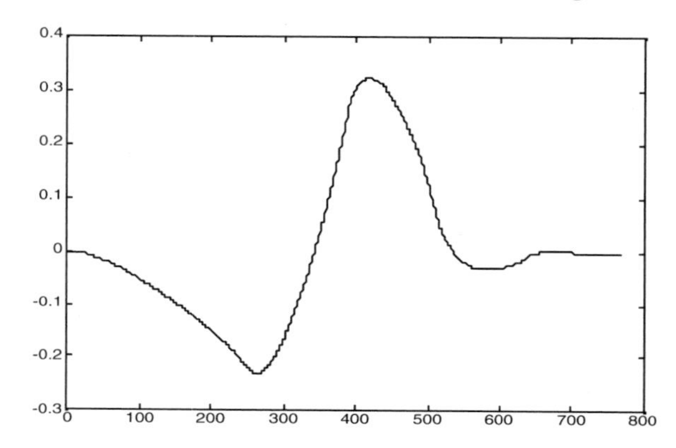

Figure 6.3–7 Numerical integration of the 766 point Db4 wavelet scaling function filter. From this integral we can now differentiate to produce filters of any desired length.

6.4 Perfect Overlay of Filter Points on the "Continuous" Wavelet Estimation

First, we superimpose the original 4 Db4 filter points that were used to build this wavelet function. As we convert our 766 point "continuous" function to a "length" of 0 to 3, the 4 "magic number" points

```
-0.1294    -0.2241    0.8365    -0.4830
```

* For numerical integration purposes, a leading and trailing zero are often added making the total length 768 points rather than 766. Haar wavelets are often estimated at 258 points rather than 256 for the same reason.

are found to be at 2/6, 5/6, 8/6 and 11/6. In other words, 1/2 integer apart starting at 1/3 and ending at 11/6. They are overplotted on the wavelet function estimation in Figure 6.4–1 (left). In actuality, there are 2 additional zero points with a value of zero. These also match perfectly the values of the 766 point estimation (mapped onto the interval 0 to 3) at 14/6 and 17/6. These are also shown at the right of Figure 6.4–1. Note that although the original Db4 filter has been lengthened to the 6 points

```
-0.1294    -0.2241    0.8365    -0.4830    0.0    0.0
```

that the filter is essentially the same (DSP engineers are familiar with appending zeros to the ends of filters or data for a variety of uses). Thus, the two additional equispaced zero points (at the end) complete the overlay.

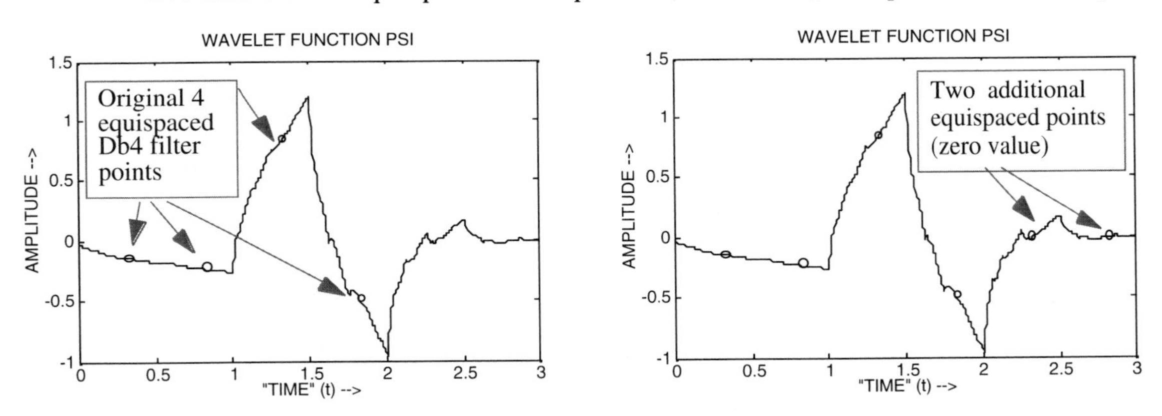

Figure 6.4–1 The original 4 wavelet filter points along with two additional trailing zeros overplotted on the 766 point estimation of a "continuous" wavelet function built from these same points. When mapped to the interval 0 to 3, the points are located 1/2 integer apart at 2/6, 5/6, 8/6, 11/6, 14/6 and 17/6.

We use the same upsampling and lowpass filtering process to interpolate estimations of other wavelet functions from a very few points. The Db6 and Db8 filters are shown in Figure 6.4–2. Note that the Db6 has 6 filter points and 4 end zeros starting at 3/7 mapped onto the interval 0 to 5. The Db8 has 8 filter points and 6 end zeros starting at 4/8 mapped onto the interval 0 to 7. As with the Db4, the Db6 and Db8 points are also equispaced 1/2 integer apart on their respective mappings.

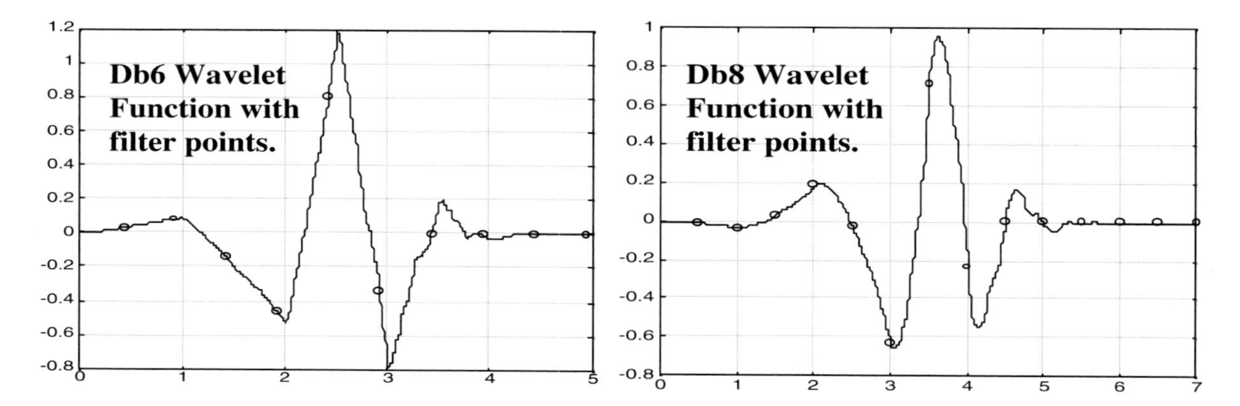

Figure 6.4–2 Db6 and Db8 estimations (approximations) built from the 6 and 8 points, respectively, in the basic wavelet filter. Note the zero valued points on the end.

We have seen that with the Db4, Db6, and Db8 we can show the filter points to be an exact fit with extra zeros on the end. All points are 1/2 integer apart. We will show later that the Db2 or Haar has 2 points (**[1 –1]**) located at 2/10 and 7/10 and no end zeros. It is interesting to note the progression from the Db2 to the Db8: Starting points are at 1/5, 2/6, 3/7, and 4/8. The number of end zeros are 0, 2, 4, and 6. The mapped intervals are 1, 3, 5, and 7. The spacing of points (including end zeros) is 1/2 integer.

We use this same process of starting with a few points from the basic wavelet filter and interpolating additional points by upsampling and lowpass filtering to produce estimations of other wavelet function filters. Figure 6.4–3 shows the Coiflet wavelet* and one of the Biorthogonal wavelets.

The Coiflet wavelet shown here is described in the MATLAB Wavelet Toolbox Users Guide as "Coif1". There are 6 values in this basic Coif1 wavelet filter. The Biorthogonal wavelet shown at the right (Fig. 6.4–3) is described as "Bior4.4". We start with the 9 points in the basic wavelet filter. A 10th zero point is used by MATLAB

* The fit (below) may not appear perfect but the Wolfram Research website cautions against small computer errors that cause this. Wolfram also shows the wavelet upside down, in agreement with our method of producing functions from filters. (The definition of the wavelet here differs from that of Daubechies by a sign).

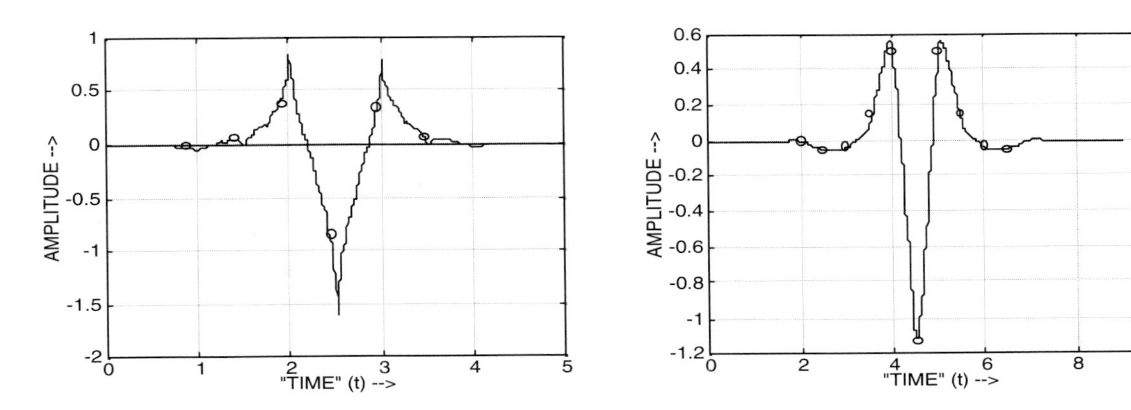

Figure 6.4–3 Estimations of the Coiflet wavelet function (left) and the Biorthogonal 4.4 0r "9/7" filter (right) with the points of the basic (mother} filter superimposed. Note the 1/2 integer spacing of the points.

6.5 Frequency Characteristics of some of the Basic Filters

Before finishing this chapter, we want to look at some additional frequency characteristics. We explored the Haar filters (section 6.3) and saw that these 2-point basic Haar filters, although excellent for detecting short-term events, have poor frequency characteristics. We compare in Figure 6.5–1 the frequency response from the basic (2-pt) Haar or Db2 filter with that of the basic (4-pt) Db4. Although still a large transition band, we see better performance (at the cost of a slightly longer filter).

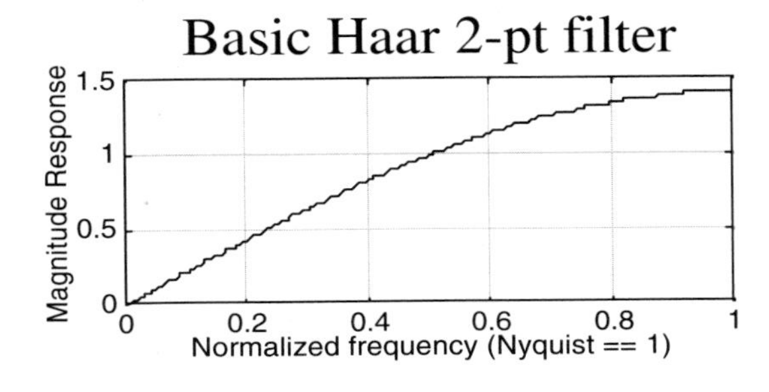

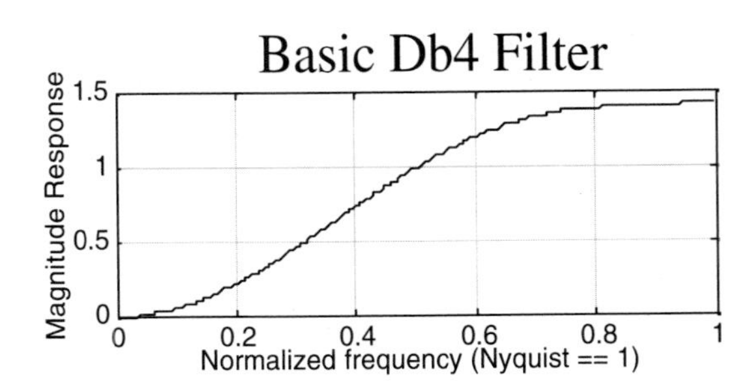

Figure 6.5–1 Frequency Response of 2-point Haar filter (**{1 –1]**) and the 4-point Db4 filter (**[–0.1294 –0.2241 0.8365 –0.4830]**)

Of course as we stretch these filters we see the bandpass nature and the relationship of bandpass center frequency to bandwidth as we did with the Haar. Figure 6.5–2 shows the frequency response for the Db4 wavelet filter stretched to 22 points (by upsampling and lowpass filtering) and then to 50 points.

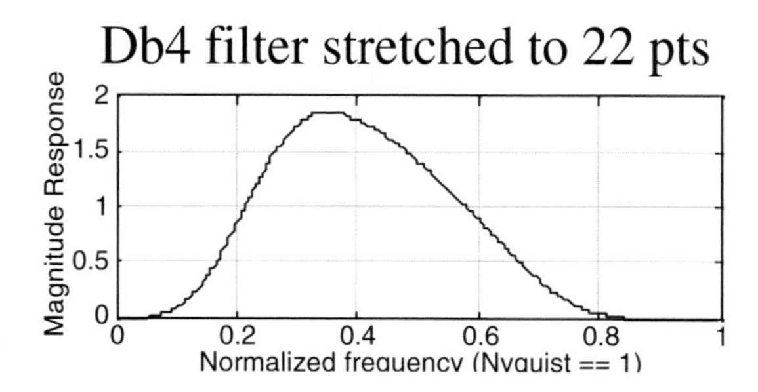

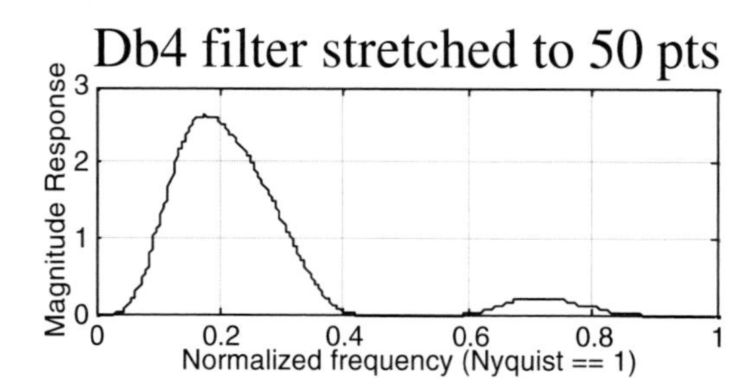

Figure 6.5–2 Frequency Response of 4-point Db4 wavelet filter stretched to 22 points and then stretched further to 50 points.

The frequency characteristics of the "Coiflet 1" (6-point) filter and the 9-point wavelet filter from the "Biorthogonal 4.4" (the 9 point filter in the 9/7 highpass filter set) are shown in Figure 6.5–3.

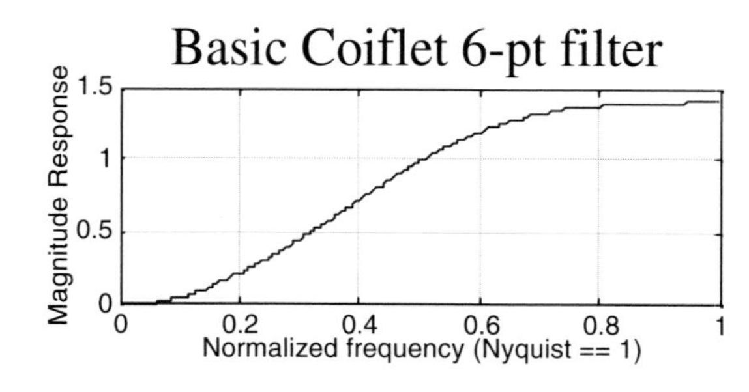

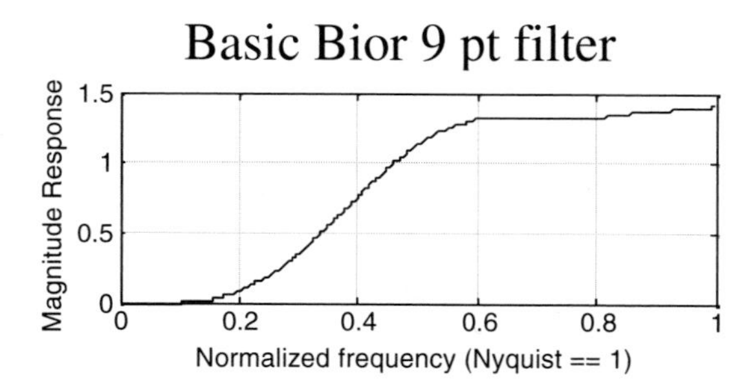

Figure 6.5–3 Frequency Response of the basic 6-point Coiflet filter and the basic Biorthogonal 9-point filter.

Comparing with Figure 6.5–1 we see that these particular basic Coiflet and Biorthogonal filters seem to produce better frequency characteristics, again at the cost of longer filters. We will discuss these filters further, along with other excellent wavelet filters, as we look at some specific properties and suggested applications later.

6.6 Summary

From DSP we learned how to design filters. We learned that a very simple *moving averager* can be approximated by the (non-recursive) Finite Impulse Response (FIR) filter having the difference equation

```
y(n) = x(n) + x(n-1)
```

which leads to the filter **[1 1]***. We have become familiar with this filter in wavelets as the *Haar scaling function filter*. Similarly a very simple *digital differentiator* can be approximated by a FIR filter having the difference equation

```
y(n) = x(n) - x(n-1)
```

which leads to the filter **[1 –1]**. We have seen this filter as the *Haar wavelet filter*.

We also learned from DSP that any time values associated with these filters depends on the data we are filtering. In other words, we simply convolve our data with the filters. This also holds true in the design of *wavelet filters*.

In the previous chapter we learned how to design wavelet filters of any desired length using an explicit equation to produce the filter points. In *this chapter* we learned to design wavelet filters by first interpolating from the very few points of the basic or *mother* wavelet filter to a very large number (256, 768, etc.) of points that approximate a continuous wavelet function. Next, we use these many points to generate an equispaced filter of any desired length. We learned to handle discontinuities (in the function and in its derivatives) by first integrating and then differentiating.

We began this chapter with a review of conventional DSP interpolation techniques, specifically the method using upsampling and the lowpass filtering. We provided insight to this process by showing how a dyadically upsampled signal can be seen as the sum of a low-frequency and a high-frequency signal. Thus filtering out the high frequency signal provides our desired interpolation.

We moved on to the basic Haar wavelet filter (**[1 –1]**) and showed how it can be interpolated by this method to first **[1 0 –1]** and then to **[1 1 –1 –1]**. We

* This filter may be multiplied by a constant (we have seen this in earlier chapters concerning perfect reconstruction to within a delay and a constant of multiplication). For example, using the FIR1 routine from MATLAB we obtain the filter **[0.5 0.5]**.

continued this process to 8, 16, . . . , 256 points. We showed an example of working with discontinuities by integrating and then differentiating. We learned earlier that this is *stretching* of filters is necessary for the Continuous Wavelet Transform (CWT). We will soon address the role of these *stretched-by-interpolation filters* in the Discrete Wavelet Transform (DWT).

We looked at the frequency characteristics of the basic and stretched Haar filters and saw that they are bandpass filters and that as the passband center frequency decreases, so does the width of the passband—while the peak value doubles (constant Q behavior).

We moved on to the basic Daubechies 4-point wavelet filter or *Db4.* As with the Haar (Db2) we demonstrated repeated upsampling and lowpass filtering to produce a 766 point approximation (estimation) of the "continuous" wavelet function. As with the Haar, we then showed how to produce a filter of any desired length for use in the CWT or DWT.

We showed how this 766-point approximation is mapped onto an interval from 0 to 3 and we were able to show a perfect fit of the 4 original points to this estimation at equispaced points 2/6, 5/6, 8/6, and 11/6. We also showed that at 14/6 and 17/6 we have 2 additional zero values. We showed a few examples of other basic wavelet filters and how they fit perfectly onto their estimations of a "continuous" wavelet.

Finally, we showed frequency characteristics of these same basic filters and saw that they were all bandpass in nature with varying passband, transition band, and stopband characteristics. As with the Haar, stretching these filters lowers the passband center frequency and reduces the passband width.

All of these filters, whether stretched by interpolation or by using an explicit equation (*crude wavelets*), can be used in the CWT. The DWT, however, cannot use crude wavelets. We now move on to a comparison of the advantages, disadvantages, requirements, and limitations of the CWT vs. the DWT.

CHAPTER 7

Comparison of the Major Types of Wavelet Transforms

In Chapter Six we moved on from crude wavelets with explicit equations to stretching a filter to any desired length by numerical techniques involving upsampling and low-pass filtering (interpolation) followed by integration and differentiation.

We now discuss the advantages and disadvantages of the Continuous Wavelet Transform (CWT) and explore the two major types of the Discrete Wavelet Transform (DWT).

7.1 Advantages and Disadvantages of the Continuous Wavelet Transform

Figure 7.1–1 (left) shows a linear chirp signal with added noise. The CWT display (right) shows the higher frequencies (at the end of the signal) to be at the lower scales (less stretching of the Db20 "comparison" wavelet filter) as can be seen at the bottom right portion of the display. The display also shows the lower frequencies (at the beginning of the signal) to be at the middle left. It is interesting to see some very low frequencies at the upper left. A close look at the darker areas of the noisy signal (left) shows a slight sinusoidal trend that may not be noticed at first.

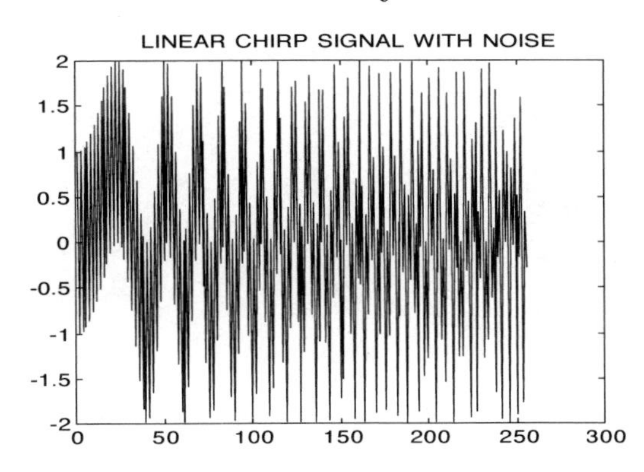

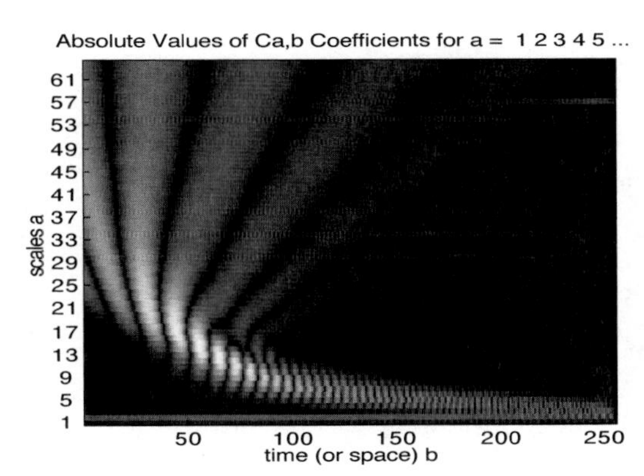

Figure 7.1–1 Linear chirp signal with noise added. The CWT indicates both the time and frequency characteristics of this signal. The excellent results shown here by using a Db20 wavelet also gives an indication of the general shape of the signal.

As we have seen in the above example and in earlier chapters, the CWT provides an excellent overview of the signal. It allows us to identify transient events and to show the time, the frequency, and the general shape of the event (by comparing the CWTs obtained by using different wavelets). We have complete control over the scales with the ability to show every possible stretching and sliding of the wavelet as it is correlated with the signal.

Because the wavelets used in the CWT are not required to conform to the stringent requirements of those used in the DWT (orthogonality, alias cancellations, etc.) we can "invent" our own wavelet. For example, Figure 7.1–2 shows a "fake" wavelet used with a sinusoid "Split Sine" signal (discussed in more detail later) that doubles its frequency at **time = 128**. The CWT using this home-made "wavelet" clearly shows the frequency characteristics of both halves of the signal.

The CWT is excellent for Signal or Image identification when the form of the desired signal is known. It can correlate a signal that has been shifted in time with the many shifted and stretched wavelets used in the CWT correlations. This can come in handy with Doppler shifts, delays, slew, chirping and other kinematic behavior. For example, we could make a "mother" wavelet that replicates the GPS signal we seek. Because of the orbital kinematics of the satellites the received signal will have noise, Doppler, delay, etc. Using the CWT algorithms we should identify the known signal at a particular shifting in time and at a particular scale.

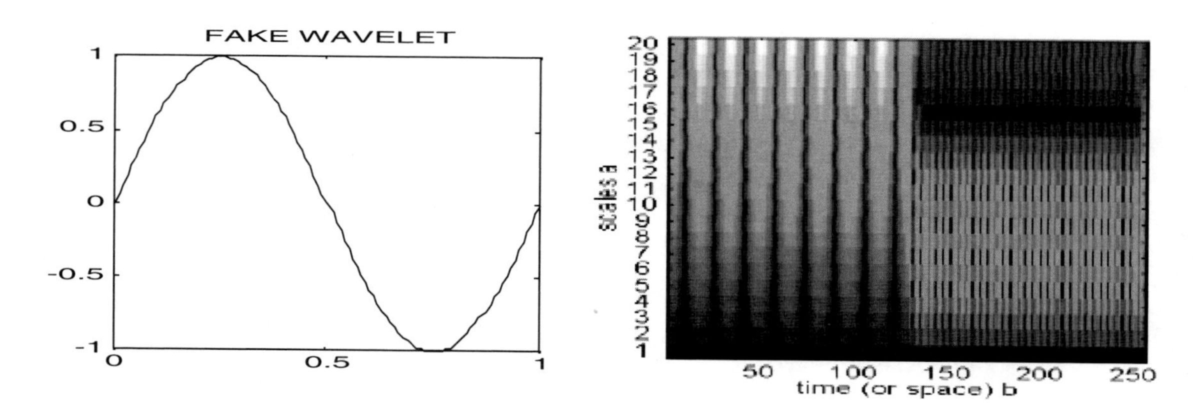

Figure 7.1–2 Arbitrary hand-crafted non-standard "fake" wavelet consisting of one cycle of a sine wave. CWT display of a "split sine" signal that changes frequency halfway through is depicted to show that the "fake" wavelet can be used (at least for a CWT).

One *disadvantage* of the CWT is that it doesn't have a viable inverse transform. We often want to transform, manipulate the data, then take an inverse transform. For example, the FFT can identify a 60-Hz "hum" but we might want to do more than locate it—we might want to filter it out in the frequency domain and then take the inverse FFT to produce a "clean" signal.

In the theoretical world of continuous wavelets an inverse CWT is possible. In practice, however, this is probably not a feasible option*. In other words, it is very easy to do the analysis, but hard to do the synthesis of signals and images. You can "see it clearly but not do much with it". The tasks of compression and denoising are thus better left to the Discrete Wavelet Transforms.

Another disadvantage to the CWT is that it is extremely redundant and can produce tremendous amounts of data. Every possible scale and time is analyzed. This is where the CWT gets the name *Continuous*—for although we are working with discrete data and discrete wavelet filters, this is as close to a "continuous" evaluation as we can get with digital computers! In other words, the analyzing wavelet is shifted smoothly over the full domain of the analyzed function for every possible scale.

In real-world problems with many scales one wonders if it is really necessary to have, for example, a correlation of many thousands of data points with a wavelet stretched to, say, 1000 points followed immediately by a correlation by the same wavelet stretched to 1001 points. In other words, is there much information to be gained by correlations with almost identical stretched wavelets? It is desirable and intuitively feasible to reduce the redundancy.

The *Discrete* Wavelet Transforms (DWT) use only those scales that are a power of 2 (*radix 2*). Thus the DWTs use what we would call a2, a4, a8, a16, etc. in CWT terminology and refer to them as *levels* 1, 2, 3, 4, etc in DWT terminology In equation form ***level* = log2(*scale*)**. For example, a *scale* of **a = 32** becomes *level* 5. Thus we see another advantage of the DWT over the CWT in being able to substantially reduce the amount of data.

When using a DWT for compression, denoising, etc. it is still a good idea to do a *CWT* first as a "sanity check" and to get an initial "feel" for the data.† If the amount of data is very large, the CWT can be specified in software to produce a subset of all the scales—every third or every tenth scale for example.

* An Inverse CWT is a *many-to-one* operation. MATLAB does not currently have an "ICWT" routine. One method requires performing a Discrete Wavelet Transform first, which defeats the purpose.
† The author likes to include a *CWT* as one of the displays for the *DWT*.

7.2 Stretching the Wavelet—The Undecimated Discrete Wavelet Transform

Before proceeding, it is a good idea to talk about the nomenclature. This first type of *Discrete Wavelet Transform* has many qualifiers—*Undecimated, Redundant, Stationary, Quasi-Continuous, Translation Invariant, Shift Invariant, Algorithme à Trous* and others . We will see why these qualifiers might apply as we proceed. The *Conventional DWT* we will look at in the next section is more complicated than the *UDWT* and includes downsampling by 2 or "decimation by 2". A single-level UDWT and Conventional DWT are compared in Figure 7.2–1.

We mentioned in Chapter One that downsampling by 2 means removing every other data point. For example, the sequence "1 2 3 4 5 6" becomes "1 3 5" or "2 4 6" depending upon choosing odd or evens. Choosing to throw away the even times will of course produce a different result than throwing away the odd times. The UDWT does not throw away data and is sometimes referred to as *time-invariant* or *shift invariant.*

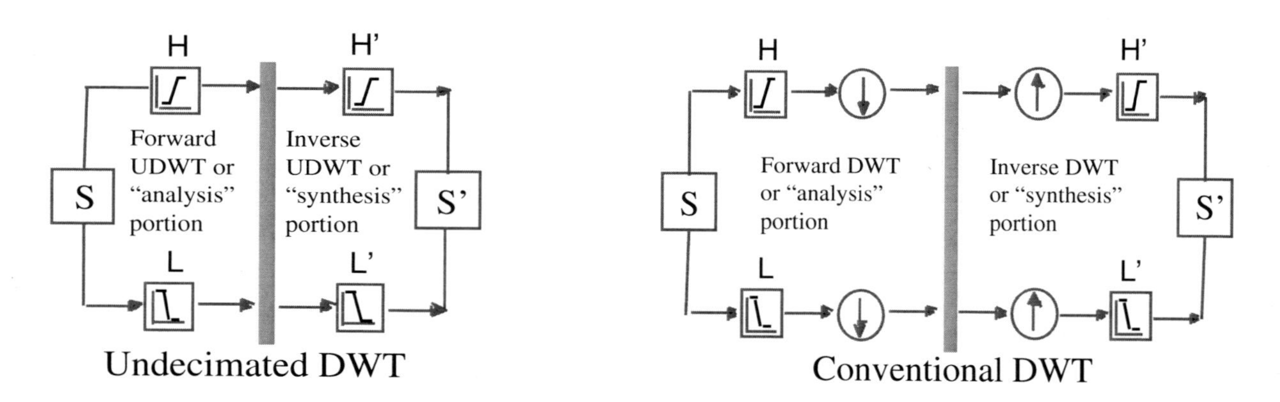

Figure 7.2–1 Comparison of the simple "undecimated" DWT at left with the Conventional (traditional) DWT at right. **H**, **H'**, **L** and **L'** are highpass and lowpass filters. **S** and **S'** are the signal before and after the transform. The circles with arrows represent downsampling by 2 ("decimation by 2") and upsampling by 2.

One wonders then why the Conventional DWT is not called a "Decimated DWT" or "Shift-Variant DWT and leave the descriptors off the simpler UDWT.

The main reason is that the Conventional DWT is in wider use A pastry or soft drink is assumed to have sugar added unless it is specified to be "sugar-free". Thus if you wanted a Cola drink without sugar and without caffeine you would have to use the longer name "Sugar-Free Caffeine-Free Cola" and reserve the simpler name "Cola" for a drink that contains both sugar and caffeine. Similarly, coffee is assumed to have caffeine unless specified as "Decaf"*. Thus we assume the DWT to have decimation unless specified as "undecimated" or by one of the other names.

In the UDWT we stretch ("*dilate*") the wavelet as we did in the CWT, but instead of every possible stretching ("*scaling*"), we stretch by factors of 2 ("*dyadically*"). We slide ("*shift*" or "*translate*") the dyadically stretched wavelet smoothly along the length of the signal as we did in the Continuous Wavelet Transform. Figure 7.2–2 illustrates this method.

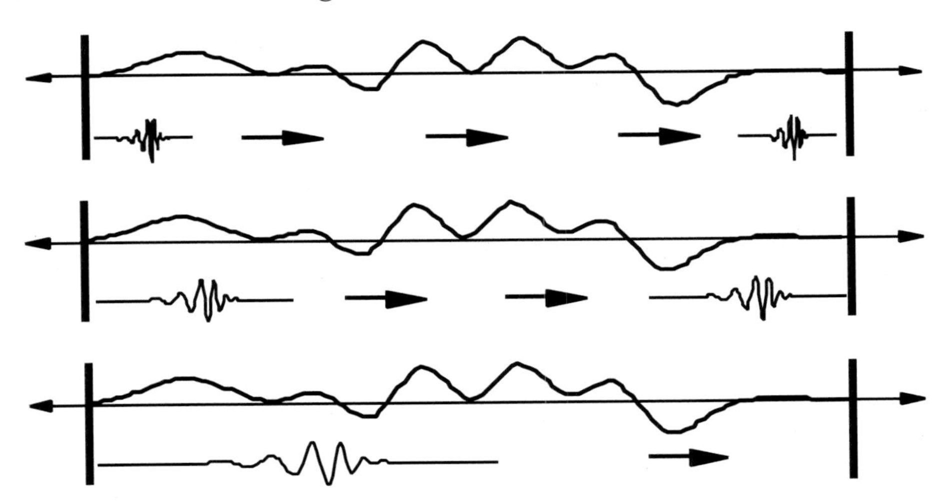

Figure 7.2–2 UDWT stretching and shifting pattern. Like the CWT, the wavelet is stretched and slid smoothly across the entire length of the signal. *Unlike* the CWT, however, The stretching is by factors of 2. The above sketches show the basic wavelet at top, the wavelet stretched by 2 in the middle, and the same wavelet stretched by 4 at the bottom. The signal remains the same at all 3 levels. These 3 levels shown here would correspond to the CWT at scales 2, 4, and 8 and would be referred to as levels 1, 2, and 3.

Notice that the one-to-one correspondence allows for an Inverse UDWT. In fact, the forward and inverse transforms are usually shown together. This is

* You would get funny looks from your waitress if you ordered "Caffeinated Coffee".

true for both the UDWT and the Conventional DWT. Further manipulations of the transformed signal (compression, denoising, etc) are usually performed between the forward and inverse transforms (the vertical gray bar as depicted at left in figure 7.2–3.

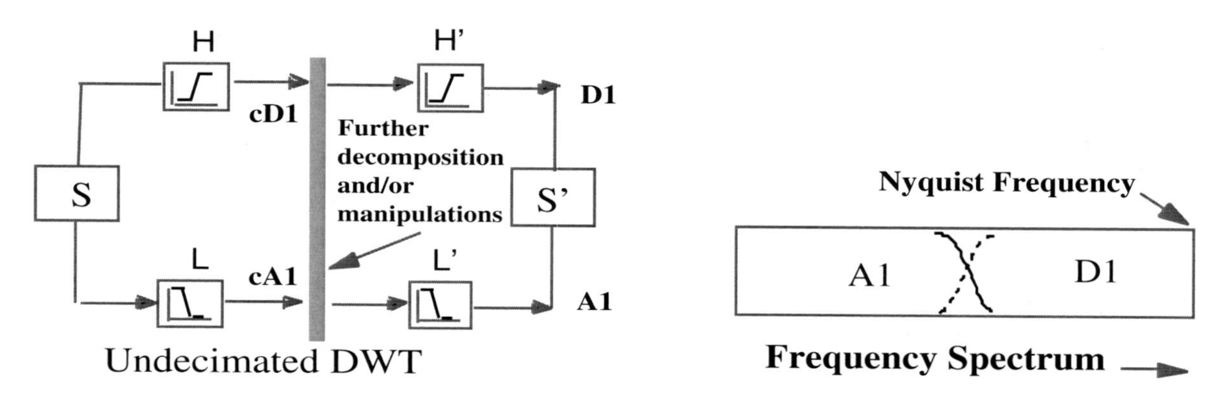

Figure 7.2–3 A single-level UDWT is shown at left. **L** and **L'** are the Lowpass decomposition and reconstruction filters (we will also refer to them as "**lod**" and "**lor**". **H** and **H'** are the Highpass decomposition and reconstruction filters ("**hid**" and "**hir**"). Details and Approximation coefficients are represented by **cD1** and **cA1**. **D1** and **A1** are the final Details and Approximation and combine to reproduce the signal. Notice how the signal bandwidth is divided into a Lowpass and Highpass frequency band as shown at right. Because the filters are imperfect there is a symmetrical overlap as depicted here.

The filters **L** and **L'** combine to make a Lowpass Halfband Filter (**Plp** in wavelet terminology) Similarly **H** and **H'** combine to make a Highpass Halfband Filter (**Php**). Summing the results of the highpass and lowpass halfband filters produces a constant in the frequency domain The final result, **S'**, is the same as the original signal, **S**, except for possible delay and/or a constant of multiplication (depending upon the filters being scaled down or the delay removed earlier).

Using the simple Haar filters we would have for this single-level UDWT

```
H = [-1 1], H' = [1 -1], L = [1 1] and L' = [1 1]
```

Php would be **H∗H'** = **[–1 2 –1]** and Plp would be **L*L'** = **[1 2 1]**. Adding these together we would have **[0 4 0]** leading to a perfect reconstruction of the signal, S, to within a delay (1) and a constant of multiplication (4).

Notice that **cD1** is the *convolution* of the signal with **H** (**[–1 1]**). This is the same as the *correlation* of the signal with **H'** (**[1 –1]**). But this is exactly what we did in the first step of the CWT—correlating the signal with **[1 –1]**. Thus **cD1** is identical with the CWT at **scale** = **2** for the Haar wavelet. In CWT terminology the "*wavelet filter*" would be **H'**. Although we don't use the other 3 filters in the CWT, **L'** is often referred to as the *scaling function filter* as first introduced in Chapter One.

We next look at a 2-level UDWT as shown in Figure 7.2–4. Notice that we have upsampled filters. We recall that upsampling and lowpass filtering is a method of interpolation and that this is how we stretch the filters. If we look in the oval dotted line we see that **L** and **Lup** function as a stretched filter. For the Haar filters, for example, **L** would be **[1 1]** and the upsampled version of **H** (**Hup**) would be **[–1 0 1]**.

Convolving **L** with **Hup** we obtain **[–1 –1 1 1]** which is **H** (**[–1 1]**) stretched by a factor of 2. Similarly we obtain stretched versions of **L** (**[1 1 1 1]**), **L'** (also **[1 1 1 1]**) and **H'** (**[1 1 –1 –1]**). We recognize **H'** as the stretched Haar wavelet at **scale** = **a** = **4** from our earlier CWT discussions. In fact, the 2nd-level Details coefficients, **cD2**, is the convolution of the signal with the stretched **H** filter (in the oval)

```
stretched H = L*Hup = [-1 -1  1  1]
```

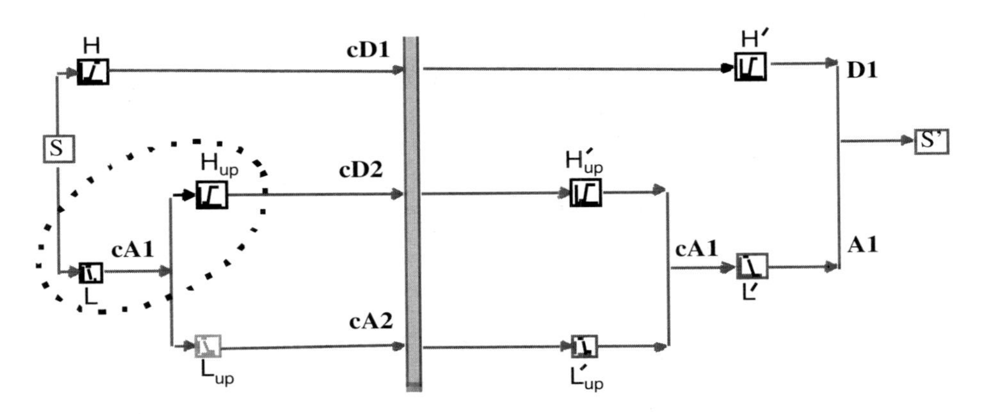

Figure 7.2–4 A 2-level UDWT. We have the 4 filters as we did in the single-level UDWT. We also have the upsampled versions of these filters. Within the dotted oval **L** and **Hup** combine to produce a stretched version of the decomposition highpass filter, **H**. Similarly **L** and **Lup** produce a stretched version of **L**. In the synthesis or "IUDWT" half at right we have the lowpass reconstruction filter **L'** combining with the upsampled versions of **H'** and **L'** (**H'up** and **L'up**) to produce a stretched version of **H'** and **L'**. The coefficients are **cD1**, **cD2** and **cA2**.

But this is the same as a *correlation* with the stretched **H'**

```
stretched H' = H'up*L' = [1  1  -1 -1]
```

Thus **cD2** in this 2-level UDWT is *exactly* the same as the CWT at **scale** = **a** = **4**. This is true not only for the Haar wavelet filters but for the Db4 and many other filters. This is important because we can relate the UDWT to the CWT in learning about these transforms. We will learn that the *Conventional* DWT can also be related to the CWT, but not in such a one-to-one manner as with the UDWT.

At this point we can see the reason for another of the several names for the UDWT. The *algorithme a' trous*, sometimes called the *a' trous* transform is French for "with holes"*. Upsampling the filters places zeros or "*holes*" in the filters. Figure 7.2–5 shows a 3-level UDWT. We now see more "holes" (zeros) as we twice upsample the filters. For the Haar wavelet filters then we would have

```
L = [1 1], Lup = [1 0 1]
Lupup = [1 0 0 0 1], Hupup = [-1 0 0 0 1]
```

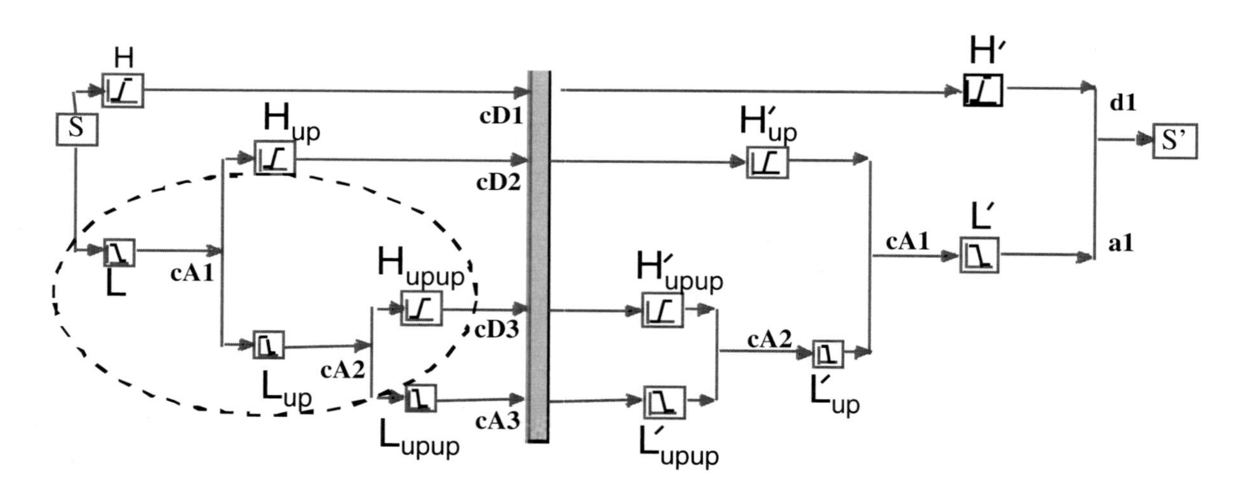

Figure 7.2–5 A 3-level UDWT. This is similar to the 2-level UDWT but now we have upsampled versions of the upsampled filters as **Lupup**, **L'upup**, **Hupup** and **H'upup**.

Within the dotted oval we then have the convolution of 3 filters

---

* *Trousers* (with holes, of course, for legs and waist) were worn by working-class males during the French revolution.

```
twice-stretched H = Hupup*Lup*L = [-1 -1 -1 -1 1 1 1 1]
```

Notice that the on the right side of diagram we have

```
twice-stretched H'= H'upup*L'up*L=[1 1 1 1 -1 -1 -1 -1]
```

But this is the stretched filter at **scale** = **a** = **8** we saw in the CWT. Again, the *convolution* of the signal with [–1 –1 –1 –1 1 1 1 1] is the same as the *correlation* of the signal with [1 1 1 1 –1 –1 –1 –1]. This means that at **cD3** we have exactly the same result as a CWT at **scale** = **a** = **8**.

Thus we can relate the UDWT to the CWT by noting that **cD1** (at the first level) is the same as the CWT at **scale** = **2**, **cD2** is the same as the CWT at **scale** = **4**, and **cD3** is the same as the CWT at **scale** = **8**. We also note the log2 relationship of the CWT scale number to the UDWT (or Conventional DWT) level number (**scale number** = **2^level number**).

Looking again at Figure 7.2–5 we see that the coefficients **cD1**, **cD2**, **cD3**, and **cA3** are all approximately the same size as the signal. We will next look at the more common method where these coefficients are smaller than the signal. This is why the UDWT (or a' trous) is sometimes referred to as the *redundant* Discrete Wavelet Transform or *RDWT*. It should be remembered, however, that levels 1, 2, 3, 4, etc. in the UDWT can be compared to scales 2, 4, 8, 16, etc. in the CWT and thus this transform is not nearly as "redundant" as the CWT. Also "redundant" doesn't mean any data is duplicated.

The UDWT is a very robust powerful transform and does not have to account for aliasing, as does the Conventional (downsampled) DWT we will now study. We will return to the UDWT later to demonstrate its capabilities.

7.3 Shrinking the Signal—The Conventional Discrete Wavelet Transform

A fair question is: "Instead of stretching all the wavelet filters, could we leave the filters alone and shrink the signal instead"? The answer is "Yes, if you know what you are doing and are careful!" This is accomplished by down-sampling the signal, usually by a factor of 2 ("*dyadic downsampling*"). This is also referred to as *decimation by 2.* In other words, by taking every other sample (downsampling by 2) we are in effect "shrinking" the signal by a factor of 2.

To an experienced DSP engineer, the terms *downsampling* and *decimation* or any other term referring to discarding all the odd or even data points in the signal should also bring to mind the possibility of *aliasing*. We will proceed to show how, if done correctly, the aliasing can be canceled out.

Figure 7.3–1 illustrates the concept of shrinking the signal while leaving the wavelet (filter) unchanged. This the method used in the Conventional DWT (usually called simply the DWT).

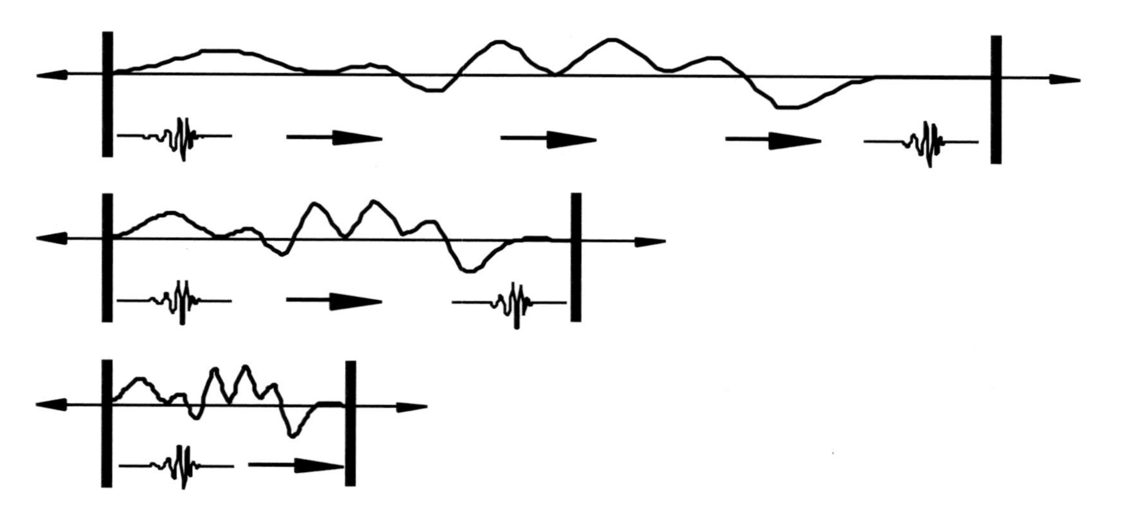

Figure 7.3–1 Conventional DWT method. Instead of *stretching the wavelet (filter)* like we did in the CWT and UDWT we *shrink the signal* instead by downsampling by 2. The above sketch shows the original signal at top, the signal downsampled by 2 in the middle graph, and the same signal downsampled by 4 at the bottom. In this transform the *wavelet* remains the same at all 3 levels.

Like the Undecimated DWT (UDWT), we change the scales by factors of 2 also. We look only at scales 1, 2, 4, 8, 16 etc. and call them “level 1, 2, 3, 4, 5 etc.”.

As the signal becomes half as long through downsampling, when we shift in time (“translate”) the wavelet forward by 1 point we are, in effect, advancing by 2 points in the original signal.

Thus the Conventional DWT works with less data then the UDWT and *far less* data than the CWT. Thus it is sometimes referred to as a *Fast Wavelet*

Transform because these factors of 2 remind one of the familiar *radix 2 Fast Fourier Transform.*

Jargon Alert: **Radix 2 means powers of 2 such as 2, 4, 8, etc.**

Like the UDWT, the one-to-one correspondence allows for an inverse DWT. We have presented the single level DWT in Chapter One and also in the previous section. We show here a 2-level Conventional DWT in Figure 7.3–2.

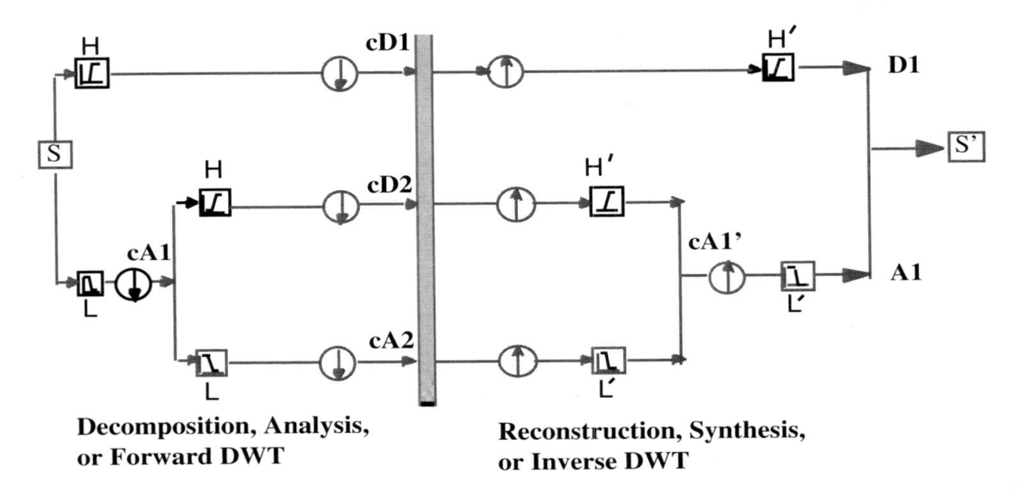

Figure 7.3–2 A 2-level Conventional DWT. Notice that only the basic 4 filters—**L**, **L'**, **H**, and **H'**—are used. Instead of stretched filters we downsample the signal as shown in the circles. The *Approximation* and *Detatils coefficients* are given by **cA1**, **cA2**, **cD1**, and **cD2**. The *final Approximation and Details* are given by **A1** and **D1**. Further manipulations such as compression and denoising are done between the Forward and Inverse DWT as depicted in the vertical bar.

If a signal, **S**, were 10,000 points long, the first level Approximation Coefficients (**cA1**) would be roughly 5000 points long due to downsampling by 2. Similarly **cD1** would be roughly 5000 points. Because of further downsampling the 2nd level Details Coefficients (**cD2**) would be roughly 2500 points long as would **cA2**. Notice that if we were to sum **cD1**, **cD2**, and **cA1** we would **2500 + 2500 + 5000 = 10,000 = original signal length** (the sum would be slightly greater due to the convolutions with the filters **H** and **L**). In other words, the computer storage required for the coefficients is roughly the same as that required for the signal.

It is tempting to think that with this decomposition that "the sum of the parts is equal to the whole". This is not the case, however. We cannot simply add the coefficients together but must first *reconstruct* them.

Figure 7.3–3 shows a functionally equivalent 2-level DWT. We have taken the last 2 steps leading to **A1** (ref. Fig. 7.3–2) and drawn them as separate paths. We are able to do this because of the *Linear Time Invariant* or *LTI* nature of these filter banks.

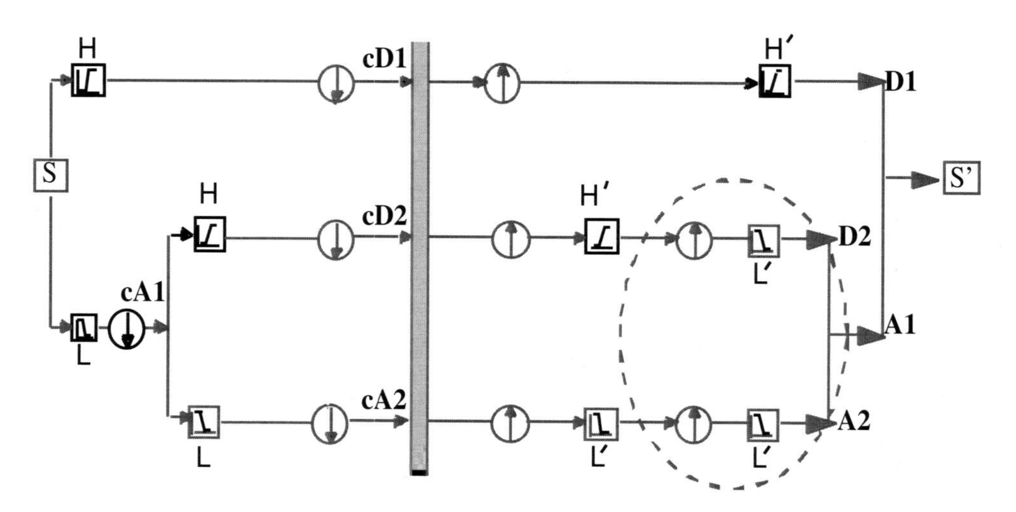

Figure 7.3–3 Functionally equivalent 2-level Conventional DWT. By replicating the upsampling by 2 and filtering by **L'** as depicted in the dotted oval, we are able to construct **D2** and **A2**. **D2** and **A2** can now be added to reproduce **A1**. Similarly, **A1** and **D1** sum to reproduce the signal. All 4 of these final Approximations and Details are now approximately the same length as the original signal. (In practice we trim these final Details and Approximations to the exact length of the signal so they can add directly.)

We can modify the *coefficients* or the *final Details and Approximations* for denoising, compression, etc. We showed examples in Chapter One of modifying the final results and in Chapter Four of modifying the coefficients.

It is important to understand that the Conventional DWT guarantees complete alias cancellation and perfect reconstruction of the signal (to within a possible delay and multiplication constant) *only when we do not modify the coefficients or final results*! Modifications must be made carefully to minimize the effects of aliasing.

When we compress or de-noise the signal we *do not wish* to have our original signal back and thus perfect reconstruction is not the goal—we willingly discard part of the data. Where engineers get into trouble is in not realizing that in doing so they are also discarding part of the alias cancellation!

It turns out that when we keep large values and discard small values (as is often done in compression) we have a minimal impact on the alias cancellation capability. When we discard zeros we have *no adverse impact* on alias cancellation. We will talk more about alias cancellation and demonstrate this concept in both the time and frequency domains later.

The frequency allocation chart for the 2-level DWT system (Fig. 7.3–3, above) is the same as that for the 2-level *UDWT* and is shown here in Figure 7.3–4.

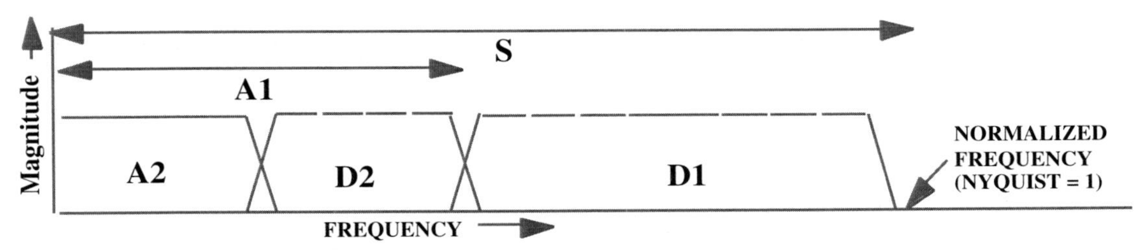

Figure 7.3–4 Frequency allocation of the 2-level Conventional DWT (and of the 2-level UDWT). Notice how **A2** and **D2** combine to reconstruct **A1**. **A1** and **D1** then can be combined to reconstruct the signal. Note the overlaps in the various subbands due to non-ideal filtering.

This figure is a simplification, of course, for tutorial purposes. There are 2 caveats that should be addressed: First, the transition bands shown above can be much wider, as we saw in Chapter Six for the Haar wavelet. Second, there may be aliasing present in **A2**, for example, and although it will be canceled by the addition of **D2** it may not lie in the frequency bandwidth depicted. Note that for the *Undecimated DWT* we have the first but not the 2nd of these warnings. In other words, with no decimation we have no aliasing problems.

There are 2 further questions that must be addressed before proceeding with the Conventional DWT and its variations: The first question would be "Why use the Conventional DWT at all when the Undecimated DWT has no aliasing concerns.

In fact, this software is available under the name Stationary Wavelet Transform or (SWT).* Furthermore, the UDWT is *time-invariant* (or *shift invariant*) and we don't have to worry about whether to discard the odd or even points. This first question takes on extra relevance as we glance again at Figure 7.3–3 and notice that **A2**, **D2**, **A1**, and **D1** have all been built up (then trimmed slightly) to be the same length as the original signal. Thus the advantage of saving storage space would be nullified. This is a correct assumption *if we actually use these final results*. In this case it may be safer to use the UDWT.

The answer to this first question is that we may be able to perform compression and/or de-noising modifying *only the coefficients* in such as way that we have minimal or zero aliasing (as is done in JPEG compression). Figure 7.3–2 showed the diagram for a Conventional DWT that produces only **A1** and **D1** as final results to be added. This does not require as much computation and storage. In fact, with the storage requirements for all the coefficients being about the same as the original signal we can perform in-place storage. If we can (safely and correctly) use only the coefficients for compression and/or denoising this can add up to a significant increase in speed and saving of storage resources for a multiple level transform.

The 2nd question, "Why not just use Finite Impulse Response (FIR) filters like we do in conventional DSP?", also deserves consideration. Looking again at Figure 7.3–4 we know there are already excellent lowpass, bandpass, and highpass digital filters to apportion the bandwidths as desired.

The answer to this 2nd question is that the wavelet filters have the ability to discriminate in both time and frequency. In other words, they not only have a limited bandwidth but they have a limited time-interval. Thus wavelets are a powerful method to examine and work with signals that also start and stop and have a limited bandwidth (transient signals).

An example was shown in Chapter One of how wavelets can be used to defeat a chirp jammer. We reproduce an excerpt here as shown below in Figure 7.3–5.

* Thus the name Stationary Wavelet Transform joins the many other descriptors of the simpler form including *Undecimated* DWT, *Redundant* DWT, *Time-Invariant* DWT, *Shift-Invariant* DWT, the *a' trous* DWT (because of the "holes") and *quasi-continuous* DWT (because of the smooth shifting of the stretched wavelets across the undecimated signal as is done in the CWT).

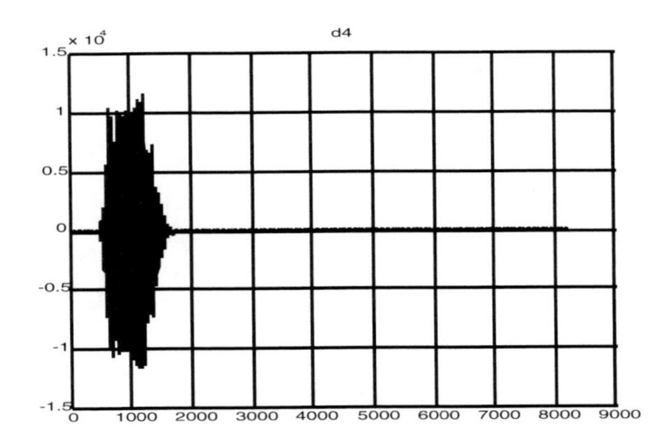

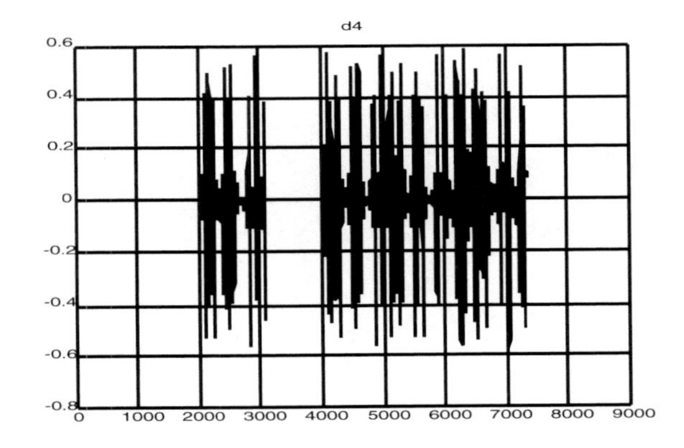

Figure 7.3–5 A signal embedded in 80 dB of noise is decomposed with a multi-level DWT using 20-point Daubechies wavelet filters (Db20). The 4th level final result, **D4**, is shown at left. The noise completely obstructs the signal for times less than 2000 as shown. With wavelet filters, however, we can selectively threshold out the very large values and keep the remnants of the signal as shown at right.

Using an automated time-dependant thresholding method we can selectively threshold out noise *for any desired time interval*. In this example we set the noisy portions of **D1**, **D2**, **D3**, **D4** (shown here), **D5**, etc. to zero at the appropriate time intervals and keep only the viable portions of the signal. We reconstructed the signal from these selectively-filtered remnants at different times and were able to successfully extract the signal from the noise using wavelet technology. This would not have been possible using conventional filtering methods.

7.4 Relating the Conventional DWT to the Continuous Wavelet Transform

In Section 7.2 we saw that the UDWT had a very direct relationship with the CWT. The Details coefficients **cD1**, **cD2**, and **cD3** (ref. Fig. 7.2–5) are exactly the same as the CWT results at scales 1, 2, and 3. There is also a somewhat direct relationship of the Conventional DWT to the CWT. We take another look at the 3-level (decimated) DWT as shown in Figure 7.4–1.

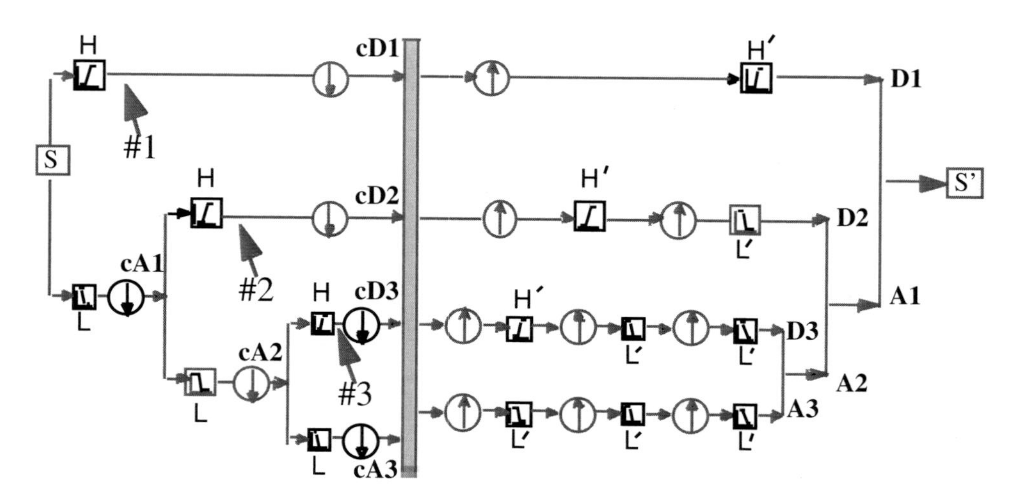

Figure 7.4–1 A 3-level Conventional DWT. The right half is drawn to produce the final Approximations and Details. Large arrows **#1**, **#2**, and **#3** indicate points at which the results can be compared to those of a CWT.

At arrow **#1** the signal has been convolved with **H** or, equivalently, *correlated* with **H'**. For the Haar wavelet filters this is exactly the same as the results from the CWT at **scale** = **2**. It is also the same at this point as the UDWT results. At arrow **#2** the results are not the same as the CWT at **scale** = **4**. However they *are* the same *if we were to downsample the CWT results.* At arrow **#3** the results are the same as if we were to *twice downsample* the CWT results at **scale** = **8**.

The comparison is valid for wavelets other than Haar. For example, using the Db4 wavelet filters, arrows **#1**, **#2**, and **#3** on the DWT (Fig. 7.4–1) correspond to scales on the CWT of "2.5", 5, and 10 (when downsampled).

Thus we see that the CWT, the UDWT and the Conventional (decimated) DWT are related in that they use comparison (correlations) of the signal with the wavelet filters. Furthermore we have seen that these comparisons are related to each other in the 3 major types of wavelet transforms. This is not a surprising result—conventional DSP filtering *convolves* the signal with a filter and this can also be viewed as a *correlation* or comparison with the flipped version of the filter.*

* Strictly speaking, a robust cross correlation zero-pads the vectors to the same size. For example the convolution of ***a*** = *[1 2]* with ***b*** = *[3 4 5]* is *[3 10 13 10]* while the cross correlation of the flipped version of ***a*** *([2 1])* with ***b*** **is** *[3 10 13 10 0]*. Ignoring leading or trailing zeros, however, the above statement still holds.

7.5 Decomposing *All* the Frequencies—The Wavelet Packet Transform

In both the Conventional and the Undecimated DWT we have decomposed the signal into Details and Approximations. In multi-level transforms we have decomposed the Approximations but have not performed any further decompositions of the Details. Figure 7.5–1 shows the bandwidths for the 3-level Conventional DWT.

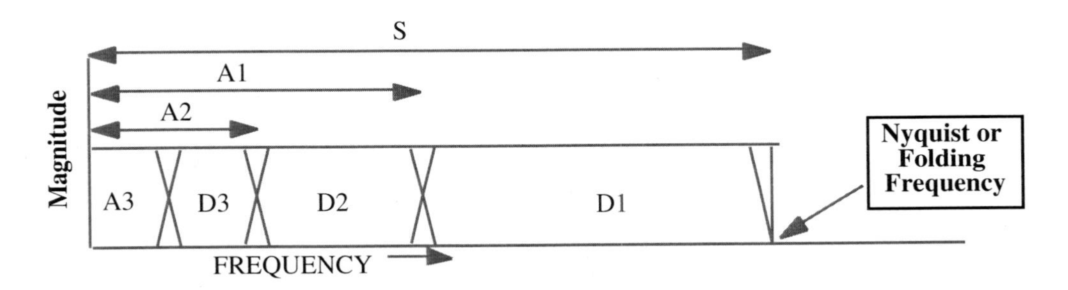

Figure 7.5–1 Frequency allocation for a 3-level Conventional DWT (or a UDWT). Imperfect filtering is indicated in this sketch by the overlapping transition bands.

In many applications we may be interested in both the higher and the lower frequencies. One of the strengths of wavelets is to provide a method to selectively remove specific unwanted frequencies for a specified period of time. In audio and speech processing the highest frequencies usually represent noise and we often want to remove them. In image processing we are often more interested in the features of the image than in the high-frequency "haze". This is the rationale for decomposing only the lower-frequency Approximations. Thus we see the wider sub-bands as shown above (Fig. 7.5–1). In other applications, however, we may be more interested in the high or the middle frequencies. For these application we use the Wavelet Packet Transform (WPT).

The WPT then is a variation of the DWT (or UDWT) that performs decomposition of both the Approximations and the Details. Figure 7.5–2 shows a 2-level WPT based on the Conventional (decimated) DWT.

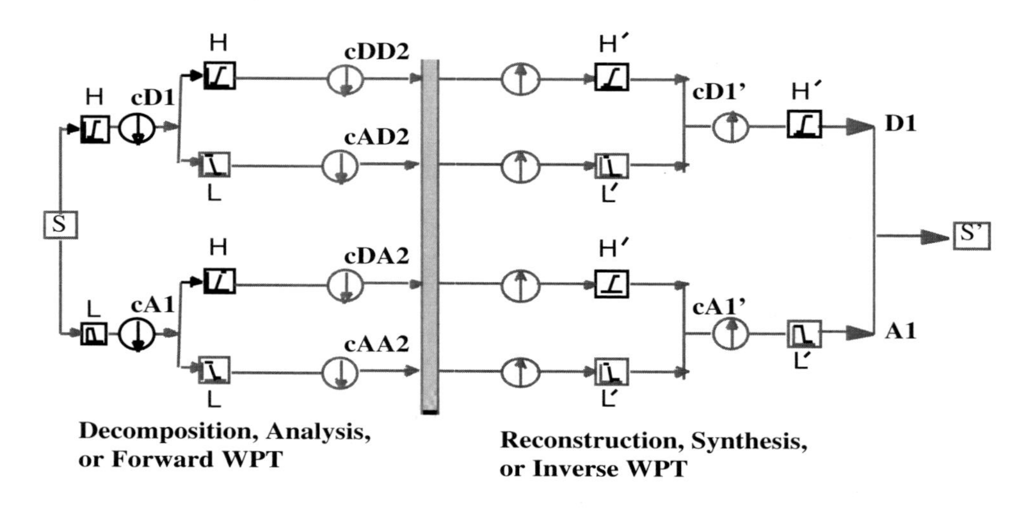

Figure 7.5–2 A 2-level (decimated) Wavelet Packet Transform. Both the high and the low frequencies are decomposed and then later reconstructed. We use the same 4 basic filters as in the DWT. Note that we have 4 sets of coefficients.

Figure 7.5–3 shows the equivalent diagram with paths drawn separately to reconstruct all the final Details and Approximations.

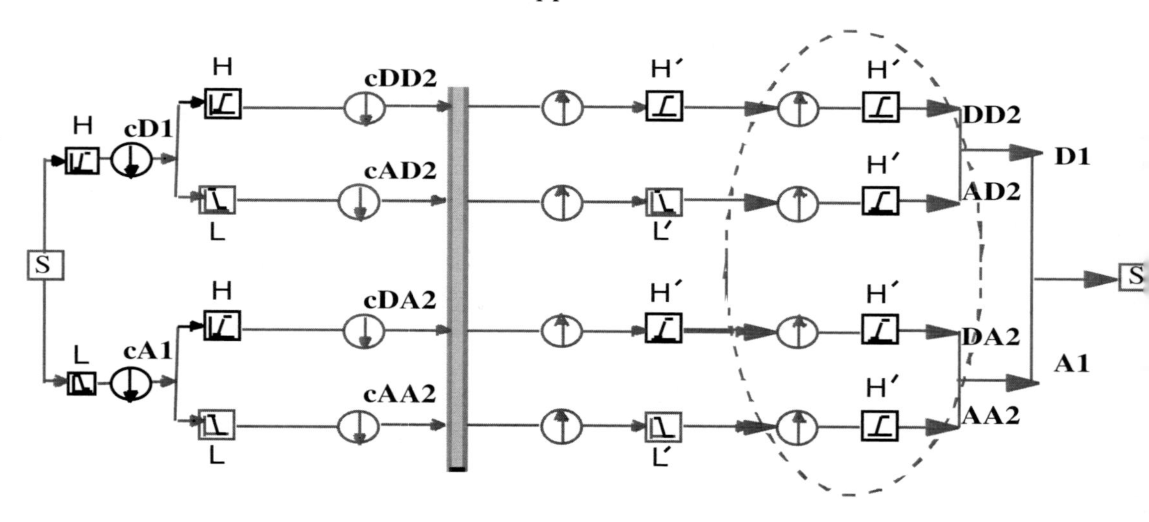

Figure 7.5–3 The 2-level (decimated) Wavelet Packet Transform redrawn (dotted oval) to produce the final Details and Approximations.

The final results are trimmed to signal length and then can be added together. In the DWT we had **D2** + **A2** = **A1** with **D1** by itself. Here we have **AA2** and **DA2** adding together to produce **A1** while **AD2** and **DD2** produce **D1**. These final results (**DA2**, **A1**, etc.) are sometime referred to as *nodes*[*] Figure 7.5–4 shows the frequency sub-bands for a 3-level WPT.

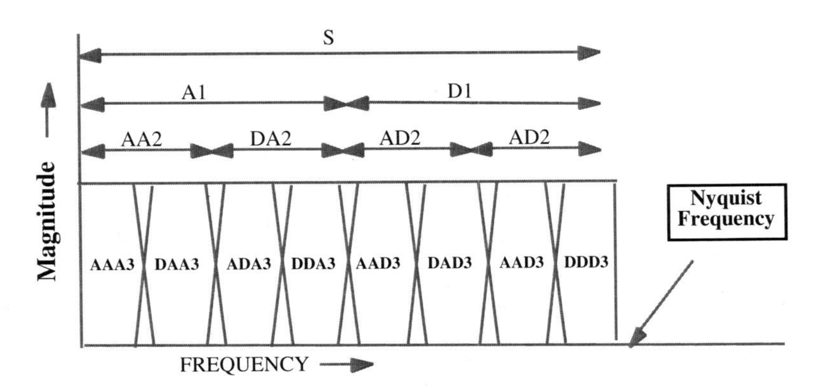

Figure 7.5–4 Frequency allocation for a 3-level WPT. Note that the sub-bands of frequency shown for level 3 here are all the same size. Imperfect filtering is indicated by the non-vertical transition bands between the frequency bands in this sketch.

We can see from the above figures that the signal, **S**, can be decomposed and reconstructed a number of ways. For example, we could use nodes **AAA3**, **DAA3**, **DA2**, and **D1** (**AAA3** + **DAA3** = AA2, **DA2** + "AA2" = **A1**, **D1** + "A1" = **S**). Wavelet packet software gives you a choice the nodes. We will talk later about "*best basis*" routines that use entropy and cost functions to help choose the optimum nodes.

Before leaving the Wavelet Packet Transform it is interesting to see what would happen if we switched the right and left halves of the transform as shown in Figure 7.5–5. This is a method used in *transmultiplexers*.[†]

[*]This same terminology of *packets* and *nodes* is found in *packet switching*, a communications scheme (and precursor to the modern internet) in which *packets* (discrete blocks of data) are routed between *nodes* over data links shared with other traffic.

[†] Transmultiplexer engineers were surprised to see their familiar filter banks used in wavelet technology.

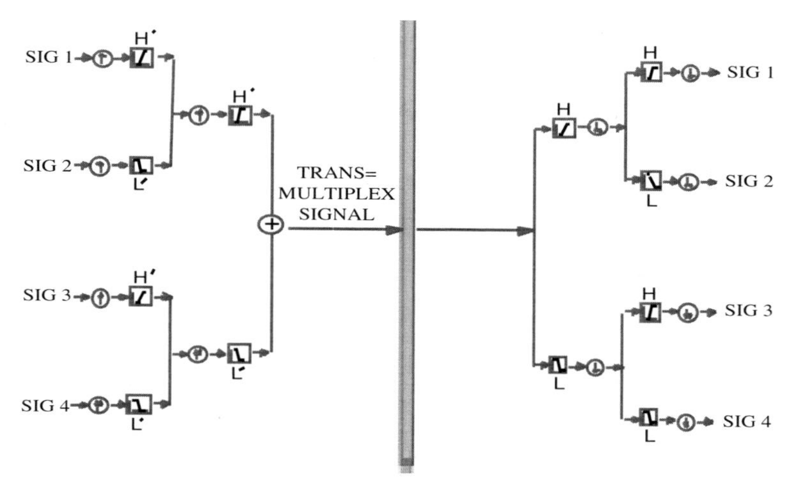

Figure 7.5–5 A 2-level transmultiplexer diagram. Four signals are input and then combined into a single path. The vertical gray bar here represents a space link or other single path. At the receiver, the transmultiplexed signal is de-multiplexed into the original 4 signals. This is a rudimentary form of a Frequency Division Multiple Access (FDMA) scheme.

7.6 Summary

In this chapter we explored and compared the 3 major types of wavelet transforms—the Continuous Wavelet Transform (CWT), the Undecimated Discrete Wavelet Transform (UDWT), and the Conventional (decimated) Discrete Wavelet Transform (DWT). We also looked briefly at the Wavelet Packet Transform (WPT) which is a variation of the DWT or UDWT which decomposes both the low and high frequencies.

We looked at the strengths of each of these major types of transforms and indicated the general types of applications of each. We also indicated vulnerabilities *which must not be ignored* such as aliasing and time-variance in the Conventional (decimated) DWT, the lack of inverse capability in the CWT and possibly problematic storage space requirements in the UDWT.

We were able to compare and relate each type of transform to the others and in doing so discovered that we are still comparing data (original signal or downsampled) with the various wavelet filters (basic or stretched).

We now proceed to look in more detail at the Perfect Reconstruction Quadrature Mirror Filters (PRQMF) that combine to build the foundation of the Discrete Wavelet Transforms—the Halfband Filter.

CHAPTER 8

PRQMF and Halfband Filters and How They Are Related

*In previous chapters we have introduced the decomposition and reconstruction highpass and lowpass wavelet filters (**L**, **L'**, **H** and **H'**) and demonstrated how they are used in the major types of wavelet transforms. We have shown how to either stretch the filters or how to (carefully) shrink the signal instead. We have also looked at some of the general strengths and weaknesses of the various filter banks that comprise these transforms.*

We will now look in more detail at the Perfect Reconstruction Quadrature Mirror Filters (PRQMF) and how the various permutations of this type of filter relate to each other. Then we will show how the Halfband Filters—the very foundation of discrete wavelet transforms—can be factored into these PRQMFs.

We will explore the concept of orthogonality and discover how a signal can be efficiently represented by orthogonal wavelets (filters) giving us a viable basis for DWTs

Finally, we will see how the Halfband Filters can be factored a different way into 2 sets of unequal length filters having limited interrelationships and modified orthogonality—but with perfect symmetry—the biorthogonal wavelets.

8.1 Perfect Reconstruction Quadrature Mirror Filters and their Inter-Relationships

The Haar filters, the Db4 filters, and many others we will study have the capability for perfect reconstruction if used correctly in a UDWT, a conventional DWT, or a Wavelet Packet Transform. We have demonstrated this capability in our earlier walk-throughs of the DWT and UDWT. This is where the Perfect Reconstruction ("PR") in PRQMF comes from.

The "M" in PRQMF indicates that these filters are *mirror* images of each other. Figure 8.1–1 sketches the frequency characteristics of the simple Haar filters. The lowpass filters (**L** and **L'**) are shown at left and the highpass filters (**H** and **H'**) are shown at right. If we were to flip horizontally either filter about a point 1/2 of Nyquist frequency they would be identical to the other.

The "Q" in PRQMF comes from using DSP terminology of 2π radians in the unit circle. Thus Nyquist frequency can be called $2\pi/2$ (instead of π). With this naming convention half the Nyquist frequency would then be $2\pi/4$ (instead of $\pi/2$) indicating one fourth of the unit circle. Thus the name "Quadrature".

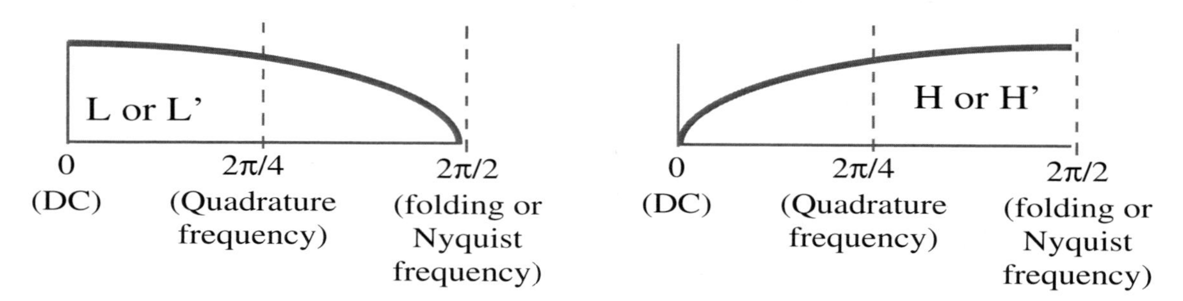

Figure 8.1–1 Rough sketch of the frequency response of the Haar PRQMF filters. Notice that these filters are the mirror image of each other and that if either filter were rotated about a point 1/2 of Nyquist ($\pi/2$ or $2\pi/4$) they would replicate the other filter.

In order to see the relationship of the filters **L**, **L'**, **H** and **H'** to each other we need to look at the Db4 filters rather than the simpler Haar filters.* The basic Db4 filters are again

```
L  = lod = [-0.1294    0.2241    0.8365    0.4830]
L' = lor = [ 0.4830    0.8365    0.2241   -0.1294]
H  = hid = [-0.4830    0.8365   -0.2241   -0.1294]
H' = hir = [-0.1294   -0.2241    0.8365   -0.4830]
```

L, **L'** **H** and **H'** all use the same 4 numbers but in different sequence and signs. Figure 8.1–2 shows the 4 basic Db4 filters and how they relate to each other.

We remember that all these values may be "pre-divided" by a constant such as 2 or sqrt(2) to avoid having to divide the reconstructed signal (S') at the end. Dividing by sqrt(2) we would have

```
L  = lod = [-0.0915    0.1585    0.5915    0.3415]
L' = lor = [ 0.3415    0.5915    0.1585   -0.0915]
H  = hid = [-0.3415    0.5915   -0.1585   -0.0915]
H' = hir = [-0.0915   -0.1585    0.5915   -0.3415]
```

* Albert Einstein said things should be made as simple as possible, but no simpler.

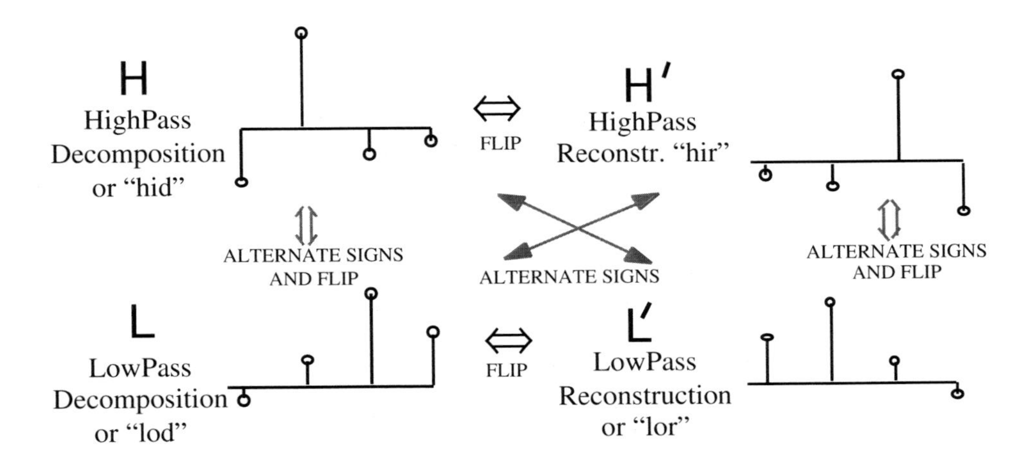

Figure 8.1–2 The four basic Db4 filters and their relationship to each other as seen in the time domain. **H'** is sometimes called "the wavelet filter" and **L'** "the scaling function filter".

The 4 basic numbers for the Db4 are not as arbitrary as they first appear. Using the sequence for **L'** or **lor** (the "wavelet filter") we have

```
(1+a)/4b, (3+a)/4b, (3-a)/4b, -(1-a)/4b
```

Where **a** = **sqrt(3)** and **b** = **sqrt(2)**. We will revisit these numbers soon.

The simpler Haar filters also obey this same relationship but it is harder to see. Again for the Haar wavelet filters we have

```
L  = lod = [ 1  1]
L' = lor = [ 1  1]
H  = hid = [-1  1]
H' = hir = [ 1 -1]
```

With the Haar wavelets, for example, **L'** is still the flipped version of **L** but the relationship is not as obvious as with the Db4 **L** and **L'** filters.

Thus, like the Haar filters, all 4 Db4 filters use the same numbers. Only the signs and the positions change. We can also see more clearly now the convolution/correlation relationships in the discrete wavelet transforms. For example, we can see that the convolution of the signal, **S**, by **L** is the same as the correlation with flipped version, **L'**. A 1-level UDWT and a conventional DWT are redrawn below in Figure. 8.1–3.

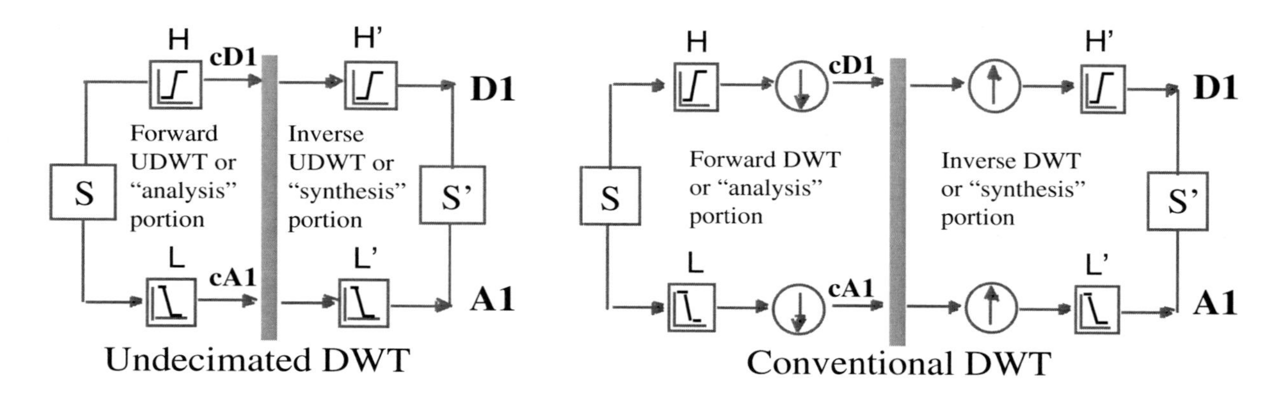

Figure 8.1–3 The four basic Db4 filters as used in a single-level UDWT and Conventional DWT.

8.2 Perfect Reconstruction Begins with the Halfband Filters

We look at a single-level UDWT as drawn in Figure 8.1–3 (left). At the end of the lower (lowpass) path we have for **A1**

```
A1 = S*L*L' = S*(L*L')
```

which means we are convolving **S** with a filter built from **L** and **L'**. If we look closer at this L*L' filter we will find it is a lowpass halfband filter. Using the above filter values from the Db4 we can compute this combined filter. We have for **L** and **L'** again

```
L  = lod = [-0.1294    0.2241    0.8365    0.4830]
L' = lor = [ 0.4830    0.8365    0.2241   -0.1294]
```

convolving the 2 we produce a lowpass halfband filter. Following popular naming convention we have ***Plp*** as

```
Plp  = [-0.0625   0   0.5625   1   0.5625   -0.0625]
```

This process is depicted in Figure 8.2–1.

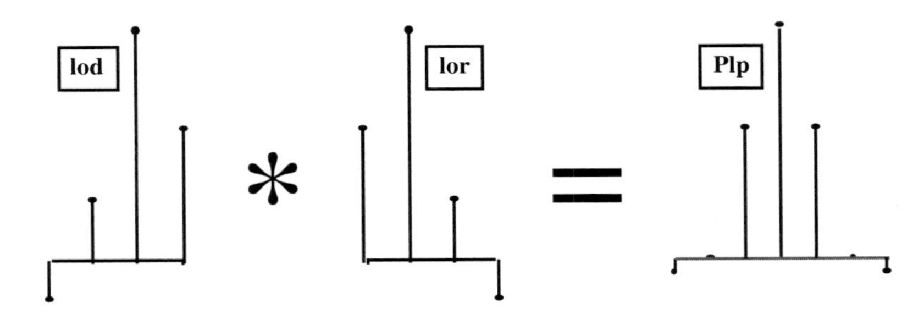

Figure 8.2–1 Convolution of the 4-point lowpass decomposition filter (L) and the 4-point lowpass reconstruction filter (L') produces the 7-point lowpass halfband filter **Plp**.

On the upper highpass path of the UDWT we have

```
D1 = S*H*H' = S*(H*H') = S*(Php)
```

where ***Php*** is the convolution of **H** and **H'**. We then have (ref. Fig. 8.2–2)

```
Php = [0.0625  0   -0.5625   1   -0.5625   0   0.0625]
```

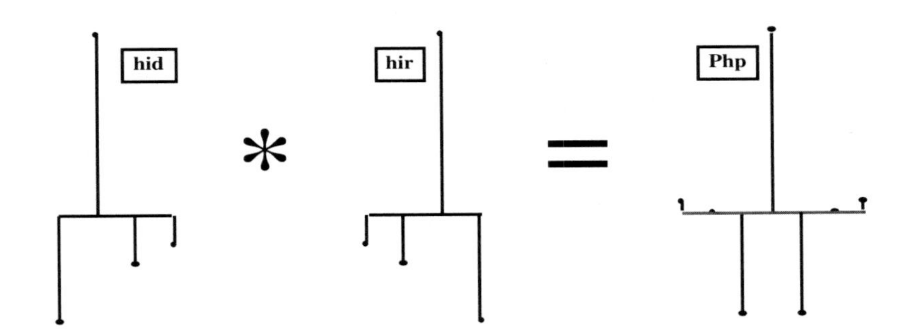

Figure 8.2–2 Convolution of the 4-point highpass decomposition filter (H) and the 4-point highpass reconstruction filter (H') produces the 7-point highpass halfband filter **Php**.

We notice that even though none of the four 4-point wavelet filters are symmetrical and thus do not have linear phase that *both* the halfband filters they produce by convolution *are* symmetrical and *will* have linear phase.

If we add **Plp** and **Php** together we have

```
Plp + Php = [0   0   0   2.0   0   0   0]
```

This means that as we convolve the signal with **Plp** and **Php** and add the results together to produce **S'** (ref. Fig. 8.1–3) we will have perfect reconstruction to within a delay (3 time increments corresponding to the 3 zeros) and a constant of multiplication (2.0 here). If we choose to "pre divide" the four Db4 filters by sqrt(2) before convolution and the subsequent addition of **Plp** and **Php*** we then have simply a delayed delta function as shown in Figure 8.2–3.

Jargon Alert: **A *delta function* (as used in DSP), sometimes called a *Kronecker delta*, has a value of 1 at time = 0 and zero at all other times. Mathematically it is represented as**

$$\delta(n) = 1 \text{ for } n = 0; \ \delta(n) = 0 \text{ for } n \neq 0$$

where δ is the symbol for the Kronecker delta.

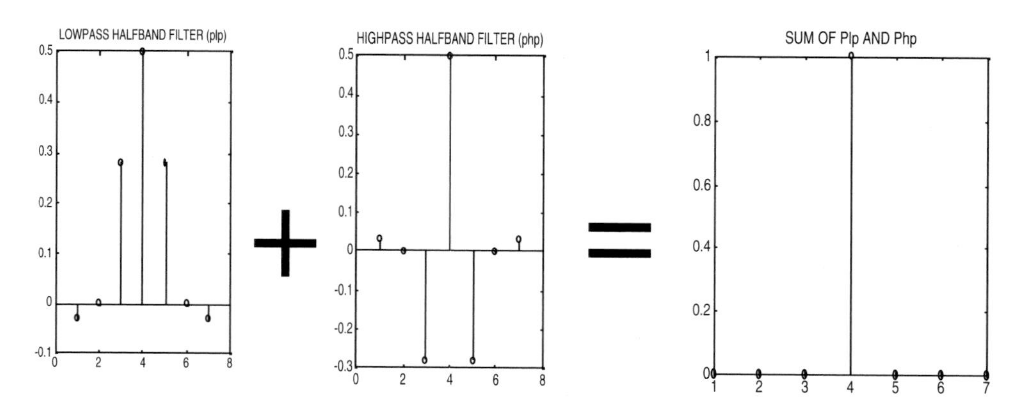

Figure 8.2–3 Addition of the lowpass halfband filter ***Plp*** and the highpass halfband filter ***Php*** produces a delayed delta function (Kronecker delta).

We notice that the odd points in **Php** are reversed in sign from those of the lowpass halfband filter. This is a familiar result to filter designers as

$$P_{hp}(n) = P_{lp}(n)(-1)^n$$

We see similar results for the Haar filters.

```
Plp = [ 1  1]*[1  1] = [ 1  2  1]
Php = [-1  1]*[1 -1] = [-1  2 -1]
```

* Some texts, including the MATLAB documentation, use 0.5 rather than 1.0 as the middle value in the half-band filter.

```
Plp + Php = [0  4  0]
```

Thus for the Haar filters we have a delay of 1 (one leading zero) and a constant of multiplication of 4. We will talk later about the role these halfband filters play in the conventional DWT and how they apply in the "stretched" filters of the multi-level UDWT.

It is not just coincidence that **H**, **H'**, **L** and **L'** combine in such a way to produce the halfband filters that, in turn, produce perfect reconstruction (once the delay and constant of multiplication have been removed). This property of being able to combine in this way is one of the several stringent requirements of these four basic wavelet filters. We will soon discuss how such filters are found.*

8.3 Properties of the Halfband Filters

We have just seen the lowpass and highpass filters in the time domain. We now look at the frequency characteristics as depicted in Figure 8.3–1 for the Db4 halfband filters.

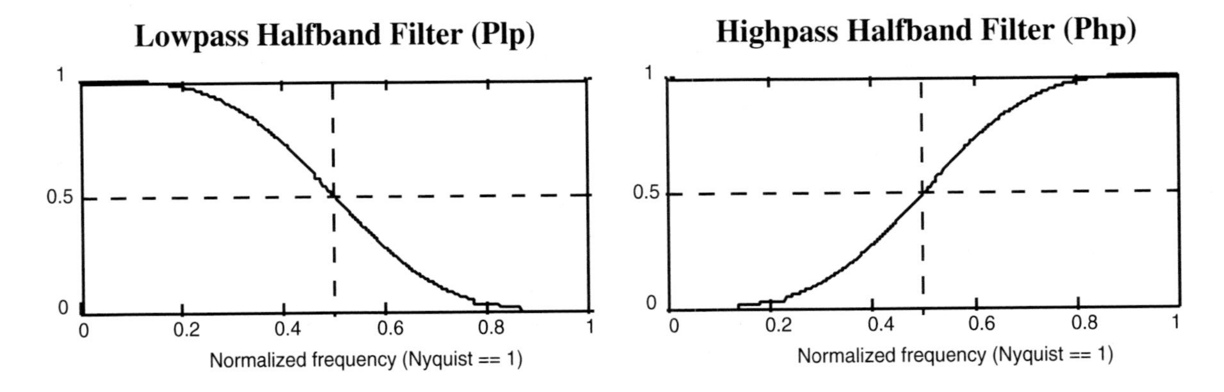

Figure 8.3–1 Frequency characteristics of the Db4 lowpass halfband filter ***Plp*** and the highpass halfband filter ***Php***. Note the symmetry about the magnitude 0.5 and about the normalized frequency 0.5 (1/2 Nyquist or $\pi/2$ on the unit circle).

---

* As discussed earlier, we can easily verify that the convolution of **H** and **H'** or **L** and **L'** produces a halfband filter. "Deconvolution" or coming up with the filters is more difficult. We will soon see that there are *biorthogonal filters* of different lengths that can also produce (by convolution) halfband filters that are *identica*l to those produced as described in this section.

These halfband filters also have linear phase. Recall that the 4-point Db4 filters themselves have *quadrature mirror* symmetry in the frequency domain with their counterparts (**H** with **H'**, **L** with **L'**) but do not have symmetry in the time domain nor do they possess linear phase.

The delayed Kronecker delta function we obtained by summing the 7-point halfband filters **Plp** and **Php** is given by

```
δ(n-4)  =  {0  0  0  1  0  0  0]
```

The frequency and phase is shown in Figure 8.3–2. It is simply a constant value for all frequencies. The phase is also linear*.

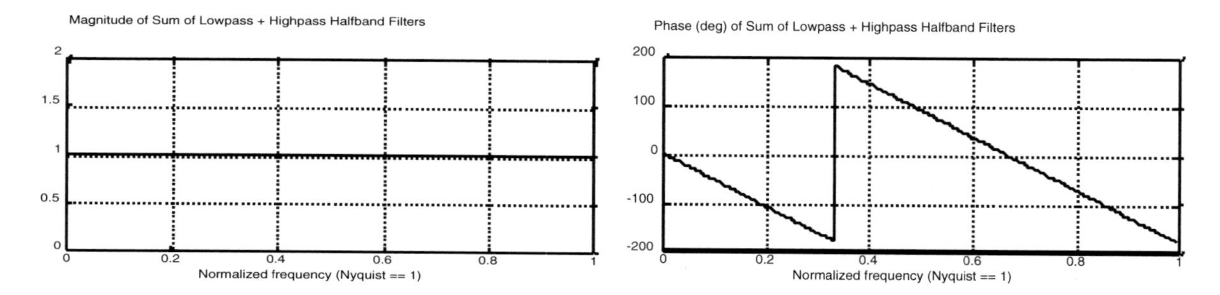

Figure 8.3–2 Frequency and phase characteristics of the delayed Kronecker delta function.

Looking again at the end of a single-level UDWT (ref. Fig. 8.1–3), when we add the lowpass and highpass paths together we can begin to see why we can achieve perfect reconstruction. In other words we have just seen how the lowpass and highpass filters combine to produce a Kronecker delta (albeit possibly delayed and multiplied by a constant).

It then seems intuitively correct that if the signal, **S**, is multiplied by **Php** to produce **D1** and the same signal is multiplied by **Plp** to produce **A1** that **A1** + **D1** should reconstruct the signal. Mathematically we can express this (for a Linear Time-Invariant or LTI system) as

```
A1+D1  =  KxS*Plp  +  KxS*Php  =  KxS*(Plp+Php)  =  KxS*δ(n-4)
```

* The phase of a Kronecker delta with no delay, **δ(n) = [1]**, is not only linear but constant at zero. In the frequency domain *multiplying* results by a *delayed* delta shifts the phase but leaves the magnitude unchanged. This is of course the counterpart to the time domain where in *convolving* the signal with the delayed delta we leave the *values* unchanged but introduce a *delay* in the signal.

where K is the constant of multiplication and δ(n-4) represents the delayed Kronecker delta for this example.

For multi-level UDWTs with stretched filters we can see the same pattern for perfect reconstruction. Figure 8.3–3 shows a 2-level UDWT. Notice that **cA1'** is supposed to be a perfect reconstruction (after removing the delay and multiplication constant) of **cA1**. We will demonstrate this using the Haar filters. We have again for the 4 basic filters

```
H  = [-1  1]
H'= [ 1 -1]
L  = [ 1  1]
L'= [ 1  1]
```

with the "prime" indicating the reconstruction or synthesis filters

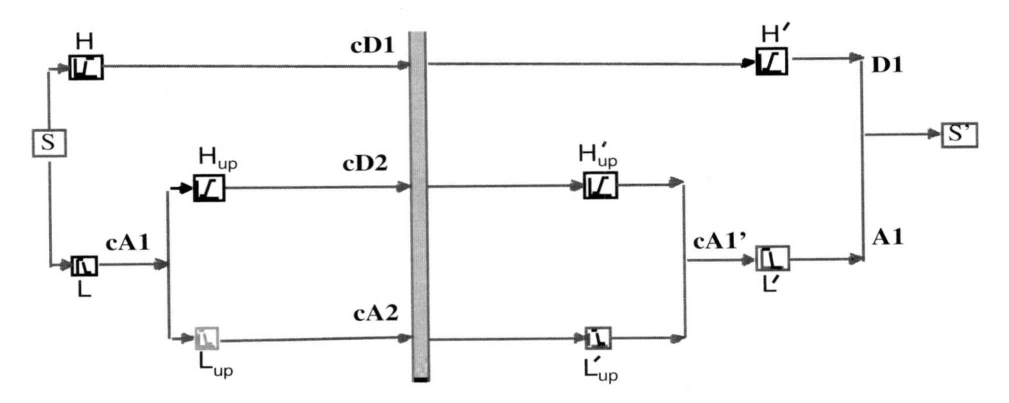

Figure 8.3–3 2-level UDWT. We can achieve perfect reconstruction for **cA1** even with upsampled filters (**Hup**, **H'up**, **Lup**, **L'up**). With **cA1'** = **cA1** we already know we can achieve perfect reconstruction for **S** with the basic filters (**H**, **H'**, **L**, **L'**). Thus after removing delays and constants of multiplication (or pre-dividing the filters) we have **S** = **S'** and perfect reconstruction

For the upsampled versions of Haar filters we then have

```
Hup   = [-1  0  1]
H'up = [ 1  0 -1]
Lup   = [ 1  0  1]
L'up = [ 1  0  1]
```

And the filters produced would be

```
Hup*H'up = [-1 0 2 0 -1]
Lup*L'up = [ 1 0 2 0  1]
```

Summing the halfband filters we obtain

```
[ 0 0 4 0 0]
```

Thus we can achieve perfect reconstruction of **cA1** with the upsampled filters. After removing the delay (2) and constant of multiplication (4) we have **cA1'** = **cA1**. But this reduces our task of showing perfect reconstruction to that of showing perfect reconstruction with a single-level UDWT (ref. Fig. 8.1–3) which we have just completed. Thus, although not a mathematical proof, we have demonstrated that for the upsampled filters found in UDWTs

```
Hup*H'up + Lup*L'up = delayed Kronecker delta
```

(a result similar to the unstretched filters) and that we can achieve perfect reconstruction for multi-level UDWTs.

For the Conventional DWT with downsampling and aliasing concerns we will later demonstrate alias cancellation in both the time and frequency domains and show how, if done correctly, we can also achieve perfect reconstruction.

8.4 "Reverse Engineering" Perfect Reconstruction to Produce the Basic Filters

We have just seen how halfband filters can combine to provide for perfect reconstruction. We now "work backwards" to see how the halfband filters are produced and, in turn, how the basic decomposition and reconstruction filters are produced.

Rather than review the concepts of Digital Filter Design as found in conventional DSP, we will simply use existing software to produce halfband filters. For example MATLAB has a linear-phase least-squares error minimization Finite Impulse Response (FIR) filter design routine FIRLS.*

* Type "help firls" in MATLAB for more information.

We simply specify the magnitudes (0 to 1) we wish at specific frequencies (0 to Nyquist with Nyquist normalized to 1). We also specify the size of filter we desire (input as "N+1"). Thus for the 7-point lowpass halfband filter ***Plp*** we have just studied **N** = **6**. We want a magnitude (**M**) of 1 at very low frequencies (**F**) and magnitude zero at very high frequencies.

Thus we have, using **.002** and **1** - **.002** = **.998** for the low and high frequencies

```
M = [1 1 0 0], F = [0  0.002  0.998  1], N = 6

FIRLS(N,F,M) =  -0.0312    0.0000    0.2812    0.5000
0.2812    0.0000    -0.0312
```

which is identical (to computer precision) to the convolution of the pre-divided Db4 **L** and **L'** filters. The highpass halfband filter can be produced using FIRLS by simply specifying zero magnitudes at low frequencies and full magnitude at the high frequencies. Thus we would have

```
M = [0 0 1 1], F = [0  0.002  0.998  1], N = 6

FIRLS(N,F,M) =  0.0312    -0.0000    -0.2812    0.5000
-0.2812    -0.0000    0.0312
```

which we could have also acquired by the DSP method of alternating signs of the lowpass filter.

For the Haar we have for the lowpass halfband filter

```
M = [1 1 0 0], F = [0  0.002  0.998  1], N = 2

FIRLS(N,F,M) =  0.2500    0.5000    0.2500
```

which is identical to the convolution of the pre-divided Haar lowpass filters

```
L = [0.5  0.5] and L' = L = [0.5  0.5].
```

Working backwards now from the halfband filters we know that the lowpass halfband filter is the convolution of **L** and **L'**. For the Haar lowpass halfband filter we have, using the pre-divided filters **L** and **L'** shown above

```
Plp = [0.25  0.5  0.25]
```

Even pretending we do not know the answer, it would still be very easy to come up with two Haar 2-point filters that, when convolved, produce this **Plp**.

The appendices go into more detail about how to perform a process called *spectral factorization*[*] to obtain the 4 basic filters but for now we will show the general concept:.

A modified z-transform of the Haar lowpass halfband filter [0.25 0.5 0.25]) is the 2nd degree polynomial

$$0.25z^2 + .5z + 0.25 = 0$$

Using the ROOTS software in MATLAB, for example, we obtain

```
ROOTS ([0.25   0.5   0.25]) = -1  -1
```

which means the above equation can be factored into

```
K(z+1)(z+1)  = 0
```

The constant K is necessary because the roots of [2 4 2], [25, 50, 25] or any number of similar "1-2-1" ratios are also -1 and -1. We can also split the constant K into 2 parts so we have

```
K1(z+1)K2(z+1)  = 0
```

Taking the inverse z transform we have the convolution

```
(K1[1 1])*(K2[1 1])  = K1K2[1  2  1]
```

One of the requirements for a wavelet filter is that the sum of the points for each of the lowpass filters (L and L') be a constant (usually 1.[†]). Thus we have **K1** + **K1** = **1** and **K2** + **K2** = **1**. Thus the constants **K1** and **K2** are both 0.5 and we have for the basic 2-point filters

```
[0.5  0.5] and [0.5  0.5]
```

which we recognize as the pre-divided Haar **L** and **L'**.

[*] For this quick review we will be dealing with several terms that are thoroughly discussed and well explained in DSP texts. "*Understanding Digital Signal Processing*" by Richard G. Lyons and "*The Scientist and Engineer's Guide to Digital Signal Processing*" by Steven W. Smith are excellent examples of explaining the concepts as well as the equations. We will thus suspend the "Jargon Alerts" for a short time until we return to wavelet terminology.

[†] Some texts use 2.0 as this constant, resulting in the more familiar **L** = **L'** = **[1 1]**. In her book "*The World According to Wavelets*", Barbara Hubbard repeatedly uses the familiar latin expression "Caveat Emptor" (Let the Buyer Beware) to highlight the different conventions used by different authors.

We have then successfully "reverse engineered" the process to first find a 3-point lowpass halfband filter and then to factor this filter into two 2-point filters using modern software. This is certainly overkill for the simple Haar filters but now that we understand the process let us proceed with a more difficult problem such as finding **L** and **L'** for the Db4 filters.

Before doing this we should make the observation that the lowpass halfband filter is the convolution of **L** and **L'** in the time domain. In the frequency domain, however, the halfband filter is the *product* of the FFTs of **L** and **L'**. In other words we *factor* the FFT of **Plp** into the FFTs of **L** and **L'**. This is why this process is called *spectral factorization* in wavelet terminology.

Following the pattern used for the 3-point halfband filter, we recall we first used FIRLS (or similar software) to design a 7-point halfband filter. We obtain as before

```
M = [1 1 0 0], F = [0  0.002  0.998  1], N = 6

FIRLS(N,F,M) = Plp = -0.0312    0.0000    0.2812
0.5000    0.2812    0.0000   -0.0312
```

Taking the z transform as we did with the Haar we can produce the 6th degree polynomial*

$$K(-0.0312z^6 + 0.0z^5 + 0.2812z^4 + 0.5z^3 + 0.2812z^2 + 0.0z - 0.0312) = 0$$

Using ROOTS or similar software we have (to computer precision)

```
ROOTS(Plp) =  3.7321   -1   -1   -1   -1   0.2679
```

Thus the polynomial can be factored into

$$K(z - 3.7321)(z+1)(z+1)(z+1)(z+1)(z - 0.2679)$$

We would like **L** and **L'** to both be 4 points in length to produce the 7-point **Plp**. Thus we will split the above factors into two 3rd order polynomials (see appendices for more mathematical details) and the constant **K** into **K1** and **K2** as we did for the Haar. Using the first 3 factors of the above 6th degree polynomial we have the 3rd degree polynomial

* For simplicity here we multiply each side of the equation by z^3 to produce this more conventional polynomial. We note in passing that the first term is 2 + sqrt(3) and the last term is 2 - sqrt(3).

```
K1(z - 3.7321)(z+1)(z+1)
=  K1(1z^3 - 1.7321z^2 - 6.4642 z - 3.7321)
```

and using the last 3 factors we have

```
K2(z+1)(z+1)(z-0.2679) =

K2(1z^3  +  1.7321z^2  +  0.4642 z - 0.2679)
```

Taking the inverse z-transform we have the filters L and L' as

```
L   = K1[1 - 1.7321 - 6.4642 - 3.7321]
L'  = K2[1  +  1.7321  +  0.4642 - 0.2679]
```

Using the wavelet condition that the coefficients of L and L' each add up to the constant 1.0 we easily obtain

```
K1 = -0.0915
K2 =  0.3415
```

Substituting we now have our pre-divided Db4 filters L and L'

```
L   = [-0.0915    0.1585    0.5915    0.3415]
L'  = [ 0.3415    0.5915    0.1585   -0.0915]
```

We can easily verify this is correct by convolving L and L'

```
conv(L,Lprime) = [-0.0312    0.0000    0.2812    0.5000
0.2812    0.0000    -0.0312] = Plp
```

which is our 7-point lowpass halfband filter. We use a similar method to determine H and H'.* Thus we see that we can start with the halfband filters and determine the constituent wavelet filters from them.

We can of course extend this process to longer filters. For example an 11-point lowpass halfband filter can be "factored" into two 6-point filters giving us the Daubechies 6 or Db6 L and L'. Similarly a 15-point lowpass halfband filter can be factored into the 2 lowpass Db8 filters.

* We can use another wavelet requirement that the sum of the squares of the coefficients is equal to a constant.

8.5 Orthogonal Vectors, Sinusoids, and Wavelets

The dictionary definition of *orthogonal* is (1) intersecting or lying at right angles; (2) having a sum of products that is zero. Cartesian Coordinates are of course an orthogonal system. Figure 8.5–1 first shows a distance vector (1, 1) in terms of the Cartesian vertical and horizontal unit basis vectors

```
Xunit = (1, 0) and Yunit = (0, 1)
```

These orthogonal basis vectors meet both dictionary definitions. In addition to being at right angles the *dot product* (inner product) of these vectors is

```
(1 0)•(0 1) = 1x0 + 0x1 = 0 + 0 = 0
```

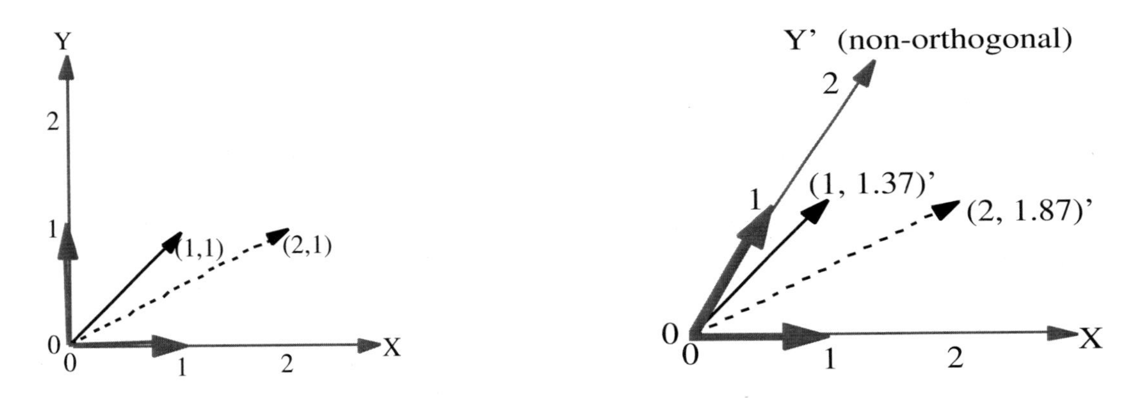

Figure 8.5–1 A distance vector (thin arrow at 1,1) can be specified in terms of orthogonal unit basis vectors (shown at left as heavy arrows). Moving the vector in the X direction (dashed arrow at 2,1) changes the X coordinate but not the Y coordinate. At right we have the same distance vector but in a non-orthogonal system with X and Y' as non-orthogonal unit basis vectors. Moving the vector in the X direction changes both the X and the Y' coordinates.

while the non-orthogonal basis vectors X and Y' at the right will produce a non-zero dot (inner) product. In other words, in a non-orthogonal system the information captured by one vector is NOT independent of the information captured by the other vector. Notice also that we are *correlating* these distance vectors with the unit basis vectors.

Jargon Alert: **Basis vectors are the "base" by which all other vectors are specified. The Cartesian coordinates unit basis vectors (Xunit and Yunit) form an *orthogonal basis* by which other vectors can be**

described. Similarly, *orthogonal sinusoids* or *orthogonal wavelet filters* form an *orthogonal basis* by which a signal can be efficiently represented.

We recall that Sines and Cosines are also *orthogonal bases* for describing a signal. The sine is 1/4 cycle or 90 degrees out of phase with the cosine and the dot products are zero. Complex numbers are thus of the form *cosine + jsine*. Figure 8.5–2 shows this graphically.

Signals can be specified by sets of basis functions, either by Fourier sinusoids or by various wavelets. We recall that the wavelet transform is a cross-correlation of a signal with a set of wavelets of various widths and shifts in time. When we have orthogonal wavelets, the information captured by one wavelet is independent of the information captured by another.

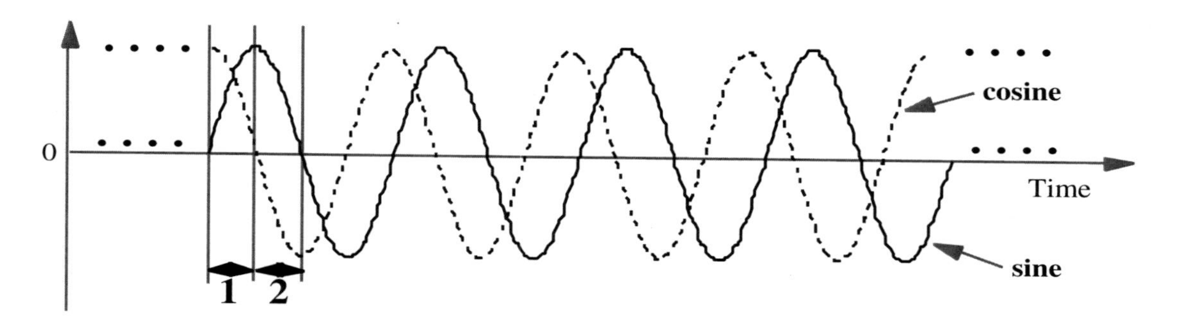

Figure 8.5–2 Sine and Cosine are 90° (π/2 radians) out of phase. Notice that the sine is symmetric in time intervals 1 and 2 while the cosine is anti-symmetric. Thus the products of the sine and cosine in interval 2 cancel the products in interval 1 and we have orthogonality.

We now look at the Haar and Db4 wavelets and show how they constitute an *orthogonal basis*. Figure 8.5–3 shows a trivial case where wavelets that do not line up in time will produce a dot product of zero. There is no correlation between the two and thus they are orthogonal.

Let's look at the basic Db4 filters for a moment. Once again we have

```
L  = lod = [-0.0915   0.1585   0.5915   0.3415]
L' = lor = [ 0.3415   0.5915   0.1585  -0.0915]
H  = hid = [-0.3415   0.5915  -0.1585  -0.0915]
H' = hir = [-0.0915  -0.1585   0.5915  -0.3415]
```

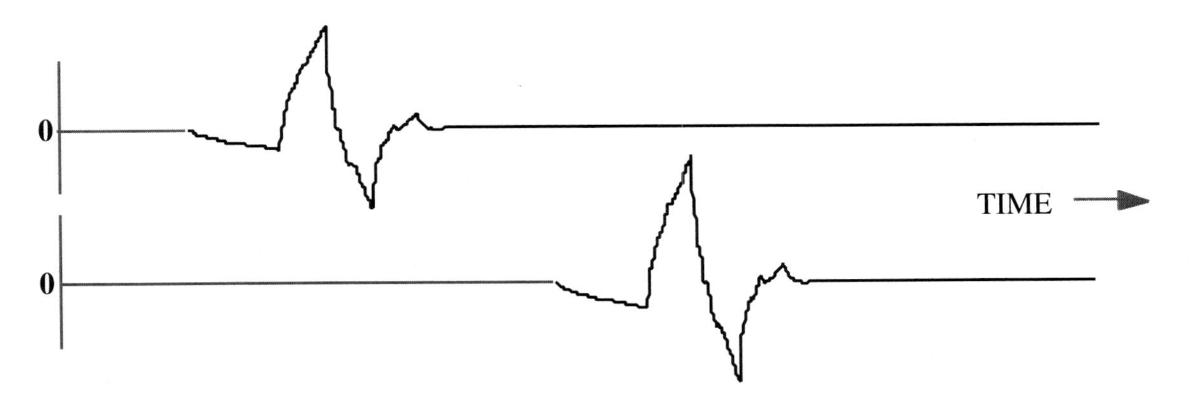

Figure 8.5–3 A trivial case where two Db4 wavelets (drawn as 766-point estimations of a "continuous" Db4 wavelet) do not overlap at all in time and the dot product will be zero. Thus these two wavelets are orthogonal. Recall that each wavelet *correlates* with the signal and thus the information captured by the first wavelet filter will be independent of the information captured by the second.

Taking the dot or inner product of **L** and **H** we have

```
sum(L.*H) = 3.4694e-18
```

or zero.*

Remembering that the 4 basic filter points are 1/2 integer apart on the mapping of the data points to the interval 0 to 3 (ref. Fig. 6.4–1), we are only interested in the dot product when we shift **H** by *2 points* (i.e. a full integer shift).

Jargon Alert: **The term *Dyadic Translation* is used in some texts to indicate this shifting by 2 to line up at a whole integer apart.**

We then have (after adding 2 trailing zeros to L)

```
Hshift = [ 0       0    -0.3415  0.5915 -0.1585 -0.0915]
L      = [-0.0915  0.1585 0.5915  0.3415 (0)  (0)   ]
```

Taking the dot product we have

```
L•Hshift = {0+0-(.3415 x .5915)+(.5915 x .3415)+0+0} = 0
```

* MATLAB uses an asterisk for multiplication. The period following "**L**" allows for computations such as the dot (inner) product. The terms are then summed resulting in zero to computer precision.

A cross correlation is a series of dot products. Remembering that the orthogonality applies at full integer shifts (including a zero shift) we have

```
xcorr(L,H) = -0.1166  0  0.2416  0  -0.1334  0  0.0084
```

in other words, **L** and **H** are *integer orthogonal* to each other. We also illustrate this in the first graphic of Figure 8.5–4 below. In addition to **L** and **H**, **L'** and **H'** are also (integer) orthogonal to each other.

All 4 basic filters are *orthonormal* to themselves (again at whole integer intervals). We show this for **L'** in the second graphic (Fig. 8.5–4). In passing, we note that the cross correlation of **L'** with **L'** is the same as the *convolution* of **L** with **L'** and thus we see the lowpass halfband filter points. Note also that as these 4-point Db4 filters are shifted by 2 or more whole integers (in either direction) they must be shifted by 4 (1/2 integer) points. Thus they do not line up at all and we revert to the trivial case orthogonality result of zero.

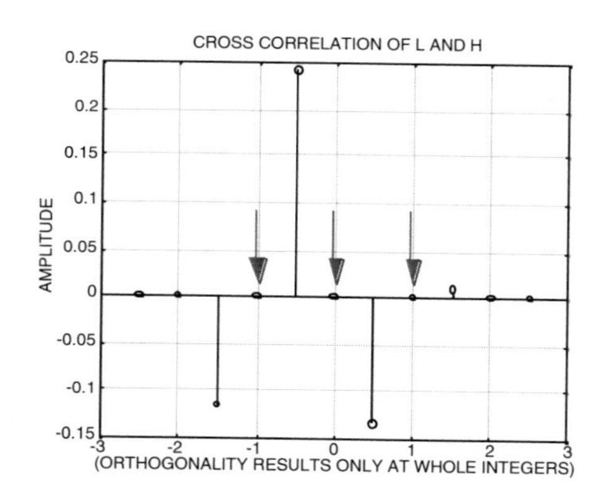

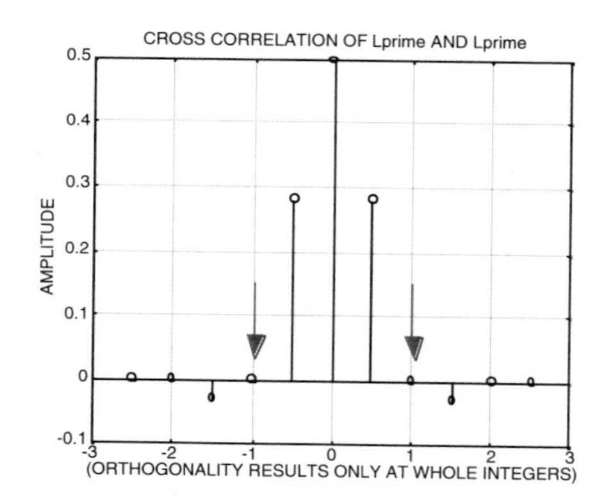

Figure 8.5–4 The cross correlation of the lowpass decomposition filter (**L** or **lod**) and the highpass decomposition filter (**H** or **hid**) is shown at left. When lined up in time they are orthogonal. They are also orthogonal at integer shifts (-1, +1) as indicated by the arrows. The right graph shows the cross correlation of **L'** and **L'** (autocorrelation). At integer shifts of -1 or +1 they are orthogonal but when aligned they produce a constant and are thus *orthonormal.* Integer shifts of 2 or more will produce zero due to alignment as can be seen in either graphic.

Jargon Alert: **Orthonormal as used here means that when aligned the filters produce a constant non-zero value. When shifted by a whole integer (two 1/2 integer points for the basic Db4 filters) the dot product is zero.**

We can also show this orthogonality for stretched filters. Figure 8.5–5 shows the 766 point estimation of the "continuous" wavelet function (built from the 4 points of **H'** or **hir**) mapped onto the interval 0 to 3. Another 766 trailing zeros are appended for this demonstration. When we shift the wavelet by an integer to the right (dotted line on left graph) and take the dot product with the unshifted wavelet we obtain zero to computer precision. Thus the wavelet function (estimation) is orthogonal to the integer-shifted version of itself. Again, **H'** and **L'** are "integer orthonormal" to themselves.

When we stretch the wavelet by a factor of 2 to 1532 points (dotted line in right graph) and take the dot product with the original unshifted 766 point "wavelet" we again obtain zero to computer precision and we have orthogonality between the wavelet and the dyadically stretched version*.

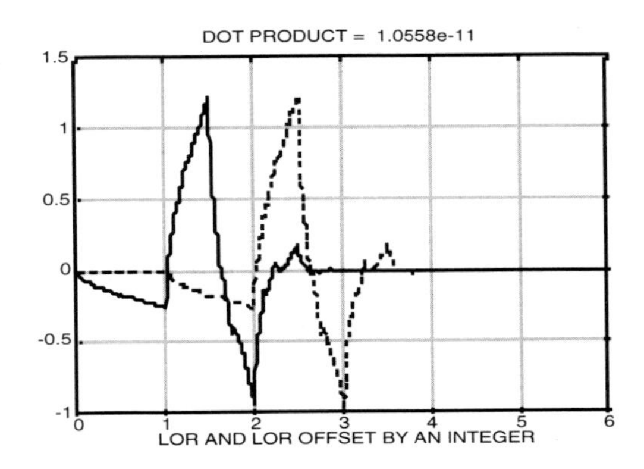

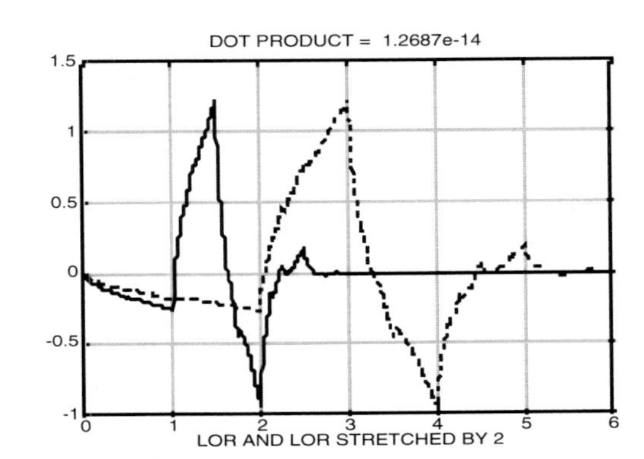

Figure 8.5–5 The first graph shows that a stretched Db4 **H'** filter (766 point estimation) of a "continuous" wavelet mapped onto the interval 0 to 3) and the same stretched filter offset by an integer (mapped on the interval 1 to 4 as shown by the dotted line) have a dot product of zero (to computer precision) and are thus orthogonal. The right graph shows that the same filter is also orthogonal to a dyadically stretched version of itself (**2 x 766 = 1532** points mapped on the interval 0 to 6 as shown by the dotted line).

We obtain similar results for the other three basic Db4 filters (**L**, **H** and **H'**) as we have demonstrated here for **L'**. They are orthonormal to themselves. Further, the decomposition filters (**L** and **H**) and the reconstruction filters (**L'** and **H'**) are orthogonal. These relationships are depicted in Fig. 8.5–6.

---

* These are important results when we recall that stretching and shifting of wavelets in the DWT is done by factors of 2 thus creating a set of orthogonal wavelet filters or an "orthogonal basis".

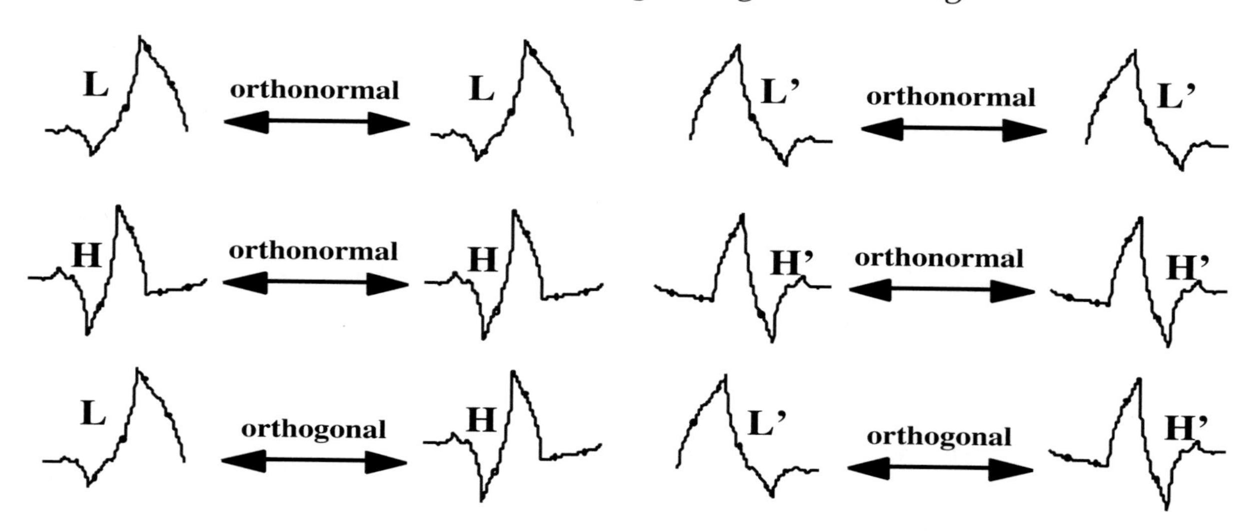

Figure 8.5–6 The orthogonality relationships between the 4-point Db4 filters. They are drawn as 766 point estimations of "continuous" functions to show the shapes with the original 4 basic filter points superimposed. We have the same relationships for the Db6, Db8, and Db2 (Haar) filters.

The Haar wavelet filters also have the same orthogonality and orthonormality but in a more trivial sense. We recall that the four basic 2-point Haar filters are also 1/2 integer apart when mapped onto their time interval 0 to 1 (ref. Fig. 6.2–4). Thus an integer shift of any of these filters reduces them to the trivial case where they are not lined up at all. For example we have for **L**

```
L•L = (1  1)•(1  1) = 2
L•Lshift = (1  1 (0)  (0))•(0  0  1  1) = 0
```

And **L** is orthogonal (orthonormal) to itself. The orthogonality relationships between different Haar filters are similar to those of the Db4. For example using **L** and **H** we have

```
L•H = (1  1)•(-1  1) = 0
```

The shifted versions do not line up at all and this also produce zero.

The stretched Haar wavelet filter **H'**(Haar), like the Db4 stretched wavelet filter **H'**(Db4), is also orthogonal to the unstretched version. We have (using trailing zeros needed to compute the dot product)

```
H'•H'stretched = (1  -1  (0)  (0) )•(1  1  -1  -1) = 0
```

Thus we see the same orthogonality relationships between the Haar filters as we did between the Db4 filters. The Db6, Db8, and many other filters have these same orthogonality relationships and are suitable for use in the UDWT or conventional DWT. Note the "crude" filters discussed earlier are suitable for the CWT but lack the orthogonality required for use in the various discrete transforms.

Remembering that the UDWT uses filters stretched by factor of 2 (dyadically) and that the conventional DWT, in effect, dyadically shifts the wavelet as the signal is dyadically downsampled we can glimpse the power of these orthogonal wavelet filters. Just as a vector can be efficiently broken down into Cartesian coordinates (X and Y), a signal can be efficiently broken down not only into orthogonal sines and cosines using Fourier techniques, but also into orthogonal wavelets in their various "dilations and translations" (stretching and sliding). They are the "constituent wavelets" that comprise a signal (ref. Fig. 1.6–1).

In other words, the wavelet techniques provide a multi-scale analysis of the signal as a sum of orthogonal signals corresponding to different time scales, allowing a kind of expedient time-scale analysis. These orthogonal wavelet techniques are sometimes referred to as a *fast wavelet transform* because of their efficiency representing the signal with the fewest possible stretched and shifted wavelet. filters.

With orthogonal unit vectors as a basis, we were able to change the X coordinate without affecting the Y coordinate (ref. Fig. 8.5–1). With orthogonal wavelets as a basis, we can de-noise or compress signals or images by removing parts of the data in *frequency and/or in time* (or space) without affecting the rest of the data. We will show some examples of this later*.

8.6 Biorthogonal Filters—Another Way to Factor the Halfband Filters

* In his short courses and seminars the author demonstrates this concept by playing his "Fugal Bugle" along with several other musical instruments in a rendition of *Yankee Doodle*—but with one of the several Middle C notes played wrong (an easy task for him, but for this demonstration done on purpose). Denoising using Fourier techniques gets rid of the wrong note but also all the other "C notes" at that *frequency*. Denoising using time-series techniques gets rid of the wrong note but also all other notes at that particular *.time.* Denoising using wavelet techniques, however, gets rid of the wrong note but leaves all the other notes untouched.

The 7 point lowpass halfband filter can be created from the convolution of the two 4-point basic Db4 **L** and **L'** filters. However, there are other filters that would produce the same result. The 7-point **Plp** filter is again given by

```
Plp = [-0.0625    0.0000    0.5625    1.0000    0.5625
0.0000    -0.0625]
```

but these numbers can be expressed in a simpler fashion as

```
Plp =  [-1  0  9  16  9  0  -1]/16
```

Ignoring the 16 divisor for moment, it is easily verifiable* that the convolution of the 3- point filter [1 2 1] and the 5-point filter [-1 2 6 2 -1] also produces [-1 0 9 16 9 0 -1]. In other words (including the divisor constants)

```
Plp = [1 2 1]/4 * [-1 2 6 2 -1]/4
```

and we could set

```
L  = [-1 2 6 2 -1]/4
L' = [1 2 1]/4
```

Figure 8.6–1 shows this convolution of the 3-point and 5-point filters.

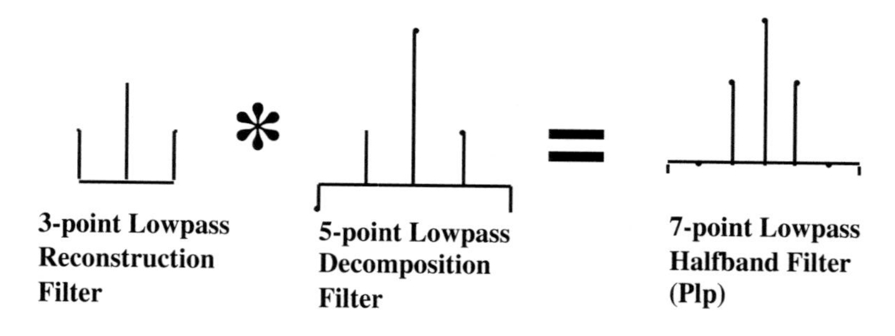

Figure 8.6–1 Convolution of the 3-pt. lowpass reconstruction filter with the 5-pt. lowpass decomposition filter produces the same 7 point lowpass halfband filter as the Db4 **L** and **L'**.

We could also "factor" the *highpass* halfband filter

```
Php =  [1  0  -9  16  -9  0  1]/16
```

* We learned that convolving the 4-point Db4 filters **L** and **L'** also produced this simpler "**[-1 0 9 16 9 0 -1]/16**" form but the 4 coefficients were in the form **±(c ±a)/4b** with **c** being 1 or 3, **a = sqrt(3)** and **b = sqrt(2)**. The Db4 convolutional relationship is thus not as easy to see as that of the 3/5 biorthogonals.

into H and H' where

```
H  = [1 -2 1]/4
H' = [1 2 -6 2 1]/4
```

To distinguish these 3/5 *biorthogonal* filters from the standard 4-point Db4 filters we will use the notation **Hb**, **Hb'**, **Lb**, **Lb'**. The relationships of these filters to each other is shown in Figure 8.6–2.

If we were to do this we would have no longer have the relationships between the equal-length filters that we saw earlier (ref. Fig. 8.1–2). For example the 5-point lowpass reconstruction filter ***L'b*** is not simply the "flipped" version of the lowpass decomposition filter ***Lb*** as we saw with the Db4 filters. In using this 3/5 scheme we are looking at 2 distinct sets of filters—each with one 3-point and one 5-point filter (and no mirror images here)—hence the name *biorthogonal*.[*] These filter sets are interchangeable for the best analysis/synthesis performance in applications such as image processing.

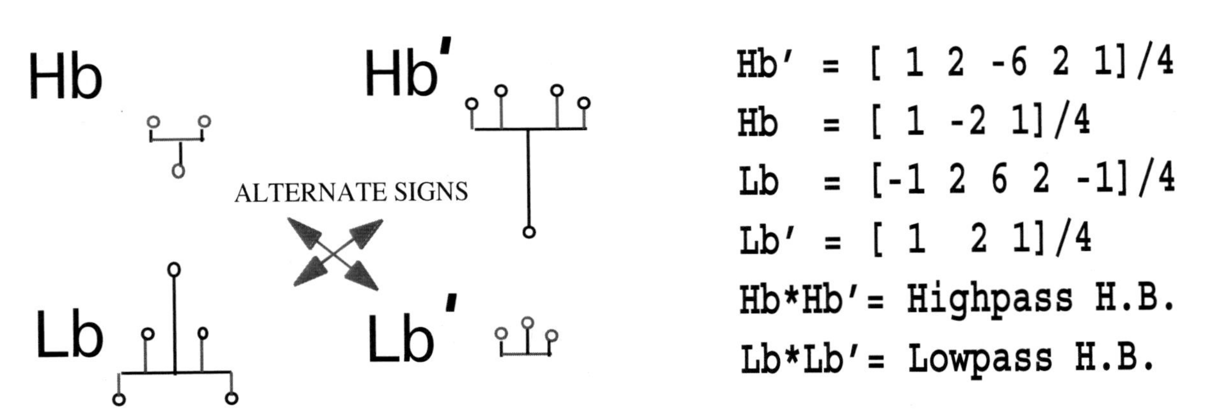

Figure 8.6–2 Depiction of the 3/5 biorthogonal filters and their interrelationships. Notice the conspicuous absence of the ability to change one filter to another by simply flipping the coefficients.

The filters are either 3 or 5 points long. However, by adding some leading or trailing zeros we can make them all 6 points long and enhance the orthogonality relationships in the process. Specifically we have (with the original 3 or 5 points underlined and again ignoring the divisor *4* for a moment)

[*] This is analogous to our American Bi-Centennial celebration in 1976 that happened only after 200 years rather than 100 years (the American Centennial was celebrated in 1876).

```
Lb       = [ 0   -1    2    6    2   -1]
Lprimeb  = [ 0    0    1    2    1    0]
Hb       = [ 0    1   -2    1    0    0]
Hprimeb  = [ 1    2   -6    2    1    0]
```

We look briefly at some of the orthogonality relationships of these equal-length zero-padded *biorthogonal* filters. We can examine the dot products at various whole integer shifts (*"dyadic translations")* as we did in the last section by looking at the *cross correlations.*

Figure 8.6–3 shows at left the orthogonality (orthonormality) relationships between the biorthogonal lowpass decomposition filter **Lb** and the biorthogonal lowpass reconstruction filter **L'b** (both zero padded as shown above). The right graph shows the orthonormality between the highpass decomposition filter, **Hb**, and the highpass reconstruction filter **H'b**.

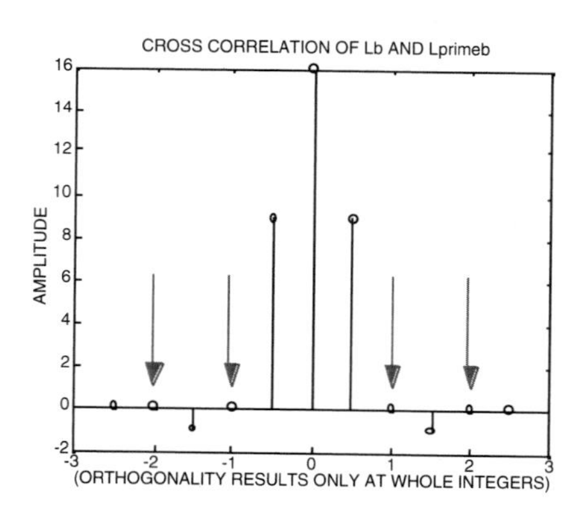

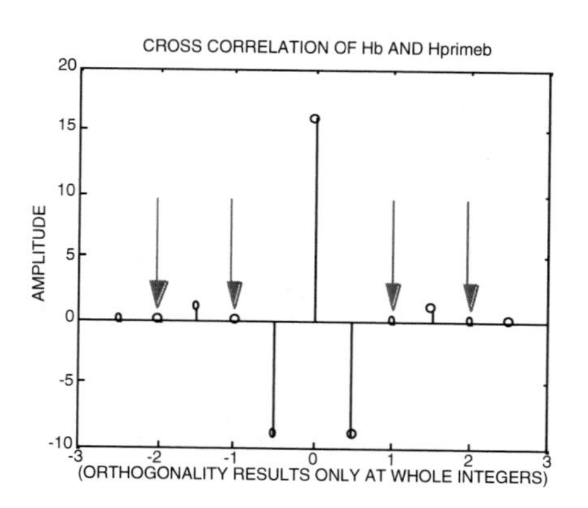

Figure 8.6–3 The cross correlation of the *biorthogonal* lowpass decomposition filter **Lb** and the *biorthogonal* lowpass reconstruction filter **L'b** is shown at left. The right graph shows the cross correlation of the *biorthogonal* highpass decomposition filter **Hb** and the *biorthogonal* highpass reconstruction filter **H'b**. The arrows show that the dot product is zero at whole integer shifts. When aligned, of course, we have a non-zero value for these *orthonormal* relationships.

We notice in the left graph above that we have produced the 7-point lowpass halfband filter (with leading and trailing zeros and constants of multiplication). This result is not surprising when we recall that because these biorthogonal filters are symmetric in the time domain (ref. Fig. 8.6–2) that the cross *correlation* is the same as the *convolution* and thus we have

```
xcorr(Lb,Lprimeb) = conv(Lb,Lprimeb) = Plp
```

We have similar results with the (zero padded) 3-point **Hb** and the 5-point **H'b** combining to form the highpass halfband filter **Php** (ref. Fig. 8.6–3, right).

In other words, our ability to factor these 7-point halfband filters into the convolution of 3 and 5 point filters (rather than two 4-point Db4 filters) gave us these biorthogonal filters in the first place! Also, because of the symmetry we have **L"b** and **Lb** being orthogonal (orthonormal) along with **H'b** and **Hb**.

Where we had self-orthogonality with the four Db4 filters (**L**, **L'**, **H**, and **H'**) we no longer have this with the biorthogonal filters. For example, the biorthogonal filter **Lb** is neither orthogonal nor orthonormal with itself (other than the trivial case where filters do not overlap at all in time). We have the same lack of self-orthogonality with the other 3 filters (**L'b**, **Hb**, and **H'b**).

Using the leading and trailing zeros as shown above, we do have other orthogonality relationships. Figure 8.6–4 shows full orthogonality between **L'b** and **Hb** and also between **Lb** and **H'b**.

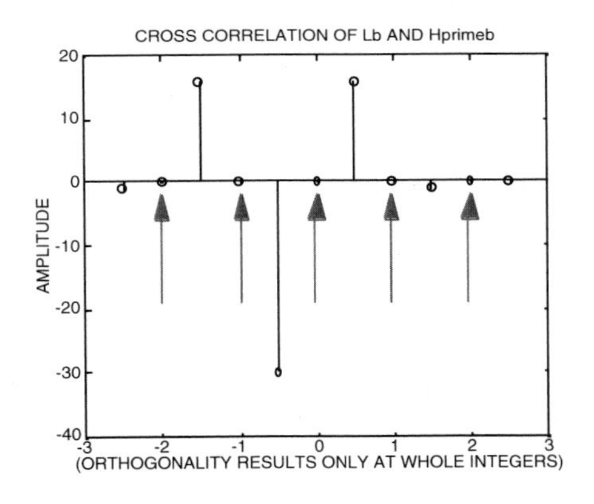

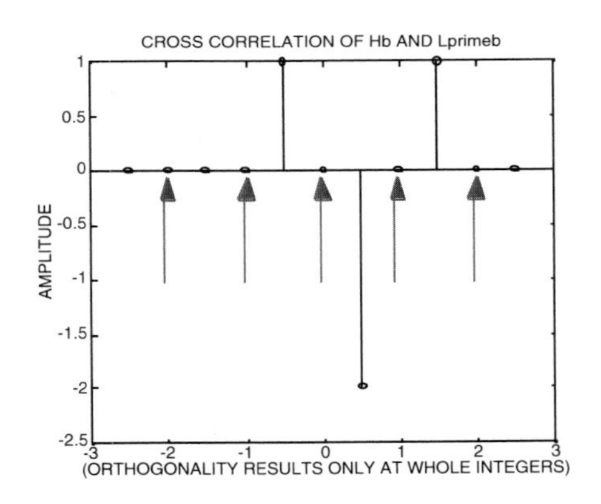

Figure 8.6–4 The cross correlation of the *biorthogonal* lowpass decomposition filter **Lb** and the *biorthogonal* highpass reconstruction filter **H'b** is shown at left. The right graph shows the cross correlation of the *biorthogonal* highpass decomposition filter **Hb** and the *biorthogonal* lowpass reconstruction filter **L'b**. The arrows show that the dot product is zero when perfectly aligned or at whole integer shifts.

Because these biorthogonal filters combine to produce the halfband filters they also allow for perfect reconstruction. Their (bi) orthogonality also allows them to be "constituent wavelets" of the signal and to form an acceptable basis to contain the information found in the signal.

The reason that we go to this extra work and put up with the loss of some interrelationships and changes in orthogonality can be seen immediately by simply looking at them—all 4 of the 3/5 biorthogonal filters have perfect symmetry in the time domain! This also means they have linear phase in the frequency domain! Symmetry and linear phase are desirable in applications such as Image Processing because human vision is more tolerant to symmetric errors. Symmetry is also handy for extending the boundaries of images.

As another example, a 15-point halfband filter can be expressed as the convolution of 2 8-point non-symmetric Daubechies filters (Db8) *or* it can be expressed as the convolution of a symmetric *7-point* filter and a symmetric *9-point filter*. This 7/9 filter set is of course called a *biorthogonal 7/9 filter*. JPEG image compression uses 7/9 biorthogonal wavelet filters. The Federal Bureau of Investigation also uses the 7/9 biorthogonal wavelet filters* for compression of fingerprints.

A sketch of the 7/9 filters showing their orthogonality relationships (same relationships as the 3/5 filters) is drawn below in Figure 8.6–5.

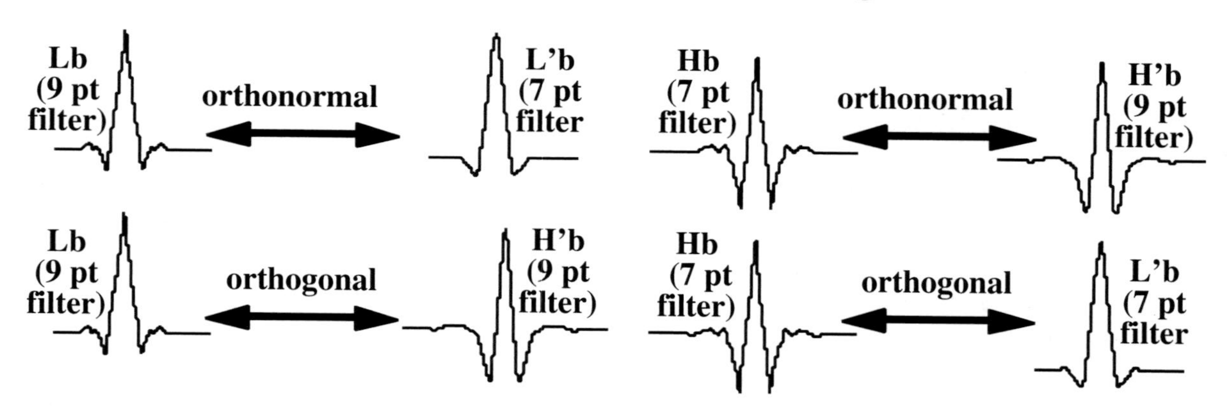

Figure 8.6–5 The orthogonal relationships between the 7/9 biorthogonal filters (drawn as estimations of "continuous" functions to show the shapes) are shown here. We have the same relationships for the 3/5 biorthogonal filters discussed earlier.

* These 7/9 filters were chosen over the 8-point Db8 filters for image processing.

8.7 Summary

In this chapter we explored each term of the expression *Perfect Reconstruction Quadrature Mirror Filter* as used in wavelet transforms and learned of their close relationships to each other.

We explored the halfband filters that are created by the UDWT and DWT and looked at some of the important properties of these filters. By looking at how the lowpass and highpass halfband filters combine we also acquired some insights into how the highpass and lowpass paths of the Discrete Wavelet Transforms add together to perfectly reconstruct the signal. We found this to be true even for the upsampled filters used in the multi-level UDWT.

We went on to "reverse engineer" or factor the halfband filters to show how the basic wavelet filters (**H**, **H'**, **L**, **L'**) are derived. We used some computer software shortcuts* such as FIRLS and ROOTS to facilitate the process.

We explored the concept of orthogonality in wavelets and showed how a signal can be represented very efficiently by orthogonal wavelet filters as a *basis* in a similar manner as a vector can be represented very efficiently by the orthogonal unit basis vectors of the familiar Cartesian Coordinate System.

We proceeded to show how these same highpass and lowpass halfband filters can be factored into *symmetric* filters *if* we are willing to relax the requirements for equal length, closer interrelationships, and some orthogonality. We derived a 3/5 set of *bi*orthogonal filters as an alternative to the 4-point Db4 filters and introduced the 7/9 biorthogonal filters used in image processing (again, because of their perfect symmetry) as an alternative to the 8-point Db8 filters.

This chapter, especially the section on determining the basic 4 wavelet filters from the 2 halfband filters, used more equations than usual (but not beyond the high-school algebra level). One of the stated goals of this book is to introduce some key equations found in conventional wavelet literature after providing an intuitive understanding of the concepts. We did this here with the concept of *spectral factorization* and will do this again in subsequent chapters when we relate concepts such as *alias cancellation* to a few key equa-

---

* Because computers are used to perform the various wavelet transforms it seems logical that computer software can be used to perform the tasks of designing the Finite Impulse Response (FIR) halfband filters and of solving for the roots in a polynomial. A rigorous treatment of these tasks, while interesting and important, falls outside this book's mission statement of understanding wavelet concepts and is better left to conventional DSP texts.

tions found in the wavelet literature. The appendices provide for a more mathematical treatment of these concepts, but we will leave the rigorous proofs to the many excellent mathematics-oriented wavelet textbooks available.*

Having explored some of the remarkable properties of the basic filters such as orthogonality (or biorthogonality) and their ability to combine to produce lowpass and highpass halfband filters and thus achieve perfect reconstruction, we now proceed to look at additional desirable properties found in the various wavelets and the wavelet filters from which they are built.

* "It is with logic that one proves, it is with intuition that one invents"—Henry Poincare

CHAPTER

9

Highlighting Additional Properties by using “Fake” Wavelets

We have talked about many of the desirable application properties of wavelets in previous chapters We learned about finite length or compact support and saw how this is found in all the wavelets that are built from the basic (finite) filters. We also learned earlier how to work with the theoretically infinite crude wavelets by using only the time interval that produces effectively non-zero values (effective support) and then using only a finite number of these values corresponding to the equispaced points in that time interval.

In the last chapter we learned about perfect reconstruction through the use of lowpass and highpass halfband filters. We saw how to factor (de-convolve) these halfband filters into orthogonal wavelet filters which can represent great amounts of data in the most compact fashion. We also discussed biorthogonal wavelet filters that provide for perfect symmetry and linear phase. We learned that these orthogonal or biorthogonal filters are suitable for use in the various Discrete Wavelet Transforms (UDWT, DWT, WPT).

In this chapter we introduce a few more desirable qualities of the various wavelets by comparing them with arbitrary or “fake” wavelets to highlight these qualities. We show how these qualities can also be used to find the 4 “magic numbers” of the Db4 filters.

9.1 Matching the Wavelet to the Signal and the Concept of Regularity

One of the principle uses of Discrete Wavelet Transforms is for denoising. This is accomplished in two general ways. The first is by correlating the finite wavelet (filter) with the transient signal and then extracting it from the noise. The other way is to correlate the wavelet with the transient noise and then subtracting it from the signal. Thus, before looking in detail at the various application properties and capabilities of the specific wavelets, we should talk about choosing wavelets by *matching* their general shapes to that of the signal (or noise). You may already be familiar with some of these techniques from DSP and/or telecommunications as *matched filtering*.

Fourier DSP techniques are about matching the signal to the sinusoids. We can design a *notch filter*, for example, that will remove a 60-Hz hum from a signal. If we are careful, we might even be able to use a *brickwall filter* (discussed in a future appendix) to first take the FFT of the signal, then remove the unwanted frequency components, and finally to take the Inverse FFT. If the signal or the noise we are interested in is *stationary* (constant frequency) then we should stick with conventional Fourier DSP techniques. In other words, the best *wavelet* would be a *wave* (sinusoid)!

Just as the Fourier sinusoid is the best "match" to a constant-frequency hum, we obtain good results when we match the *wavelet* to the signal (or noise) of interest. If the signal or noise is *transient* (starts then stops) wavelet technology is indicated.

We have learned that as the wavelet filters are *stretched* and *shifted* thus they will (at some "dilation" and "translation"), be the same length and at the same position as the signal or noise (ref. Fig. 1.6–2 as an example). The strength of the correlation now depends on how well the wavelet shape matches that of the transient signal or noise and some wavelets will match better than others.

For example, if the signal or noise is a *chirp* then a wavelet such as a Db20 might be a good match. If the signal is binary in nature a binary-looking wavelet such as the Haar might be the wavelet of choice. Figure 9.1–1 shows these 2 wavelets with possible applications (based on similarity of shape).

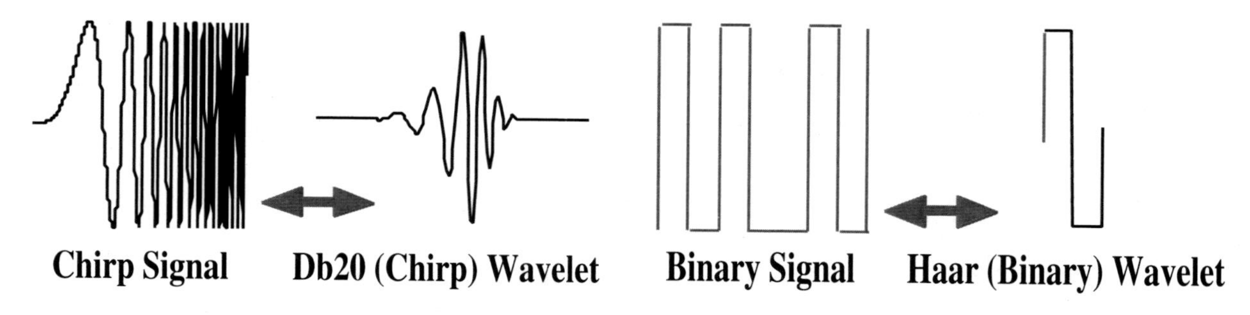

Figure 9.1–1 Choosing a wavelet that looks like the signal (or noise) is a good start. A 20-point Db20 wavelet filter would be a good starting point for a chirp signal as shown at left. A Haar wavelet would be a good starting point for a binary signal. Wavelets are drawn here as multiple point estimations of "continuous" wavelets. In the CWT and UDWT the wavelets filters are stretched and begin to look like this. In the conventional DWT the signal is downsampled instead but the principle of "matching" is still valid.

With Wavelet Technology being a "child of the digital computer age", we can often find the optimum wavelet by trying several candidate wavelets. In wavelet software the command to perform a CWT or DWT is often a single line of code with the wavelet name as one of the input parameters. Evaluation a different wavelet is as easy as changing the wavelet name and executing again. It is possible to evaluate many wavelets in a short time this way.

We discussed earlier the possibility of *creating* a *customized wavelet* (or *template* in matched filtering terminology) to match the transient waveform we are looking for. We might know the pattern of a GPS signal, for example, but with the slew, chirp, and other kinematic distortions arising from the relative motion of the satellite and receiver the pattern may be shifted or stretched. Using this customized "GPS wavelet" in a CWT we should be able to determine the amount of stretching and shifting required for a good match and in turn determine the kinematics.

We will now demonstrate this general process. Suppose we have another "Split Sine": signal as shown in Figure 9.1–2. An FFT would tell us the *frequencies* but not the *time* when the signal changed frequencies. A likely candidate for a custom wavelet to match this changing-frequency signal might be one cycle of a sine wave. The CWT process would then stretch and shift this customized or "fake" wavelet until it matched each of the 2 different sinusoids in the split sine signal. We would expect to be able to see on the CWT display the point in time where the frequency changed.

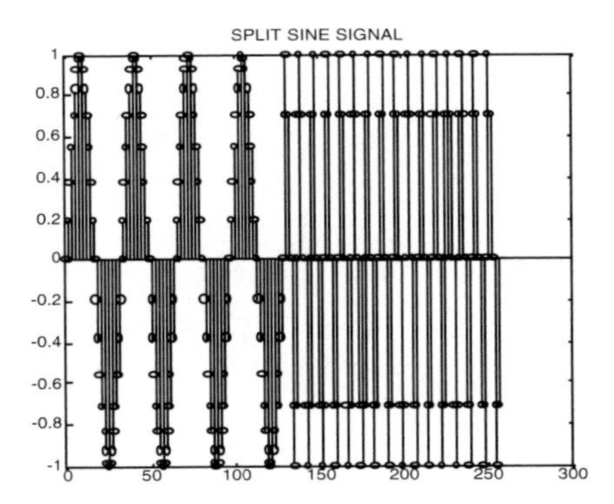

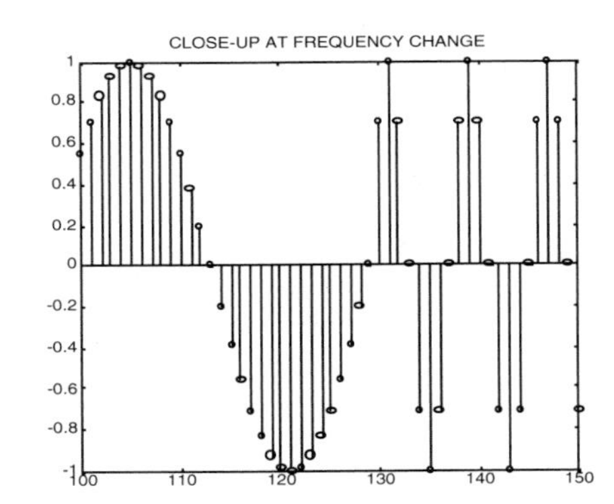

Figure 9.1–2 256-point Split Sine signal. The frequency increases by a factor of 4 halfway through as is seen in the close-up at right near the halfway point (128).

Figure 9.1–3 shows this "fake" wavelet first with 5 points and then with 9 points. Although not an established wavelet, this sine "wavelet" does meet some of the conditions. For example the sum of the filter points is zero. We also know the integral of a sine wave (cosine), This wavelet also has a beginning and an end. We might think that this would be an ideal wavelet to use for this particular signal. One of the problems with this particular homemade "fake" wavelet, however is it's lack of *regularity.*

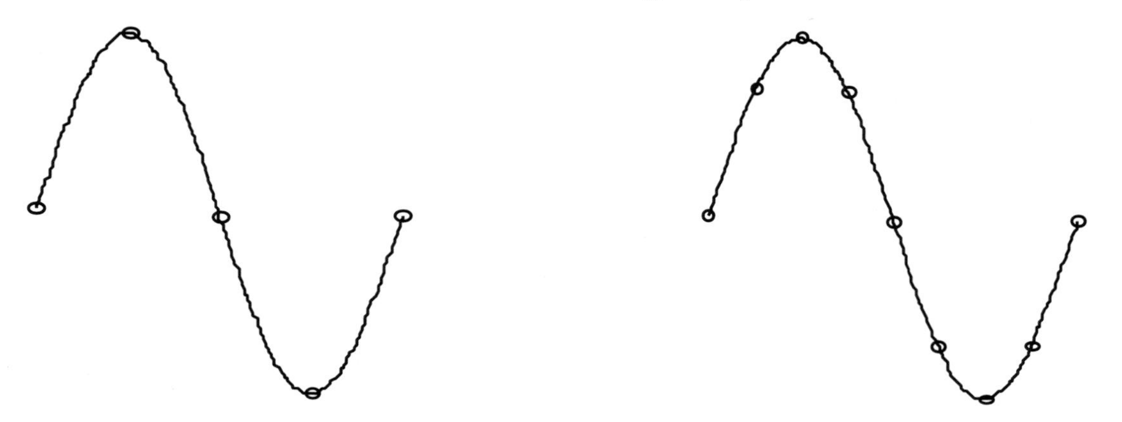

Figure 9.1–3 An arbitrary wavelet constructed from one cycle of a sine wave. A 5 point filter is shown at left and a stretched 9-point filter is shown at right.

***Jargon Alert: Regularity* as used here has to do with smoothness or lack of local rapid variations. The fractal-looking Db4, for example, has poor regularity. The Db2 or Haar has outright discontinuities while the Db10 is fairly smooth. For the DbN wavelets, regularity (and smoothness) increases with N. *Infinitely Regular* means that it is smooth everywhere in time—the slope or first derivative has no discontinuities.**

At first glance, this "fake" wavelet seems very smooth. However we defined this arbitrary wavelet as having a start and stop time. This implies that its value outside these times would be zero. Figure 9.1–4 shows this wavelet (at left) drawn over a longer time interval. Notice the sharp changes in the slope (first derivative) at the beginning and end of the sine wave cycle. This "fake" wavelet is closer to the Db4 in appearance (center). Contrast this single-cycle sine wave and the Db4 with the Morlet wavelet (right). The Morlet wavelet is *infinitely regular* and can be seen to be very smooth with no discontinuities in the waveform or slope (first derivative). Of the 3 wavelets, the Morlet appears to be the best match to the portions of the split sine signal.

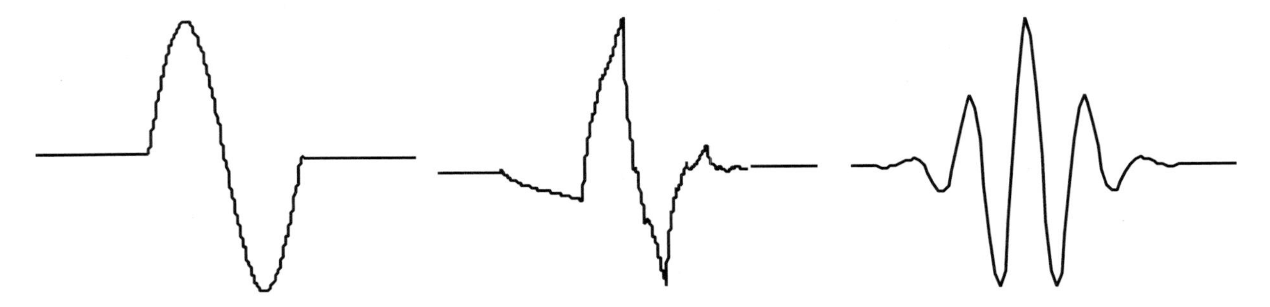

Figure 9.1–4 The single-cycle sine wave "fake wavelet" at left has abrupt changes in slope. The Db4 wavelet (center) also has abrupt changes in slope and other discontinuities in higher derivatives. The Morlet wavelet has infinite *regularity* (smoothness everywhere in time).

We now simply "plug in" these wavelets to the CWT software to obtain displays as shown in Figure 9.1–5. The Morlet wavelet gives the best discrimination between the low-frequency and high-frequency halves of the split-sine signal as seen in the CWT display at right. Notice at level 6 the strong correlation between the Morlet wavelet and the high-frequency right half of the signal while there is little or no correlation with the low-frequency left half. This is not true with either the Db4 or the "fake" sine wavelet. It is most interesting, however, that both the Db4 and the "fake" sine wavelet "did the job" in determining the time that the frequency change occurred.

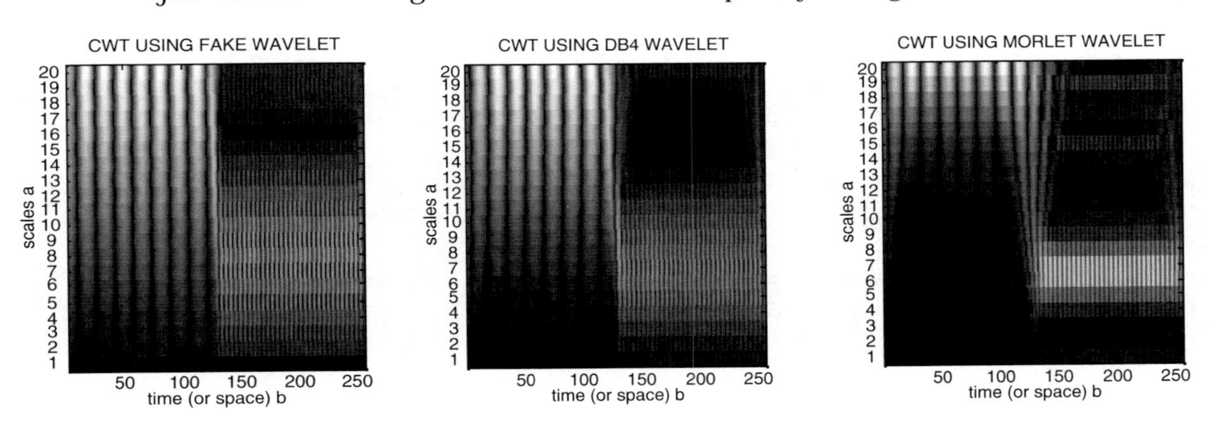

Figure 9.1–5 The displays of Continuous Wavelet Transforms (CWT) using the "fake" sine wavelet, a Db4 wavelet, and a Morlet wavelet. Notice that all 3 wavelets show where the Split Sine signal changes frequency from low to high. The Morlet CWT at right (that best matches the signal) provides the best discrimination.

9.2 Customized Wavelets, Best Basis, and the "Sport of Basis Hunting"

The above example illustrates the choice of wavelet need not be perfect. Inexperienced wavelet transform users sometimes miss the point that with all the shifting and stretching involved in the CWT and the various DWTs that wavelets are extremely adaptable. They are amoeba-like in their ability to change shape and location. Thus one extreme is to spend an inordinate amount of time searching for the "perfect" wavelet for each application. This has been referred to as "*The Sport of Basis Hunting*". Some even try to construct a new wavelet each time. While this is not difficult for the CWT, it can be very tricky for the Discrete Transforms with their requirements for perfect reconstruction, orthogonality, and alias cancellation (DWT).

At the other extreme some users will stick with one wavelet and trust to its adaptability. While it is not a bad idea to start with a general-purpose wavelet like the Db4, the ease of simply substituting another wavelet name in the software allows for rapid comparisons to choose a better wavelet for that particular application. The reader is to be commended for his/her interest in understanding and using wavelets. Otherwise with only Fourier analyses he/she is trying to get by with only one basis—an unchanging, infinite ("stationary") sinusoid!

Software exists with names such as "Best Basis" or "Matching Pursuit" that help chose an optimal wavelet from a library or an optimal wavelet transform configuration using cost functions such as energy or entropy.

Jargon Alert: Cost Functions **deal with return on investment (e.g. most energy or best entropy using a particular wavelet).**

Although this software can be helpful you still need to understand what is going on (one of the main objectives of this book) rather than putting blind faith in equations or software. For example, with this type of software you can inadvertently match the noise or other unwanted artifact, rather than the signal.

One recent method of building a wavelet is known as *The Lifting Scheme.* While a full explanation is beyond the scope of this book we present a very brief overview. This process takes a trivial wavelet and gradually improves ("lifts") properties such as smoothness and vanishing moments (more on vanishing moments later in this chapter).

***Jargon Alert: Lifting* is used here in the sense of raising in rank or condition rather than the more common meaning of raising vertically from a lower to a higher physical position. Think *uplifting*.**

With this method a basis function can be constructed from simpler basis functions. Orthogonal or biorthogonal wavelets can be built with the lifting scheme, but only after the fact and only from existing wavelets. This method does not built wavelet filters from scratch.

Figure 9.2–1 illustrates the methodology. The lifting scheme splits, predicts, updates, then repeats the process to refine ("lift"). Advantages of using this scheme include in-place computation, integer-to-integer transforms (used in lossless coding), zero-delay filter banks, and faster processing.

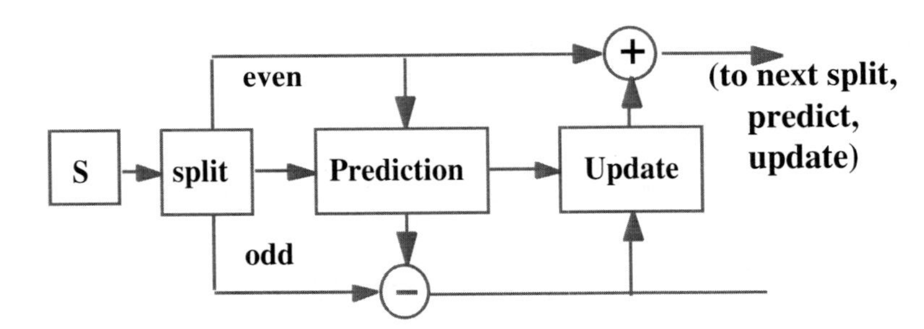

Figure 9.2–1 The first step in the *lifting scheme*. To begin with, the signal **S** is first split into odd and even components. A prediction is then made and an update is performed. Note that the minus means signal from left minus signal from top This is an iterative process producing a more refined (*lifted*) result.

9.3 Vanishing Moments and another Fake Wavelet

***Jargon Alert: Vanishing Moments* means that the wavelet can correlate with a linear, quadratic, or higher order polynomial and obtain small or zero correlation coefficients.**

A signal, a transient pulse or event within the signal, or perhaps noise may be represented by a polynomial—a constant bias, a linear slope, a quadratic, or higher degree. It these cases it is possible to separate the signal from the noise using wavelet technology. Sometimes the noise can be suppressed directly if all or part of the noise can be represented by a polynomial. The

Daubechies family of wavelets (Haar or Db2, Db4, Db6 etc.) is excellent for *vanishing moments.* *

We will demonstrate vanishing moments by creating another "fake wavelet". This time we will *round off* the numbers that comprise the Db4 wavelet filters. Thus for the pre-divided (by sqrt(2)) Db4 filters we have

```
L  = lod = [-0.0915    0.1585    0.5915    0.3415]
L' = lor = [ 0.3415    0.5915    0.1585   -0.0915]
H  = hid = [-0.3415    0.5915   -0.1585   -0.0915]
H' = hir = [-0.0915   -0.1585    0.5915   -0.3415]
```

and the rounded-off ("r") fake wavelet versions become

```
Lr  = [-0.1   0.2   0.6   0.3]
L'r = [ 0.3   0.6   0.2  -0.1]
Hr  = [-0.3   0.6  -0.2  -0.1]
H'r = [-0.1  -0.2   0.6  -0.3]
```

These fake wavelet filters still have some of the desirable qualities of wavelet filters. For example the sum of the coefficients is 1.0 for **Lr** and **L'r** and 0.0 for **Hr** and **H'r**. Further, if we convolve **Lr** and **L'r** we have a lowpass halfband filter and convolving **Hr** and **H'r** produces a highpass halfband filter. The sum of these two halfband filters produces a delayed delta function as shown in Figure 9.3–1. Thus these "fake wavelets" can produce perfect reconstruction. The results to this point are very similar to using the actual Db4 filters (ref. Fig. 8.2–3).

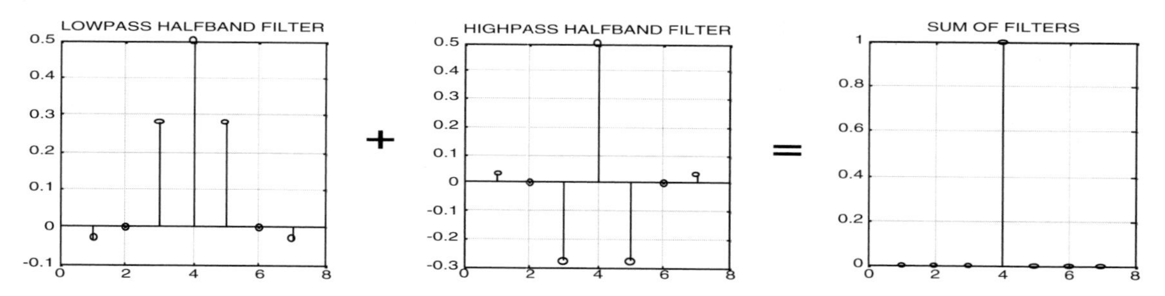

Figure 9.3–1 The rounded-off Db4 or "fake" wavelets produce lowpass and highpass halfband filters. These filters add together to produce a delayed Kronecker delta.

* The author has simulated Traveling Wave Tube Amplifiers (TWTAs) that had output power related to input power by a Volterra Series involving a 45th degree polynomial!

These "rounded Db4" filters can even produce so-called "continuous" wavelets and scaling functions by upsampling and lowpass filtering. Figure 9.3–2 compares the familiar Db4 scaling function (estimation) to the scaling function estimation of the "fake" scaling function. These rounded filters also produced an estimation of the "fake" wavelet function.*

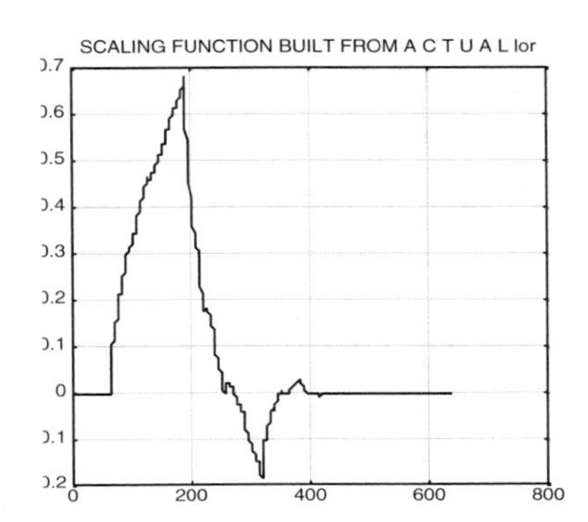

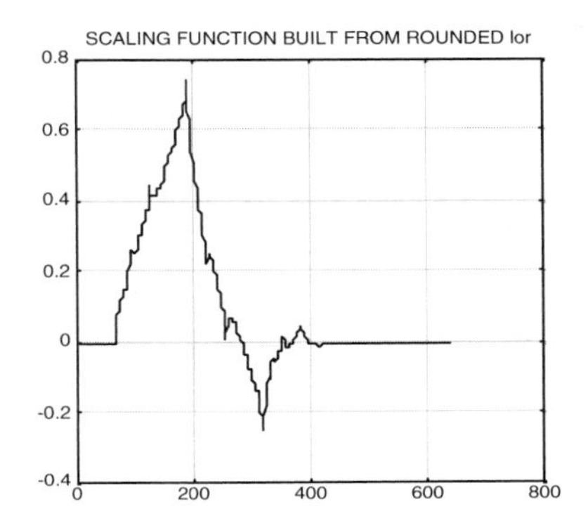

Figure 9.3–2 At left is the familiar scaling function from the Db4 filters. At right is a scaling function from the "rounded" filters.

It is in the capability of *vanishing moments* that the superiority of the Db4 is seen. The Db4 basic wavelet filter has the capability to suppress noise (or a signal) that is in the form of a constant or a linear slope (first degree polynomial)

The dot or inner product (1-point correlation) of the Db4 filter **H'** with a constant (e.g. 1.0) is

$$(-.0915 \times 1)+(-.1585 \times 1)+(.5915 \times 1)+(-.3415 \times 1) = 0$$

Similarly, the dot product of the rounded filter **H'r** with the constant 1.0 is

$$-0.1 -0.2 + 0.6 - 0.3 = 0$$

Looking at a slope = **[1 2 3 4]** we would have for the dot product with the Db4 **H'** values

$$(-.0915 \times 1)+(-.1585 \times 2)+(.5915 \times 3)+(-.3415 \times 4) = 0$$

* At this point the author was excited that perhaps he had invented a new wavelet with the same properties as the famous Db4. His elation was short-lived, however, as he soon saw a lack of *vanishing moments* capability.

But for the rounded filter **H'r** we have

```
(-.1 x 1)+(-0.2 x 2)+(0.6 x 3)+(- 0.3 x 4) = 0.1 ≠ 0
```

and thus the rounded "fake" filter has only one vanishing moment while the Db4 has two. This process is illustrated in Figure 9.3–3.

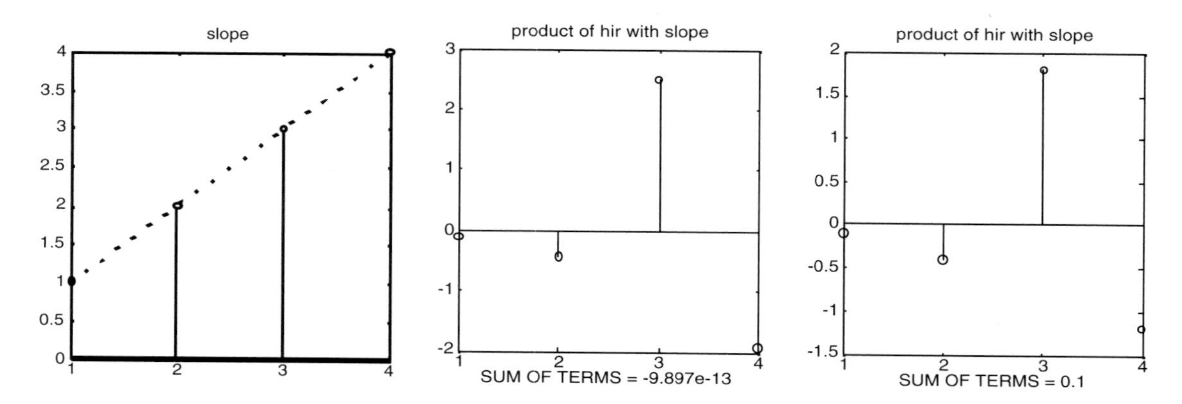

Figure 9.3–3 At left a linear slope **[1 2 3 4]**. The dot product or 1-point cross correlation of this slope with the Db4 filter **H'** yields a sum of terms of 9.897e–13 or zero to computer precision as shown in the middle graph. The dot product of the same slope with the "fake" rounded off wavelet **H'r**, however, yields a non-zero value of 0.1 as shown at right.

9.4 Examples of Use of Vanishing Moments

Consider a very small arbitrary event given by **[1 2 –2]**. We now add this to a 1000-point parabola

$$y(n) = n^2$$

at **n** = **400** and obtain the combined signal as shown at the right of Figure 9.4–1. At **n** = **400**, **y(n)** = **160,000** and the small event would be lost among these large numbers.

We now pretend that we don't know where the event is located or what it looks like as we try various methods to find and examine these small numbers (integers between –2 and 2) overwhelmed by the large polynomial (y values up to 1,000,000 as n goes from zero to 1000)

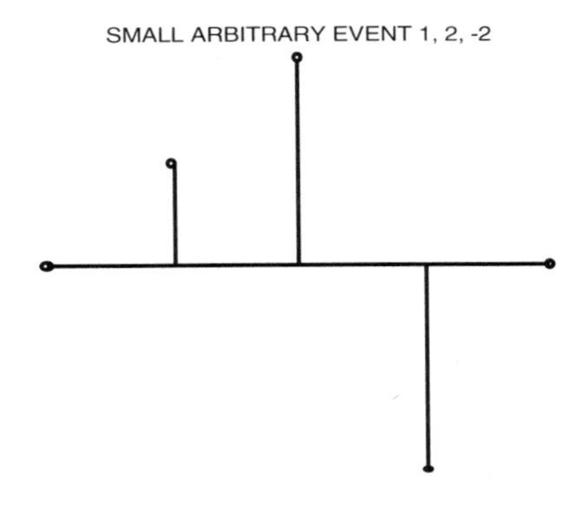

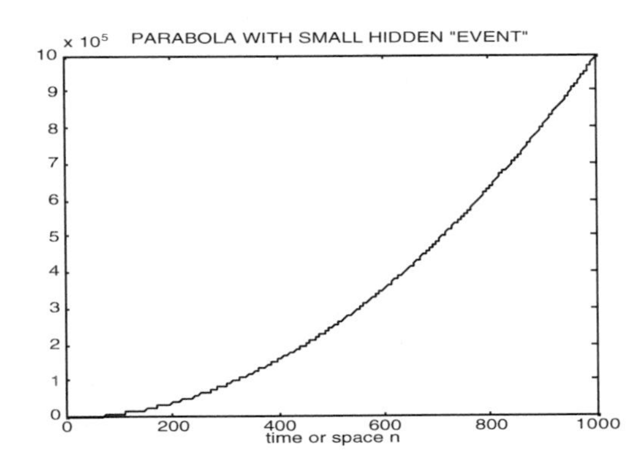

Figure 9.4–1 A small event ranging from –2 to +2 (left) is added to a parabola that ranges from 1 to 1,000,000. The combined signal is shown at right and, without foreknowledge of the location, the event would be impossible to detect using conventional methods.

Using an FFT, as shown in the left graph of Figure 9.4–2, indicates only the very low "frequency" of the parabola and does not show the high frequency of the event. Also, we recall that an FFT does not indicate the location of an event. Even using a standard CWT (right) we see only some edge effects.

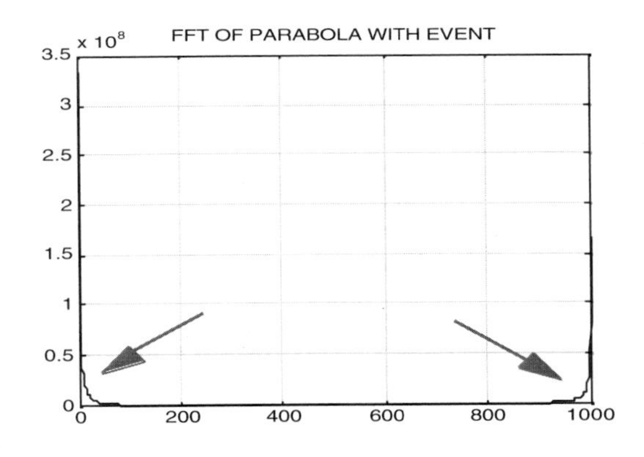

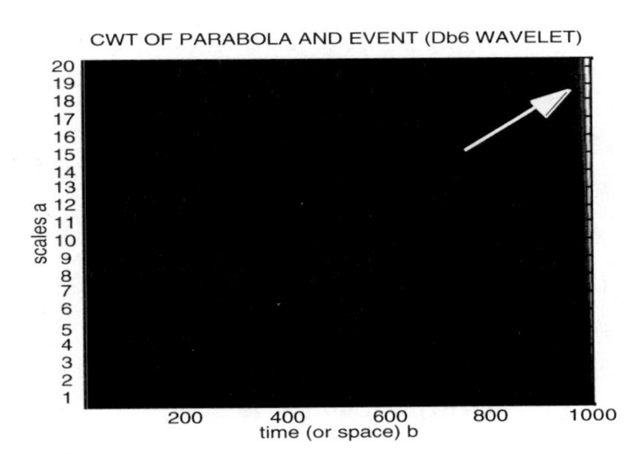

Figure 9.4–2 The FFT of the combined signal (left graph) shows only the parabola components as indicated by the arrows. The CWT of the combined signal (without special image enhancement) shows only some edge effects as indicated by the arrow at right.

Using a conventional DWT with a Db6 basic wavelet, however, we obtain very interesting and useful results. The Db6 6-point wavelet filter when correlated with a constant line, a linear slope, or a parabola will produce zero results. Figure 9.4–3 shows the conventional DWT display.

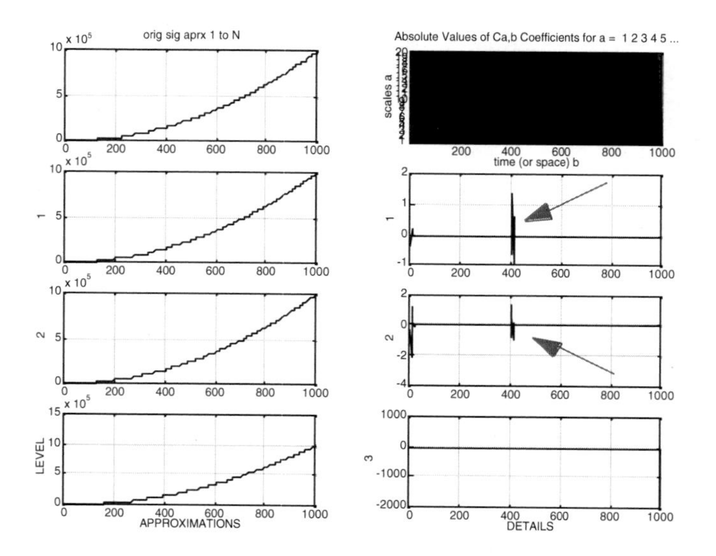

Figure 9.4–3 Three-level conventional DWT display for the combined signal using a Db6 wavelet. Note in Details **D1** and **D2** zero correlation everywhere but the hidden event (arrows).

Using the results from **D1** and/or **D2** we could locate the event. **D3** (bottom right graph) also should show the event, but when compared with some end effects, the event is too small. Compare the scale on this **D3** graph (–2000 to +1000) with the **D1** and **D2** graphs (integers less than 10).

We can produce even more remarkable results. One of the strengths of wavelets is that we can work with specific sub-intervals of time. Seeing from the DWT display that the event is in the neighborhood of **n** = **400**, we focus on the range from **300** to **500** and run the same DWT but down to level 6. Figure 9.4–4 shows **D1**, **D2**, **D3**, and **D4** in close-up (**D5** and **D6** not shown). Notice that the size of these details is decreasing. Also notice that **D3** and **D5** look like the **Db6** basic wavelet (ref. Fig. 6.4–2). We'll show you why later.

We recall that the signal **S'** is reconstructed from the level 1 Approximations and Details. Specifically,

```
S' = A1 + D1
```

A1 is reconstructed from **A2** and **D2**, **A2** is reconstructed from **A3** and **D3** and so on. Thus the signal can be reconstructed, for this 6-level DWT, as

```
S' = A6 + D6 + D5 + D4 + D3 + D2 + D1
```

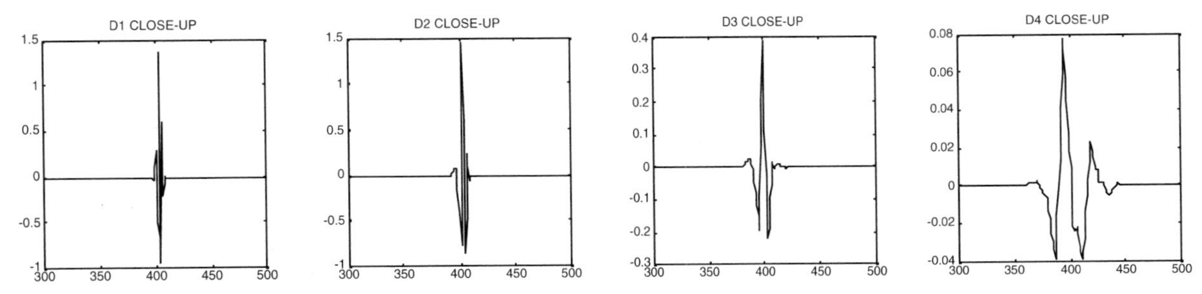

Figure 9.4–4 Close-ups in the range **n** = **300** to **500** of the Details **D1**, **D2**, **D3**, and **D4**. At each level, including **D5** and **D6** (not shown), the correlation of the Db6 basic wavelet filter with the parabola is zero (vanishing moments) while the correlation with the event produces non-zero results.

But because of the vanishing moments property we saw that **D1** through **D6** were all zero except for the event. Thus if we set all the details to zero and then reconstruct we should be able to isolate the polynomial noise portion (parabola) of the combined signal. In other words

```
S = A6 + 0 + 0 +0 + 0 + 0 + 0 = A6
```

Figure 9.4–5 shows the result of subtracting **A6** from the combined signal. At left we can see the event at **n** = **400**. In a further close-up at right we can see that we have isolated the event from the combined signal very well.

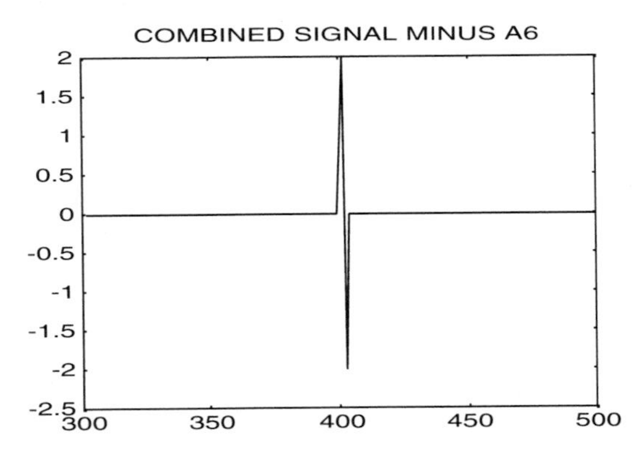

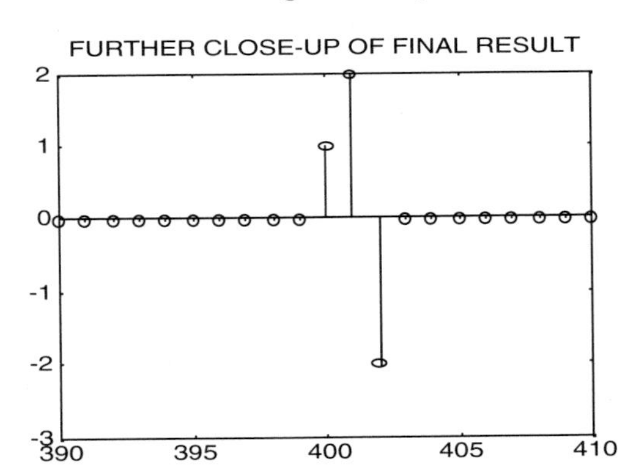

Figure 9.4–5 Extracting the event by subtracting the parabola from the combined signal. The further close-up at right shows the original event [1 2 –2] located at 400.

To recap, we added an event ranging from –2 to +2 to a polynomial ranging from zero to one million. We pretended we didn't know the event or where it was located in the combined signal. Using DWT vanishing moments, however, we were able to isolate it and show the exact location and the (almost) exact values for the event*.

Most texts use the number of points in the Daubechies basic filter to name the wavelet (e.g., 4 points in the Db4, 6 points in the Db6, etc.). However a few, including the MATLAB documentation, use the number of vanishing moments to describe the wavelet. Thus the 4-point wavelet filter we call the Db4 has 2 vanishing moments† (and thus can isolate up to a 2nd degree polynomial) and is designated "*Db2*" in MATLAB. Similarly, the 6-point Db6 filter has 3 vanishing moments (able to discern a constant, linear slope, and a parabola as we have seen) and is designated "*Db3*".

There is one more interesting example that may serve to give the reader a "feel" for vanishing moments. Consider a view of a city skyline that might have been designed by *I. M. Pei*‡ and notice the slopes and higher order polynomials. Treating the skyline as a "signal", figure 9.4–6 shows a 2-level conventional Db6 DWT display for this "signal".

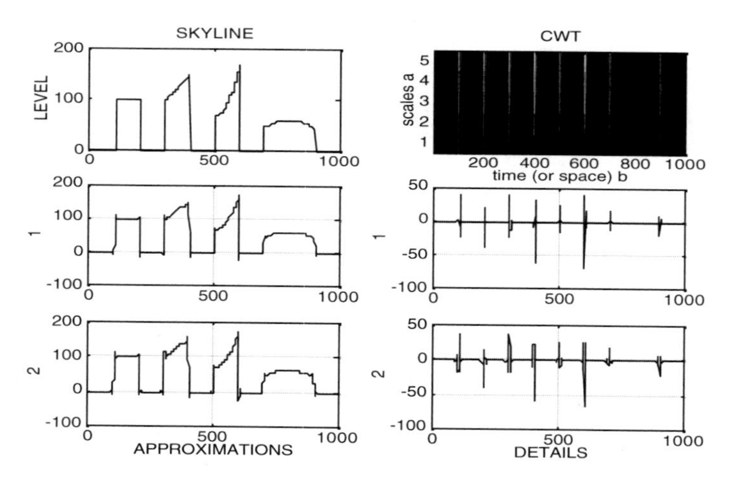

Figure 9.4–6 DWT display of an *I. M. Pei*-type skyline. The skyline "signal" is shown in the upper left graph. The CWT display in the upper right graph shows some high frequency (small scales) that indicate the building edges. This can be seen much more clearly, however in the **D1** display (middle right graph) that clearly shows the building edges.

---

* Starting at **n = 399** the perfect values would be **[0 1 2 –2 0]**. The actual values are very close to perfect with **[-0.0181 0.9825 1.9831 –2.0163 –0.0160]**

† We discussed earlier how the Spectral Factorization process has an exponent of 2 for what we call the "Db4".

‡ I. M. Pei is a prize-winning architect, known for his high modernist architecture that often involves mathematical shapes such as linear slopes and parabolas.

As with the "event" in the parabola, the correlation of the Db6 wavelet with the flat roofs, the sloped roofs, or the stylized parabolic roof produces zeros and only the building edges are shown.

9.5 Finding the "Magic Numbers" of Basic Db4 Filters using Wavelet Properties

In the last chapter we demonstrated how to find the four "magic numbers" used to create each of the Db4 wavelet filters by "reverse engineering" using a computer software and the z-transform. Now that we understand additional concepts such as *vanishing moments* we'll show you an even easier way.

We'll start with the basic "wavelet filter" which is the highpass reconstruction filter **hir** or H' (if we upsample and lowpass filter these 4 numbers repeatedly we have an estimation of the wavelet function—ref. Fig. 6.4–1). We will designate the 4 values as

```
H' = [ c0,  c1,  c2,  c3]
```

Using the relationships of the four filters to each other (ref. Fig 8.1–2) we have

```
H' = [ c0,  c1,  c2,  c3]
H  = [ c3,  c2,  c1,  c0]
L  = [ c0, -c1,  c2, -c3]
L' = [-c3,  c2, -c1,  c0]
```

We learned in this chapter that the dot product of H' with a constant is zero (1st vanishing moment). Using **[1 1 1 1]** as the constant bias we have for our first equation (using "x" as the multiplication indicator)

```
c0x1 + c1x1 + c2x1 + c3x1 = c0 + c1 + c2 + c3 = 0
```

We desire that the dot product of H' with a linear slope also be zero (2nd vanishing moment). Using **[1 2 3 4]** as the slope we have for our 2nd equation

```
c0x1 + c1x2 + c2x3 + c3x4 = c0 + 2c1 + 3c2 + 4c3 = 0
```

We learned that each of the wavelet filters is *integer orthogonal* (ref. Fig. 8.5–6). With the 4 filter values spaced 1/2 integer apart in time we would have a zero value for the dot product of H' and H' shifted by 2. In other words the dot product of

```
H'          = [c0  c1  c2  c3  (0) (0]
```

with the integer-shifted version of itself

$$H'_{shifted} = [0 \quad 0 \quad c0 \quad c1 \quad c2 \quad c3]$$

gives us our 3rd equation

```
c0c2 + c1c3 = 0
```

Another requirement for these filters is that the coefficients of the scaling function filter, L', add to a constant.* This gives us our fourth equation

```
-c3 + c2 -c1 + c0 = sqrt(2)
```

We now have four equations in four unknowns. With simple substitution we obtain

```
c0 = -0.1294, c1 = -0.2241, c2 = 0.8365, c3 = -0.4830
```

and we have for our filters the familiar results

```
L   = lod = [-0.1294    0.2241    0.8365    0.4830]
L'  = lor = [ 0.4830    0.8365    0.2241   -0.1294]
H =   hid = [-0.4830    0.8365   -0.2241   -0.1294]
H'  = hir = [-0.1294   -0.2241    0.8365   -0.4830]
```

9.6 Summary

In this chapter we talked about *matching* the signal (or the noise) to a wavelet with the same general shape. The stretching and shifting of this wavelet that occurs in the wavelet transforms will then allow for a strong correlation and the ability to segregate the noise from the signal.

We demonstrated how we could use a customized wavelet in a CWT and introduced the concept of *regularity*. We saw that an *infinitely regular* (smooth) wavelet like the Morlet out-performed our "fake" wavelet.

* If we use 1.0 instead of sqrt(2),.as is found in some texts, we obtain the "pre-divided" (by sqrt(2)) filters.

We noticed, however, that either the Morlet or the "fake" wavelet would suffice to show where the frequency changed in the split-sine test signal. Thus we saw that the "Sport of Basis Hunting" to find (or build) the "perfect" wavelet may be counter-productive and ignores one of the major strengths of wavelet technology—the ability to stretch and shift the wavelet in time or space to line up with the event. The other extreme is to try to get by with only one wavelet as we have with the sinusoidal "wave" used as a basis for Fourier analysis. A good rule of thumb then is to pick a wavelet that looks somewhat like the signal or noise we want to isolate and let the wavelet transforms do their magic.

We touched on the concept of *Best Basis* or *Matching Pursuit* where software is used the choose the "best" wavelet based on *cost functions* using criteria such as *entropy* and added a caveat that the software could chose a wavelet that matched an artifact or noise instead of what we wanted. We then gave a very brief overview of the *lifting scheme* where a *trivial* wavelet can be recursively improved (*lifted*) to create a more suitable wavelet for a particular use.

Next, we introduced the concept of *vanishing moments* and demonstrated their utility to isolate the signal from the noise when either one can be represented by a polynomial. We generated another "fake wavelet" similar to the Db4 by rounding off the numbers. This fake wavelet emulated the Db4 properties very well until we tried to use it to isolate a slope (first degree polynomial). The Db4 classic wavelet filter performed flawlessly producing a zero dot product (vanishing moment) while the fake wavelet fell short. We showed some examples of the use of vanishing moments.

Armed with our knowledge of vanishing moments, orthogonality, and some other basic desired attributes of wavelet filters, we were able to easily derive the filter coefficients for the Db4.

Having added to our knowledge of desirable qualities in wavelets, we are now ready to introduce the various wavelet families and describe the strengths, weaknesses, and suggested applications of each.

"The roots of education are bitter, but the fruit is sweet."

—Aristotle

CHAPTER 10

Specific Properties and Applications of Wavelet Families

By this point, we have learned much about the properties of the various wavelets and wavelet filters and how they can (or cannot) be used in the various wavelet transforms. We are now in a position to look at the major wavelet families and how the specific properties lead to some practical applications.

It has been said that one of the biggest contributions of wavelets has been to bring various disciplines together and to provide material for research papers. Engineers in one discipline are surprised to see the same methods used by other disciplines but labeled differently. A wavelet transform to one person is seen as multiresolution analysis, a multirate system or a filter bank by others. Add to this Tower-of-Babel-like diversity in nomenclature the proliferation of the exponentially-increasing number of papers describing new wavelets or variations on existing wavelets and it is not surprising that we have a great number of wavelets and wavelet families out there.

Some words of encouragement are in order. In the first place, the increasing number of wavelets is really an "embarrassment of riches". We don't have to understand completely the mechanics and be able to use every wavelet in existence any more than we have to understand the mechanics and personally drive every make and model of car, truck, or bus on the highway. We learn how to "drive" the general families of wavelets and enough mechanics to know where to look (or who to ask) if we encounter problems. Trying out additional new wavelets is then not much different than test-driving a new car. *

If conventional techniques such as the FFT now seem easy by comparison to Wavelet Technology this is understandable because in the "good old days" we were looking only at frequency. Wavelets add literally an entirely new dimension by simultaneously working with both time and frequency. Recall that the CWT needs three dimensions to show the results. We have time or space as the x-axis, frequency or scale as the y-axis,

* When first learning to fly a small plane, the author was overwhelmed by the number of gauges, displays, and meters found in modern cockpit avionics and felt he was not up to the task of operating such a complicated machine with such an abundance of information! The wise flight instructor covered up all the gauges, etc. except the compass, altimeter, airspeed indicator, and fuel gauge. On a nice day, with the instructor by his side, the author could now fly! The additional gauges were then uncovered one-by-one and they were then seen to be helps rather than hindrances. The addition highly useful information "uncovered" by using wavelet technology on your data is analogous to the additional information found by uncovering modern avionics gauges and displays. Hang in there!

and magnitude at that particular time and frequency as the z-axis (indicated by brightness or color in the CWT display).

Being able to discern the time, the frequency and even the general shape of a transitory event (by comparisons to the candidate wavelets), is a powerful technique. And although it is by necessity more complicated, the extra effort in learning to understand and use these amazing tools is more than offset by the vastly improved ability to fully exploit any digital signal containing events that start and stop and/or change frequency (the most interesting kind!).

10.1 (Real) Crude Wavelets

The properties we have previously studied can now be used to classify the various types of wavelets. The first type is the *crude* wavelets. We have already learned about *crude wavelets* and how the basic (real) filter coefficients are produced by evaluating the explicit mathematical equation at equispaced points in time. Stretched filters are easily obtained by evaluating the equation at more points.*. Crude filters are *bandpass* with the center frequency decreasing for the stretched wavelet (more interpolated points).

We have recently learned that this type of wavelet has no orthogonality, no vanishing moments and no additional filters. In other words, we have a (stretchable) "*wavelet filter*" (**H'**) from the explicit equation for use in the CWT but not the "*scaling function* filter" (**L'**) or the other 2 filters (**H** and **L**). Thus crude filters can be used in a CWT but not a DWT.

Because these wavelets (and filters) come from an explicit mathematical equation, they are smooth and are thus *regular*. They are also by design *symmetrical.*

These wavelets are theoretically infinite in time but the values are essentially zero outside a small range (e.g. from **t** = –5 to **t** = +5) and thus they have "*effective support*" in that range (rather than "*compact support*" where the wavelet is a finite length). Some of the better-known crude wavelets with real coefficients are now discussed.

---

* Some authors refer to these as "*continuous wavelets*" because of the continuous nature of the filter-coefficient-generating explicit mathematical equation. Most avoid this term, however, stressing that we never actually use a continuous wavelet in Digital Signal Processing but rely on the various discrete filters instead.

MEXICAN HAT WAVELET—This wavelet was used as our first example of a crude wavelet. The estimation for a continuous Mexican Hat wavelet is shown in Figure 10.1–1 along with the three dimensional version.

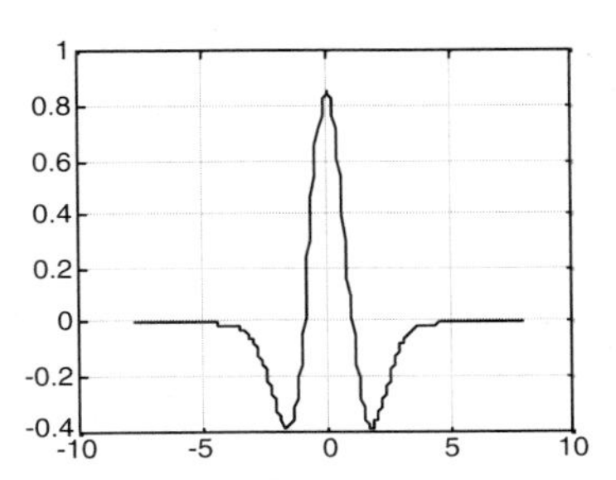

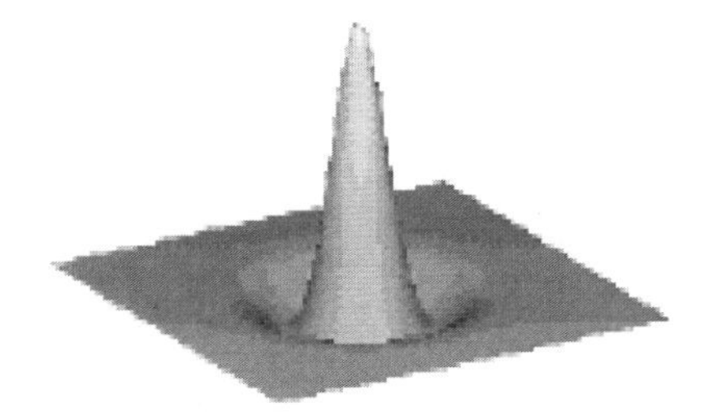

Figure 10.1–1 The Mexican Hat wavelet as drawn from its explicit mathematical equation. The 3-D version is shown at right. It looks even more like a Sombrero, especially if the square corners were rounded.

We can observe that this wavelet is symmetrical in time.* It also looks very smooth. In fact, the Mexican Hat is a good example of an *infinitely regular* wavelet as discussed earlier. The explicit mathematical equation for the Mexican Hat Wavelet is given by (using the MATLAB format of "*" to indicate multiplication)

```
mexh(t) = (2/(sqrt(3))*pi^0.25)* exp(-t^2/2))*(1-t^2)
```

The effective support is from –5 to +5 (we mentioned earlier that the value at **t** = **5.1** was less than 0.000004). The process of "stretching" this filter to obtain more points was shown in Figures 5.5–3 and 5.5–4. We also demonstrated the *constant Q bandpass* nature of the Mexican Hat wavelet (with the center frequency decreasing as it is stretched) in Figures 5.7–1 and 5.7–2.

Having the traits of smoothness (regularity), symmetry, and a very rapid decay, the Mexican Hat wavelet is often used in vision analysis because these traits are similar to those of the human eye†. The 3-D version is used in earthquake analysis by placing these smooth, rapidly decaying, symmetrical "sombreros" along the fault lines in the computer simulations.

* The Mexican Hat wavelet is in fact derived from a function that is proportional to the 2nd derivative of the Gaussian Probability Density Function—thus the symmetry.

† Neurons in the human eye work somewhat like a Mexican Hat wavelet. Hold 2 fingers together near your eye and look through the slit. As you squeeze the slit together most people will see little black lines that are not really there, nor are they diffraction patterns. See **williamcalvin.com/bk7/bk7ch13.htm** for a discussion.

MORLET WAVELET—Named for Jean Morlet, a geophysicist and one of the early inventors of wavelets, this was the first wavelet (after the Haar) to be developed and is another example of a crude wavelet. We used this wavelet earlier to demonstrate regularity by showing its improved performance over a "fake sine wavelet" in correlating with a "split sine" signal (ref. Fig. 9.1–4 and 9.1–5).

The explicit mathematical equation for the Morlet Wavelet is given by

```
morlet(t) = exp(-t^2/2 x cos(5t)
```

The Morlet is sometimes described as a modulated Gaussian (see next subsection). The above equation used to generate points equispaced in time is graphed below in Figure 10.1–2.

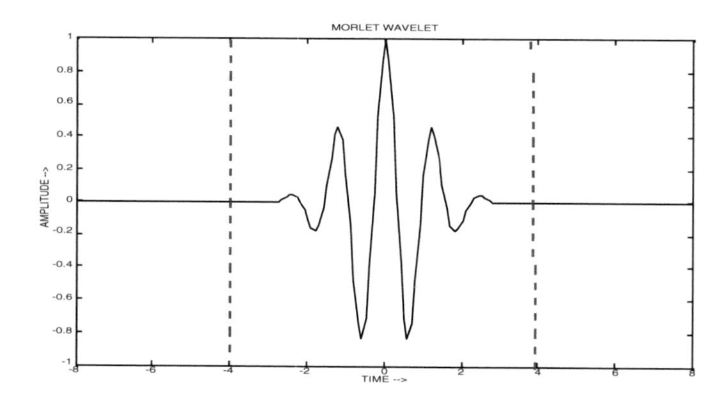

Figure 10.1–2 The real Morlet wavelet as drawn from its explicit mathematical equation. Effective support is from –4 to +4 as indicated by the dotted lines.

The Morlet Wavelet shares most of the same traits as the Mexican Hat. Points are produced from its explicit mathematical equation (ref. Fig. 5.6–3), it is symmetrical, infinitely regular, and has an effective support from –4 to +4. It has only the one stretchable filter and is not orthogonal, thus it can be used in a CWT, but not in a DWT. We saw earlier that more points are required to simulate this shape than that of the Mexican Hat (ref. Fig. 5.63) but that the frequency resolution was better.

With smoothness and periodicity, we saw that the Morlet is good for periodic or continuously varying data. Examples of use of the real Morlet wavelet include sinusoidal pulses and atmospheric indices (such as cyclical changes in air pressure and in storm tracks).

GAUSSIAN WAVELETS—The familiar Gaussian probability density function, shown at the left in Figure 10.1—3, does not sum to zero and is not a wavelet, crude or otherwise. However its derivitives can produce wavelets. The well-known equation for the Gaussian function is simply

```
gaussian(t) = exp(-t^2)
```

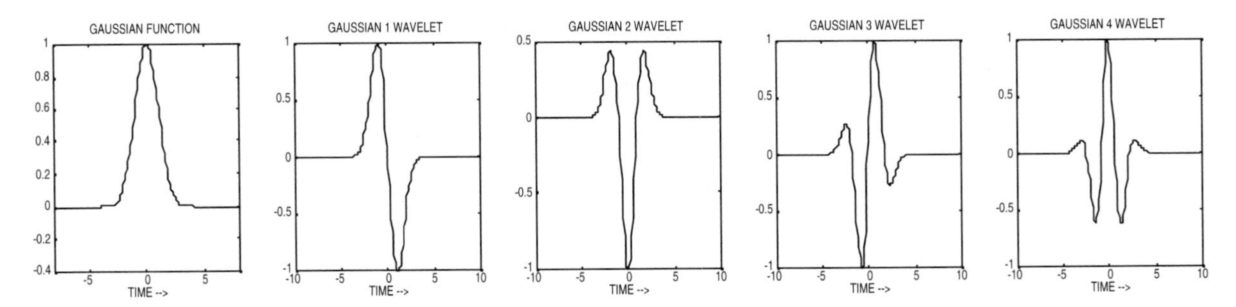

Figure 10.1–3 The Gaussian function (not a wavelet) is drawn in the first graph. The first through 4th derivitives of the Gaussian function (normalized) are drawn in the next 4 graphs. The final graph, for example, can be referred to as a *Gaussian 4* wavelet.

If we look at the slope of the Gaussian function we see that it starts at zero, becomes positive, goes to zero at the top, becomes negative, and goes to zero again. This is exactly what we see in the 2nd graph. This waveform does indeed have components that sum to zero and we can use this to generate points for the wavelet filters at equispaced time intervals.* This wavelet can be called a "*Gaussian 1 wavelet*" indicating the first derivative of the Gaussian. The effective support is from –5 to +5.

Looking at the slope of the 2nd graph we can see the pattern for the 3rd graph (the 2nd derivative of the Gaussian). This graph should look familiar as an upside-down (negative of the values) Mexican Hat wavelet. The final 2 graphs are the 3rd and 4th derivitives of the Gaussian and are called the *Gaussian 3* and *Gaussian 4 wavelets.* This process can be continued with more derivatives to produce additional Gaussian wavelets. They are of course normalized as required.

These wavelets are either symmetric (Gaussian 2, 4, 6, etc.) or anti-symmetric (Gaussian 1, 3, 5, etc.). Like the other crude wavelets they are

* This is possible, of course, because we use numerical differentiation and are thus working in the world of Digital Computers.

suitable for a CWT but not a DWT and they do not have orthogonality, vanishing moments, or scaling function filters.

As can be seen, they appear smooth and have good regularity. A low-number Gaussian wavelet (Gaussian 1 or 2—the inverted Mexican Hat) can be represented by a small number of points and can provide good time resolution. A high-number Gaussian wavelet requires more points but provides better frequency resolution. This flexibility makes the Gaussian family a good choice for applications such as Music, Speech and Tomography.

MEYER WAVELETS—Named for wavelets pioneer Yves Meyer, these wavelets are defined in the frequency domain rather than by the time domain equations we have seen earlier in this chapter. A (non-standard) Inverse FFT is then used to produce the filter points at equispaced time intervals. Unlike the other (real) crude wavelets we have discussed, however, an estimation of the scaling function is also produced. Although theoretically infinite in the time domain, an effective support from –8 to +8 is used. The 256-point estimations of the Meyer wavelet function and scaling function, mapped onto an interval from –8 to +8 are shown below in Figure 10.1–4.

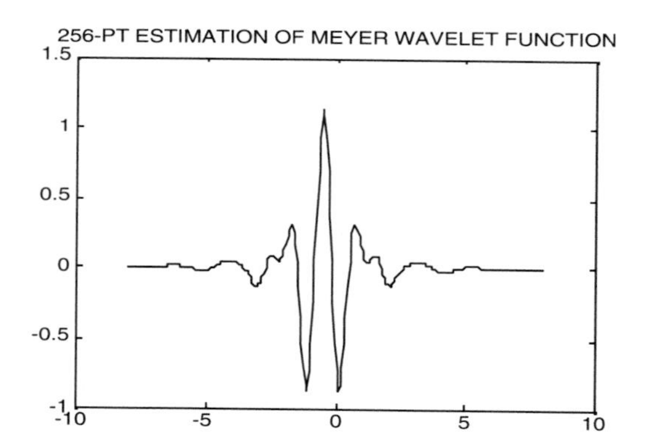

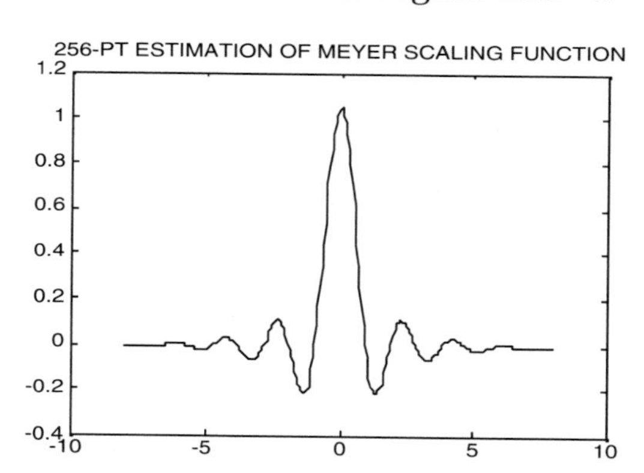

Figure 10.1–4 The 256-point estimation of the Meyer wavelet function and scaling function are drawn here. Like the other crude wavelets the sum of the *wavelet* function points is zero.

Meyer wavelets share the same properties as the other crude wavelets. Like the Mexican Hat, Morlet and Gaussian wavelets they are symmetrical. Also like the other crude wavelets, they are not suitable for use in the DWTs.

The Meyer highpass reconstruction filter (wavelet filter) stretched to various lengths can be used in a CWT. The CWT, however, is not concerned with scaling function filters thus the most general use of the Meyer wavelets is, like the other crude wavelets, stretched filters at equispaced time intervals that come from an explicit equation (in this case from the IFFT of a frequency-domain equation). We start with 32 equispaced points on the interval –8 to +8 as shown in Figure 10.1–5.

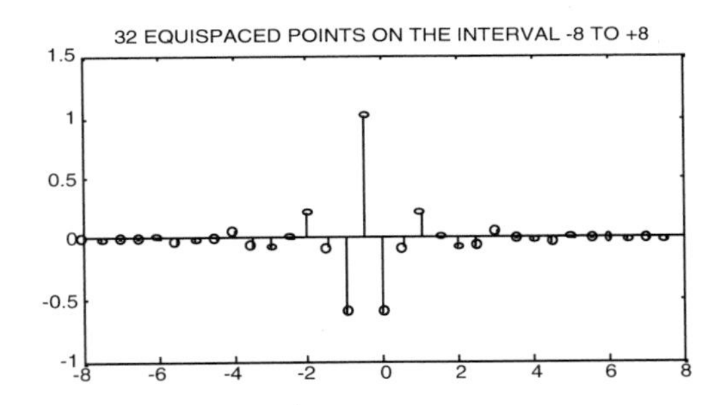

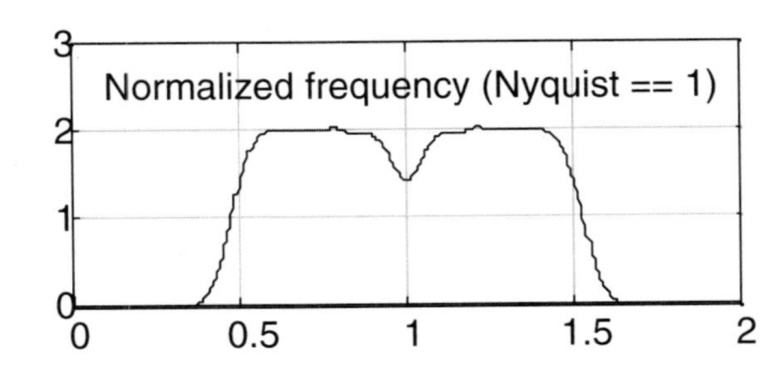

Figure 10.1–5 Meyer wavelet 32 points on the interval –8 to +8. Note the uneven frequency response being nowhere near *flat* in the passband.

We notice on the left that the points are symmetric but that the peak is slightly to the left of zero. We also notice that this 32-point filter is not a good bandpass filter. If it were a good, flat, highpass filter we could allow this as the first of several bandpass filters, but it is not.

We will discuss these issues soon in the *Discrete Meyer Wavelets*. For now, however, we simply stretch the filter by specifying 64 points instead of 32 on the same interval. The first graph of Figure 10.1–6 shows this. On the 2nd graph we see that we now have some very good bandpass filters. The 3rd graph shows a stretching to 128 points on the same interval. Notice on the last graph the *constant Q* behavior in that the center frequency and bandwidth are both halved while the magnitude of the peak is doubled.

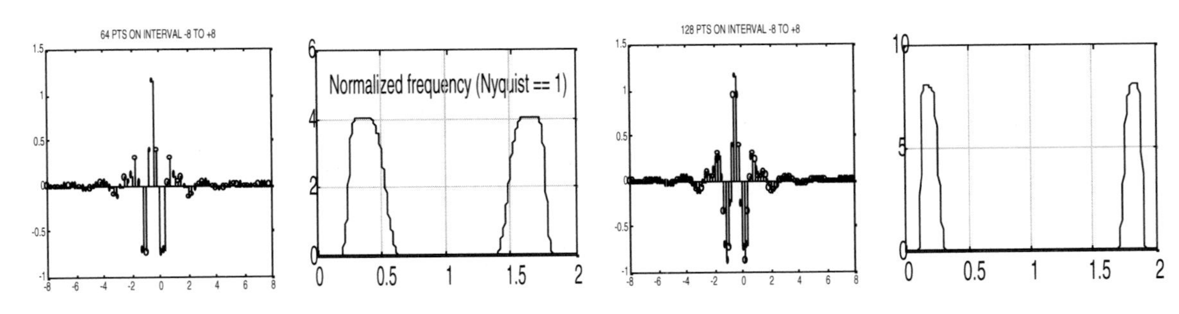

Figure 10.1–6 Meyer wavelet filter stretched to 64 points and then to 128 points. Notice the constant Q behavior.

When used in a CWT, some software begins with as few as 17 points to represent the Meyer wavelet, however we don't get good bandpass filters till we stretch to about 64 points. This is part of the rationale for creating a 62-point *Discrete Meyer* approximation that we will look at a little later.

Because of the outstanding frequency discrimination of Meyer wavelets they are an excellent choice to isolate events by frequency. As with all wavelets, being limited in both the time and frequency domains, the "cost" of using Meyer wavelets is a longer filter.

Another advantage of Meyer wavelets is that they have real filter coefficients and can be used with real signals. We will now look at some complex crude wavelets that also provide excellent frequency discrimination.

10.2 Complex Crude Wavelets

There are two reasons to learn about working with *complex wavelets*. The more obvious reason is that we can then use them with *complex signals*. The other, more subtle reason, is that some waveforms *must be complex* in order to function at all as a wavelet. We will look first at a "*complex-only*" wavelet.

SHANNON ("SINC") WAVELET—We discussed the familiar truncated Sinc function as a possible wavelet (ref. Fig. 5.3–1 and 5.3–2). Although this classic DSP waveform would be desirable as a wavelet because of its relatively sharp frequency cutoff, it has a major problem—It is a lowpass filter, not bandpass. For example, as we stretched the Mexican Hat and Morlet crude wavelet filters (by interpolating more points), we lowered the bandpass center frequency. With the Shannon "wavelet" we lowered the cutoff frequency but it was still a lowpass filter.

The solution here is to make the Shannon filter complex and then "end-around" shift the filter in frequency by *bandshifting*. A MATLAB statement to produce this complex wavelet is given by (where "*" indicates multiplication)

```
Shan = (FB^0.5)*sinc(FB *tlin)*exp(2*j*pi*FC*tlin))
```

where *FB* is the bandpass width, *FC* is the center frequency, and *tlin* is the number of equispaced points in time (another reminder that the time points used in the explicit equation for a crude wavelet are only to generate filter values)

For a tutorial example we start with 1/2 integer spaced points from –10 to 10 which gives us 41 time points for the *tlin* and then 41 corresponding values from the explicit equation. For simplicity we let **FB** = **1** and **FC** = **0**. The first and 3rd terms in the above equation go to zero and we have the familiar real Shannon wavelet

```
Shan = sinc(FB x tlin) =

sin(π x FB x tlin)/(π x FB x tlin)
```

(remembering that the sine function uses radians for compatibility with most computer software). Figure 10.2–1 shows this real wavelet in the time and frequency domains. With the center frequency of this bandpass filter set to zero, however, it is still really a *lowpass* filter and not acceptable as a wavelet filter.

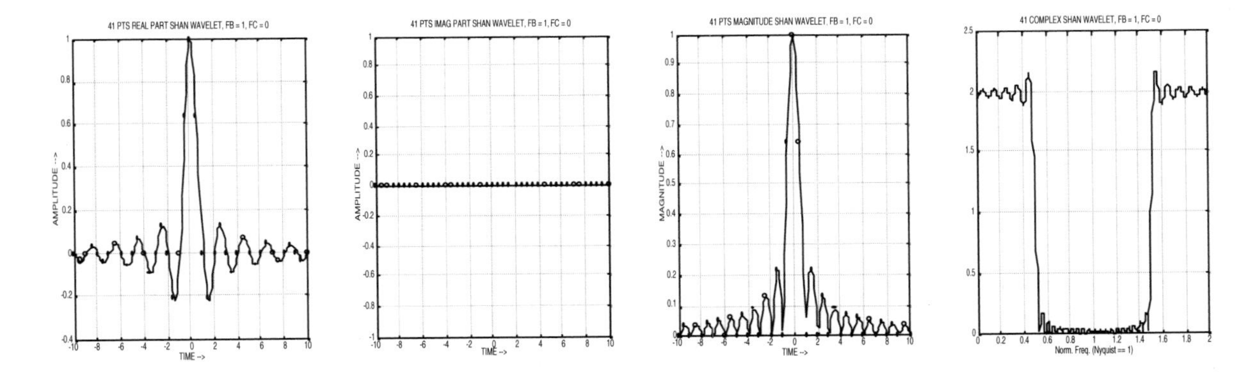

Figure 10.2–1 With the center frequency = 0, we have the real Shannon "wavelet". Notice the familiar form in the first graphic, the imaginary part is zero as shown in the 2nd graphic. The modulus or magnitude is shown in the 3rd graphic. The 41 filter points are shown overplotted on the curves produced by the above equation. The frequency characteristics are shown at right for this familiar lowpass filter. We see that the frequency bandwidth is indeed 1—but only if we count both the left and right sides of the passband (0 to 0.5 and then 1.5 to 2 with Nyquist = 1).

If we change the center frequency from 0 to 1 we now have the same width passband (1.0) but "end-around-shifted" in frequency to be centered at Nyquist (π radians) as seen in the last graph of Figure 10.2–2. This is of course a *highpass filter* but still technically *bandpass* with the passband from 0.5 to 1.5.

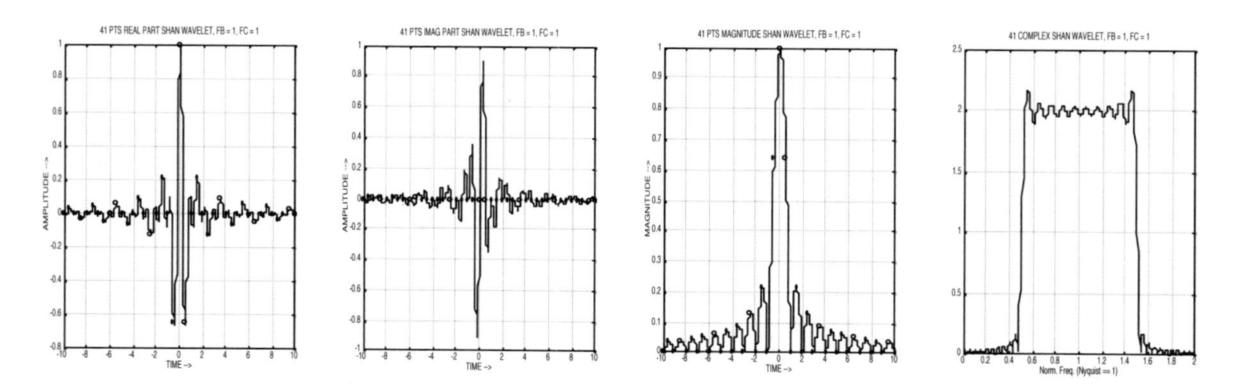

Figure 10.2–2 With the same bandwidth but the center frequency moved to 1.0 (Nyquist) *we have bandshifting* (circular shifting of frequency) as seen in the last graph. The magnitude or modulus (3rd graphic) , remains the same as in the previous figure (ref. Fig. !0.2–1). Note the real and imaginary values in the first 2 graphs.

Using this same equation we now stretch (dilate) the filter by adding more interpolated points. We stretch by a factor of 2 and have points at 1/4 integers from –10 to +10 giving us 81 time points and thus 81 filter values from the equation.

Figure 10.2–3 (left) shows this wavelet in the frequency domain. The wavelet is now complex. It also follows the *constant Q* criteria discussed for the Mexican Hat and Morlet wavelets (ref. Fig. 5.7–3 and 5.7–4). In other words, by stretching this wavelet (filter) by a roughly a factor of 2 from 41 to 81 points the center of the bandpass filter has changed from 1.0 to 0.5 and the height of the bandpass filter (in the frequency domain) has changed from 2 to 4 as shown in the figure below at left.

Further stretching from 81 to 161 points moves the passband center frequency from 0.5 to 0.25 and the height from 4 to 8 as shown in the figure below at right.

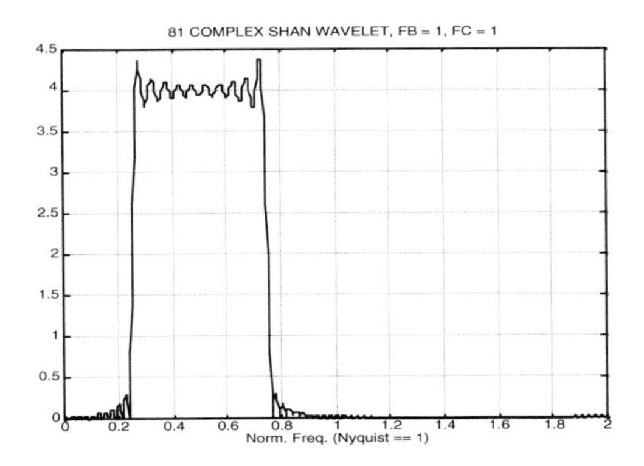

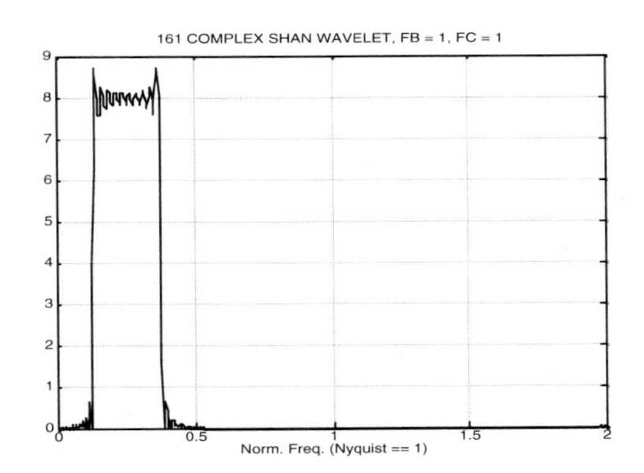

Figure 10.2–3 Interpolating from 81 to 161 points (stretching the filter) causes both the passband width and the passband center frequency to change from 0.5 to 0.25. It also shows constant Q behavior as the magnitude changes from 4 to 8.

If then we are willing and able to work with complex signals we can use the Shannon (Sinc) Wavelet. We have talked about *matching* the wavelet to the signal or noise. This wavelet is excellent for finding specific frequencies in an event. In other words, the *match* is in the frequency domain.

We can better see now why the Sinc wavelet is called the "dual" of the Haar wavelet. The 2-point Haar wavelet ([1 –1]) is excellent for finding very short events but has poor frequency cutoff. The Sinc, as we just saw, has very good frequency cutoff (short transition band) but requires a large number of points and would not be well suited for short-time events. This is an example of the *Heisenberg Uncertainty Principle* as applied to wavelets.

Jargon Alert: **The Heisenberg Uncertainty Principle for Quantum Physics says you can't know the exact position and the exact momentum of a particle simultaneously* ($\Delta X \Delta P \geq h/2$ where h is Planck's Constant). In time/frequency analysis this principle refers to the fact that you can't know the exact time and the exact frequency of a signal simultaneously ($\Delta T \Delta F \geq$ non-zero constant). We will discuss this principle as applied to wavelets, including *Heisenberg Boxes*, further in the appendices.**

* This fundamental uncertainty of position often leads to graffiti on the Physics Lab blackboard that says something like "Heisenberg was *probably* here."

The other major difference between the Haar and the Shannon (Sinc) wavelet is of course that the Shannon is a *crude* wavelet and can be used only in a (complex) CWT while the Haar can be used in a CWT *or* DWT.

The applications of the Complex Sinc Wavelet are many and varied. Those familiar with DSP know that the classic (real) Sinc function is a mainstay in filtering. The ability to create a wavelet with some of the powerful frequency discrimination characteristics of the Sinc function is well worth the trouble of working with the complex numbers.

The Shannon Wavelets, like the other complex wavelets we will study, share the same attributes as their *real* counterparts. The filter points are generated from an explicit mathematical equation, they are smooth (regular) and symmetrical* They can be used in a (complex) CWT but not a DWT. They have no scaling function, vanishing moments, or additional filters. They have constant Q behavior.

The Shannon Wavelet is theoretically infinite in length and does not have *compact support per se*, but can be made finite the same way we make the *real* Sinc function finite—either by direct truncation (boxcar window) or with more sophisticated windows such as Hamming, Hanning, Blackman, etc.

COMPLEX FREQUENCY B-SPLINE WAVELETS—The equation for the Shannon Wavelet just studied was

```
Shan = (FB^0.5)x(sinc(FB x t)x exp(2 x j x π x FC x t))
```

which could be looked at as *specifying the wavelet in the frequency domain.* In other words we chose the Center Frequency (FC) and the Bandwidth (FB). Then after choosing the finite interval for time (we used t from −10 to +10) we had our wavelet filter.

As the name implies, the *complex frequency b-spline wavelets* are also specified in the frequency domain. The explicit equation for this crude wavelet (in MATLAB format) to generate the filter points is given by

```
Fbsp = (FB^0.5)*(sinc(FB* t/M)^M)* exp(2*j*pi*FC*t))
```

which is the same as for the Shannon except for the factor "M". If **M** = **1**, in fact, the equations are identical. For **M** = **2** we have a "sinc squared" or sinc function times itself (with M also used in the denominator for normalization).

* Although the real and the imaginary parts by themselves may not be symmetrical, the magnitude or *modulus* is symmetrical.

With **M** = **3** we would have a "sinc cubed" function. Figure 10.2–4 shows the *frequency domain* representation of the "Fbsp" wavelets for **M** = **1**, **2**, and **3**.

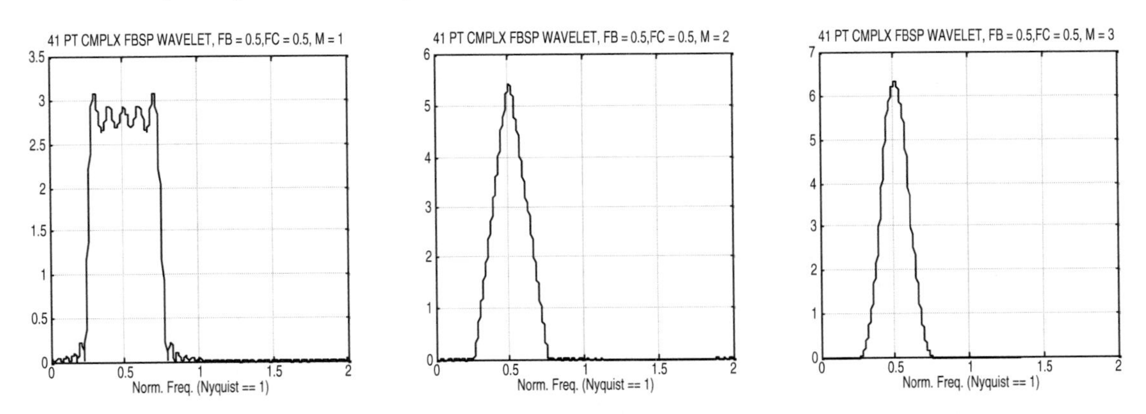

Figure 10.2–4 Frequency domain representation for the Complex Frequency B-Spline Wavelets of orders 1, 2 and 3 (**M** = **1**, **2** and **3**). 41 equispaced points were obtained from the above equation on the time interval −10 to +10. Note the Boxcar, Triangle, and Gaussian shapes here in the frequency domain.

Like the Shannon (Sinc) wavelet, these Frequency B-spline wavelets are good for isolating desired frequencies. Furthermore, by matching the wavelet shapes in the frequency domain to the frequency domain shapes of the event (e.g. Gaussian noise), we can better determine its nature.

The reason for the rectangular, triangular and Gaussian shapes can found with a short DSP review. Recall from DSP that multiplication in the *time domain* is equivalent to *convolution* in the frequency domain. In the above equation, with **M** = **2**, we are multiplying the sinc function by itself in the *time* domain. This is equivalent to convolving the boxcar shape with itself in the *frequency* domain which produces a triangle shape. With **M** = **3** we are convolving the triangle shape with a box which gives a Gaussian (*bell-curve*) shape. The process is illustrated below in Figure 10.2–5 for a 9-point "box".

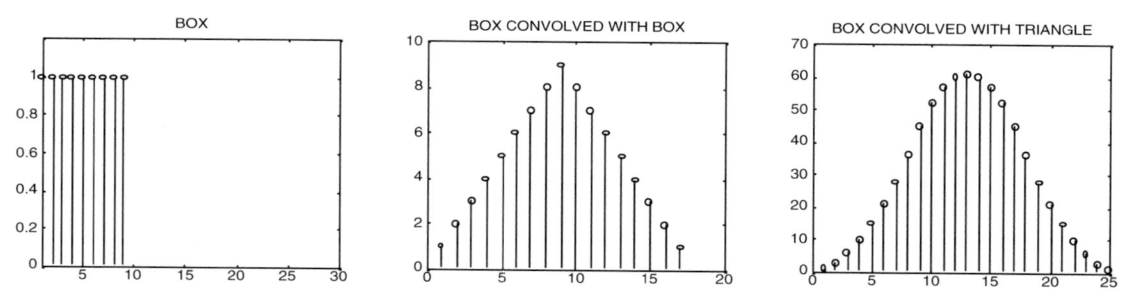

Figure 10.2–5 A 9-point "box" (left graph) is convolved with itself to produce the triangle shown in the middle graph. When this "box convolved with box" triangle is convolved with the same box again we obtain the Gaussian as shown at right. The horizontal and vertical axes are simply *x* and *y* for this figure.

***Jargon Alert:* The term B-Spline means Basic polynomials that are connected (splines are connected functions). The first graph in Fig. 10.2–5 is a Constant B-spline. The polynomial here is simply y = 1 on the interval 1 ≤ x ≤ 9. The second graph is a Linear B-Spline—it connects two linear polynomials. We have y = x on the interval 1 ≤ x ≤ 9 and y = 18 - x on the interval 9 ≤ x ≤ 17. A reminder that these graphs are used here in the freq. domain—hence the name *Frequency B-Splines*.**

Like the Complex Shannon wavelet, the *constant Q* behavior is also found as we stretch the wavelet. For **M** = **1** (order 1) we saw this for the Shannon wavelet in Fig. 10.2–3. For **M** = **2** we show this behavior for the original 41 points, 81 points, and 161 points in Figure 10.2–6.

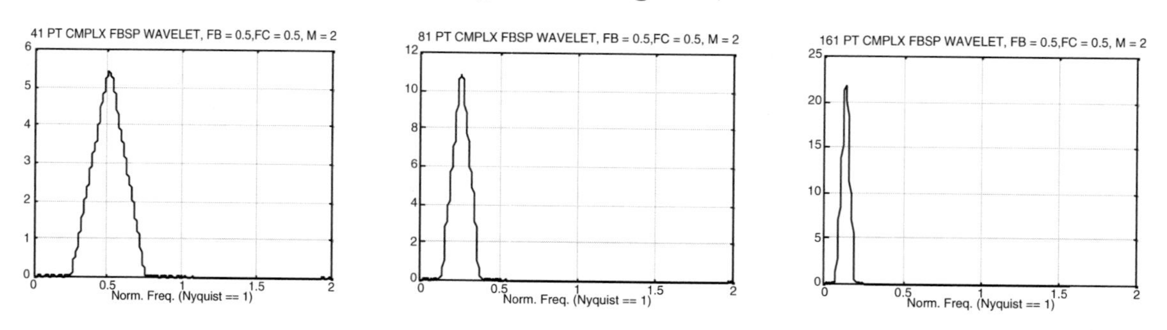

Figure 10.2–6 Using the 2nd-order Frequency B-Spline (**M** = 2) equation with 41 points on the interval from −10 to +10 (points 1/2 integer apart), we obtain the frequency response shown at left. Interpolating to 81 points and then to 161 points on the same time interval (points 1/4 and then 1/8 integer apart) stretches the wavelet in the time domain and produces the constant-Q behavior seen in the frequency domain. Notice the Center Frequency and Bandpass Frequency are decreased by roughly factors of 2 while the height is increased by factors of 2.

COMPLEX MORLET WAVELET—The Morlet wavelet also has a complex version as shown below in Figure 10.2–7. In this form we can use the modulus (right graphic) for edge detection. In other words, as the Complex Gaussian is shifted (translated) smoothly along the signal or image, the location of a change in the signal will show up on the CWT without the "ringing" we would get sliding the real (sinusoidal) version past the discontinuity.

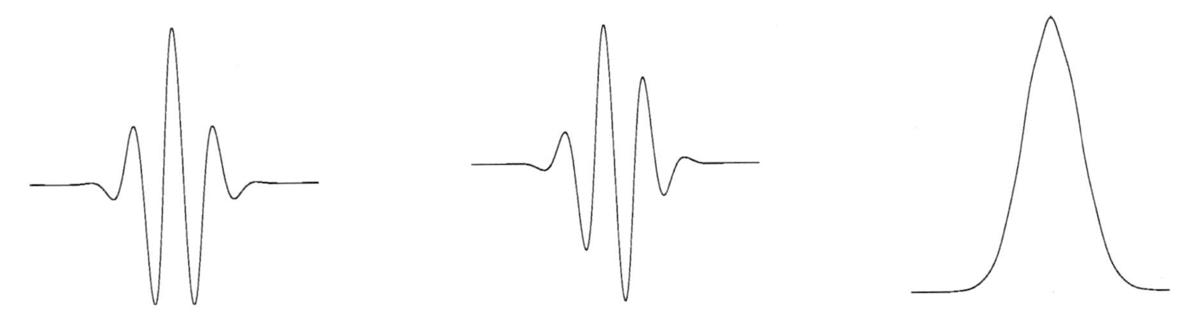

Figure 10.2–7 The Morlet Wavelet in its complex form. The real portion at left is familiar now as the real Morlet Wavelet. The imaginary portion is shown in the middle. Combining the real and imaginary portions produces the Gaussian-looking modulus as shown at right.

The explicit (MATLAB) equation for the complex Morlet wavelet is given by

```
cmor(t)=((pi*Fb)^(-0.5))*exp(-t^2/Fb)* exp(j*2*pi*Fc*t)
```

where Fb is the bandwidth parameter and Fc is the center frequency.

COMPLEX GAUSSIAN WAVELETS—Like the Morlet, the Gaussian wavelets also have a complex version. The explicit equation for this crude wavelet is given by

```
cgau(t) = exp(jt) x exp(-t^2)
```

which is the same as the equation for the real Gaussian wavelet multiplied by the complex exponential. Using the Euler identity we can re-write this as

```
cgau(t)= cos(t) x exp(-t^2) + jsin(t) x exp(-t^2)
```

Figure 10.2–8 (a) shows the exp(-t^2) term (the ordinary Gaussian probability density function we saw earlier). Figs. (b) and (c) show the cosine and sine terms, and (d) and (e) show the real and imaginary parts of the Complex Gaussian function itself. Notice that the Cosine (b) is symmetrical about **time** = **0** and thus the real part of the complex Gaussian (d) is also symmetrical. With the Sine (c) anti-symmetrical about **time** = **0**, the imaginary part of the Complex Gaussian Function will also be anti-symmetrical (e).

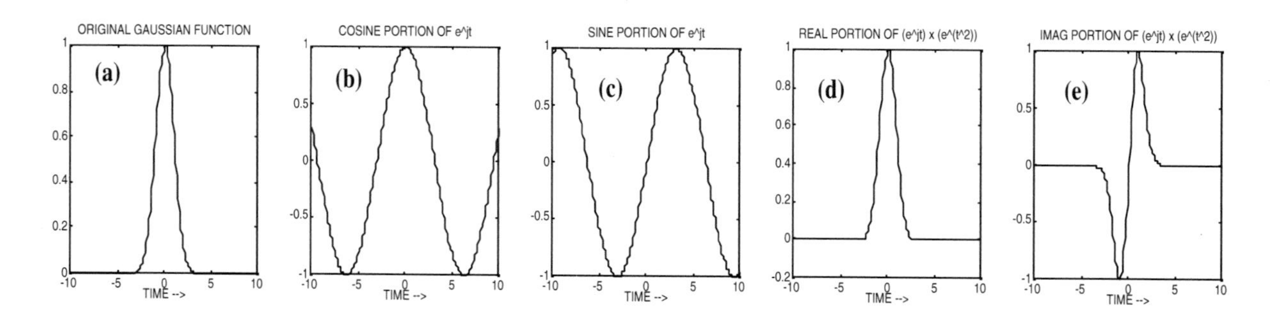

Figure 10.2–8 The familiar Gaussian function $\mathbf{e}^{-t^2}$ is drawn in the first graph. The 2nd and 3rd graphs show the cosine and sine components of $\mathbf{e}^{jt}$. Multiplying (a) by (b) gives the real component (d) while multiplying (a) by (c) gives the imaginary component (e).

As with the non-complex Gaussian wavelets in the last section, it is the *derivitives* that are wavelets (i.e. produce viable wavelet filters). Figure 10.2–9 shows the 1st and 2nd derivitives of the *real* and the *imaginary* portions of the complex Gaussian function. If we look at the slope of the real portion of the complex Gaussian (Fig. 10.2–8 (d)) we see this in Figure 10.2–9 (a). Similarly we see the slope of the imaginary portion of the complex Gaussian (ref. Fig. 10.2–8 (e)) plotted in (b) below.

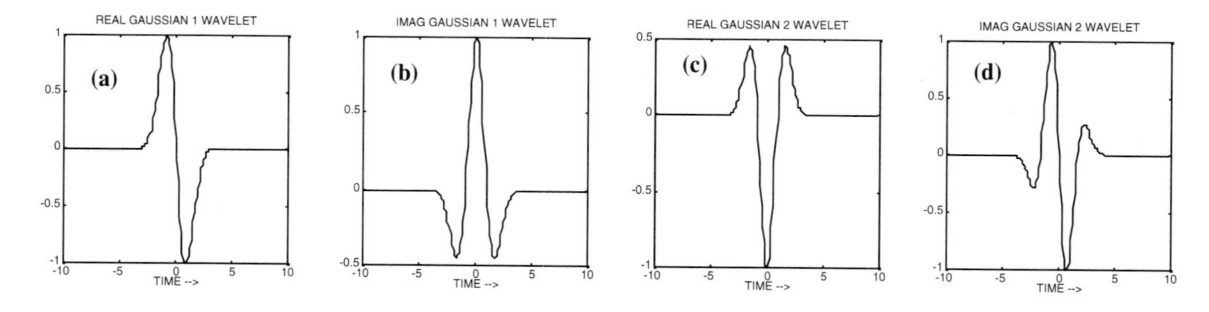

Figure 10.2–9 Complex Gaussian wavelets. The wavelets made from the real and imaginary first derivatives are shown in (a) and (b). Those made from the second derivitives are shown in (c) and (d). The effective support of these wavelets is usually from –5 to +5.

Continuing (Figure 10.2–9), the slope of the first derivative of the real portion of the complex Gaussian (a) is plotted in (c). For the imaginary portions, the slope is plotted in (d)

As with the non-complex Gaussian wavelets, taking further derivitives allows for better frequency resolution. In other words, the number of "cycles" in these wavelets is increased and they become a better match to some hidden sinusoidal-looking event.

Also notice that with the real and imaginary wavelets at each order or level of differentiation we have a combination of symmetric and anti-symmetric wavelets. This gives us further flexibility in matching events with symmetrical and anti-symmetrical components.

Jargon Alert: The nth derivative of a Gaussian Wavelet is designated as *order n*. The wavelets in Fig. 10.2–9 are thus "*order 1*" and "*order 2*"

Because of this flexibility, the Complex Gaussian Wavelet is used in a variety of applications including vibration analysis and medicine (e.g. ECG analysis).

10.3 Orthogonal Wavelets

We learned that a great deal of data can represented by a very few orthogonal wavelet filters just as any point in a city can be referenced by an (orthogonal) North-East-South-West Coordinate System* (ref. Section 8.5—Orthogonal Vectors, Sinusoids, and Wavelets).

Because of this compact form of representation, it is possible to perform a Discrete Wavelet Transform (DWT) in addition to the "Continuous" Wavelet Transforms (CWT) we have just discussed.

Just as the *crude* wavelets shared certain capabilities and characteristics, the *orthogonal* wavelets also have certain properties in common. These include

- Ability to be used in a conventional DWT, UDWT, or Wavelet Packet (Discrete) DWT due to the capability for *perfect reconstruction*. This allows for not only identification (as we saw in the CWT) but also for compression and de-noising.
- Orthogonal wavelet filters also have *alias cancellation* capabilities when used in the conventional DWT. In other words they "clean up their own mess" of aliasing caused by the downsampling in the conventional DWT. However, we have to be careful about throwing out some alias cancella-

* Salt Lake City, Utah is laid out this way. To travel from "1st North, 3rd West" to "21st South, 13th East" you simply travel 22 blocks South and 17 blocks East—in almost any sequence you desire.

tion capability when we compress or de-noise using the conventional DWT.

- These wavelets include not only a "*wavelet function filter*" (the highpass reconstruction filter ***hir*** or H') but also the "*scaling function filter*" (the lowpass reconstruction filter ***lor*** or L') and 2 additional filters (highpass decomposition filter ***hid*** or H and lowpass decomposition filter ***lod*** or L).

- These basic orthogonal filters have *compact support*—they begin and end.

We now examine some of the major families of orthogonal wavelets in more detail.

HAAR WAVELETS—Although we have discussed Haar wavelets before, we can now classify them better. They are the simplest, shortest, and the first to be used. Although a "continuous" Haar wavelet doesn't exist in the real world of digital computers, a good estimation can be produced by upsampling and lowpass filtering (interpolating) the simple H' filter (**[1 –1]**) to produce a multiple-point (e.g. 258 points) estimation. As shown in Figure 10.3–1 they have a support width of 1.0

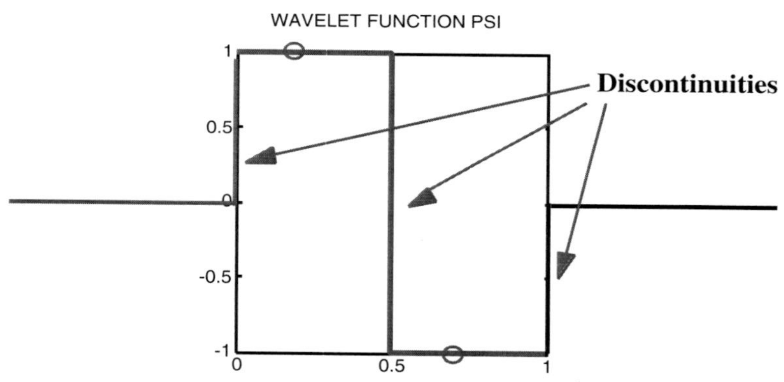

Figure 10.3–1 A 266 point estimation of a "continuous" Haar wavelet generated from the basic 2 points spaced 1/2 integer apart and located at 0.2 and 0.7 (shown superimposed here).

Haar wavelets (filters) have only one vanishing moment. This can be seen by observing that the dot product of **H' = [1 –1]** and a constant **[1 1]** is **(1 x 1) + (–1 x 1) = 1 – 1 = 0** while the dot product with a slope like **[1 2]** will be non-zero. Stretched versions of the Haar wavelet will of course also have one vanishing moment (e.g. the dot product of **[1 1 –1 –1]** and the constant **[1 1 1 1]** is also zero.

Haar wavelets have outright discontinuities and are thus not smooth or regular. They are not strictly symmetric, but are *anti-symmetric* and they do have linear phase. With a basic filter length of only 2 points they are excellent for time resolution but poor for resolution in frequency.

Haar wavelets can be used with CWT, UDWT and DWT (Perfect Reconstruction/Alias Cancellation capability). As we have demonstrated, they are good for edge detection, for matching binary pulses, and for very short phenomenon. The humble Haar wavelet is actually a mainstay in wavelet technology.

DAUBECHIES WAVELETS—We are also familiar with these through our extensive study of the Db4 wavelets. The left graph of Figure 10.3–2 shows the Db4 wavelet (768 point estimation of "continuous") with the 4 original points superimposed 1/2 integer apart at 2/6, 5/6, 8/6 and 11/6. Two additional trailing zero points are found at 14/6 and 17/6. The center graph shows the Db6 estimation with the 6 points (and 4 end zeros) superimposed starting at 3/7 and spaced 1/2 integer apart. The 3rd graph shows the Db8 with 8 points and 6 end zeros starting at 4/8 and spaced 1/2 integer apart.

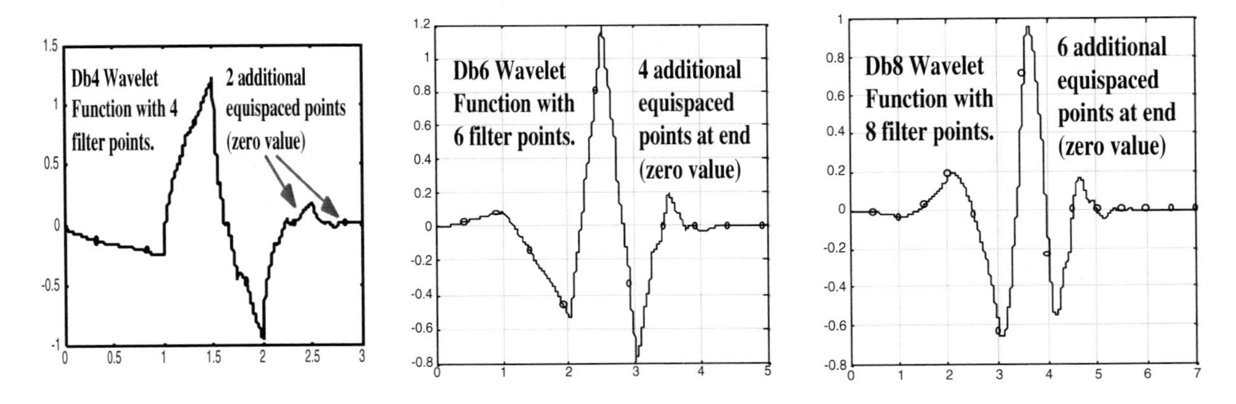

Figure 10.3–2 Estimation of a continuous Db4 wavelet on the interval 0 to 3 followed by the Db6 wavelet on the interval 0 to 5 and then the Db8 wavelet on the interval 0 to 7. They are all zero outside these intervals and thus have a *support width* of 3, 5, and 7, respectively.

The Haar wavelet is also a member of the Daubechies family and is sometimes referred to as a "Db2" wavelet. When mapped onto an interval from 0 to 1, we saw that the Db2 (Haar) had 2 points starting at 1/5, spaced 1/2 integer apart with no end zeros.

Looking back at Figures 10.3–1 and 10.3–2 we notice that the Daubechies wavelets are by no means smooth, however we can see the regularity increases as the "order" (N in a DbN wavelet) increases. A Db20 wavelet looks somewhat smooth, for example (ref. Fig. 9.1–1). As expected, the longer Daubechies wavelets provide better frequency resolution at the expense of decreased time resolution.

The number of vanishing moments in a Daubechies wavelet is N/2 (i.e. half the number of filter points). For example, a Db8 wavelet has 4 vanishing moments).* The Daubechies wavelets have the most vanishing moments for their size of any wavelet.

With the exception of the Haar, Daubechies wavelets do not have linear phase. Figure 10.3–3 shows this for the Daubechies 8 (Db8) highpass filter.

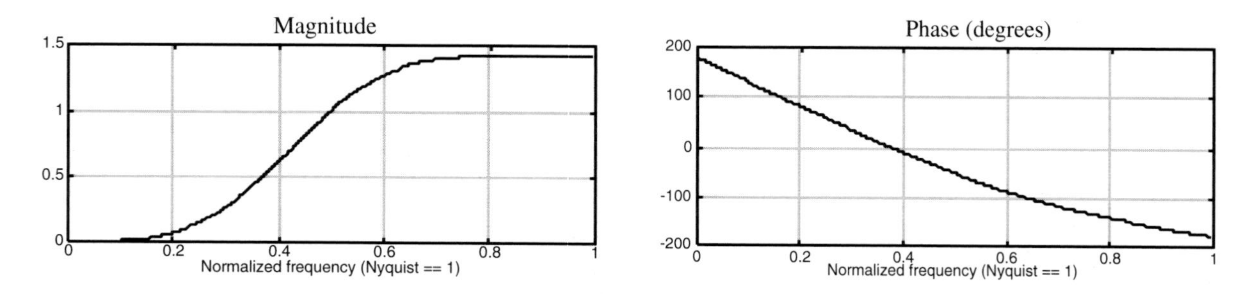

Figure 10.3–3 Magnitude and phase of a Db8 (8-point) highpass filter. Note the phase is non-linear.

All four wavelet filters (**H**, **H'**, **L**, and **L'**) are orthonormal to themselves and are orthogonal their counterparts (**H** with **L**, **H'** with **L'**—ref. Fig. 8.5–6).

These wavelets (filters) are robust, fast, and adaptable. They are in wide use for identifying signals with both time and frequency characteristics (we use longer filters for better frequency resolution). They are especially well suited to speech, fractals, and non-symmetrical transients. We have used the Db20 in this book to match chirp signals and chirp noise (ref. Figs. 7.1–1 and 9.1–1).

Being non-symmetric, Daubechies wavelets may be passed by in favor of some symmetric wavelets we have discussed (and will discuss) for image processing because the human eye is more tolerant of symmetric errors.

* A reminder that a few authors, including the Mathworks, use the number of vanishing moments rather than number of basic filter points to describe the Daubechies wavelets. Thus the Haar is referenced in MATLAB code as "Db1", the 4-point Db4 is referenced as "Db2" and so on.

However, there are instances in image processing such as edge detection where we want to "jar" the human eye with unusual patterns and shapes. Airport luggage screening is an example. Thus we see the Haar and Daubechies wavelets used in this form of image processing.*

SYMLETS—As the name implies, Symlet Wavelets or "Symlets" are more symmetrical than the Daubechies wavelets. As we have discussed (ref. Fig. 8.6–1) the halfband filters can be "factored" in more than one way into a decomposition filter and reconstruction filter that, when convolved, produce the same halfband filter. For the longer halfband filters (23 points or more), we can produce alternative filters to the Daubechies that are more symmetrical. For example, a 31 point highpass halfband filter can be factored into the two 16-point Db16 filters **H** and **H'** as shown in Figure 10.3–4.

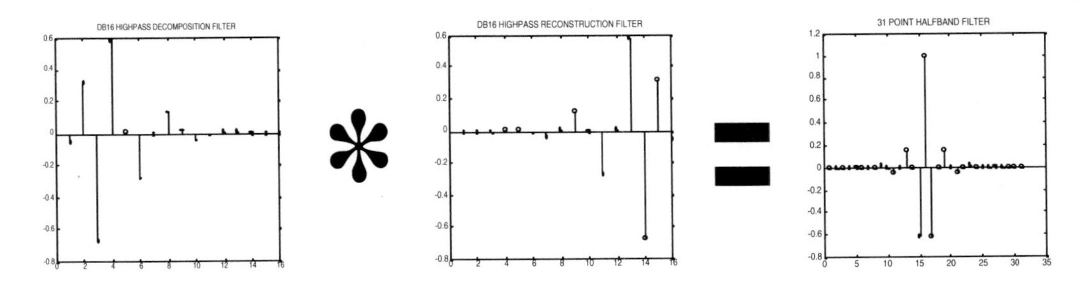

Figure 10.3–4 The convolution of the Db16 highpass decomposition filter with the Db16 highpass reconstruction filter produces the 31-point highpass halfband filter.

This same 31-point highpass halfband filter can also be factored into two other 16 point filters that are more symmetrical (but not perfectly so). This is shown in Figure 10.3–5.

* The Db8 wavelet was a finalist in the choice for JPEG image compression.

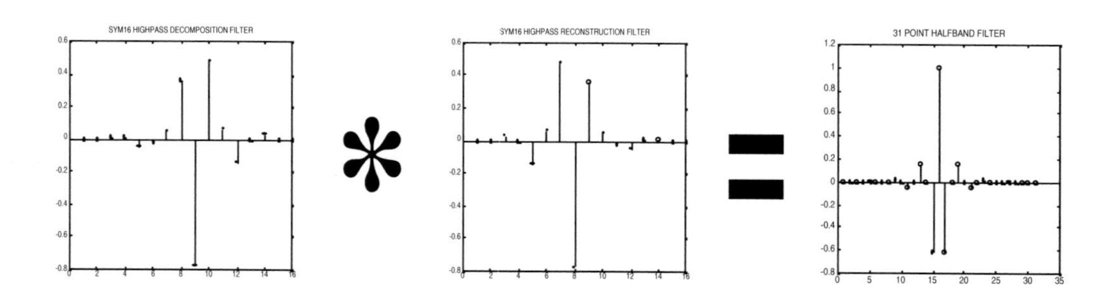

Figure 10.3–5 The convolution of the 16 point Symlet (Sym16) highpass decomposition filter with the Sym16 highpass reconstruction filter produces the same 31-point highpass halfband filter as the highpass Db16 filters **H** and **H'**.

The lowpass 31-point halfband filter can also be factored* into either two Db16 lowpass filters or two Sym16 lowpass filters As with the Daubechies filters, we can create an estimation of a "continuous" Symlet by repeatedly upsampling and lowpass filtering. The Symlet waveforms are shown in Figure 10.3–6.

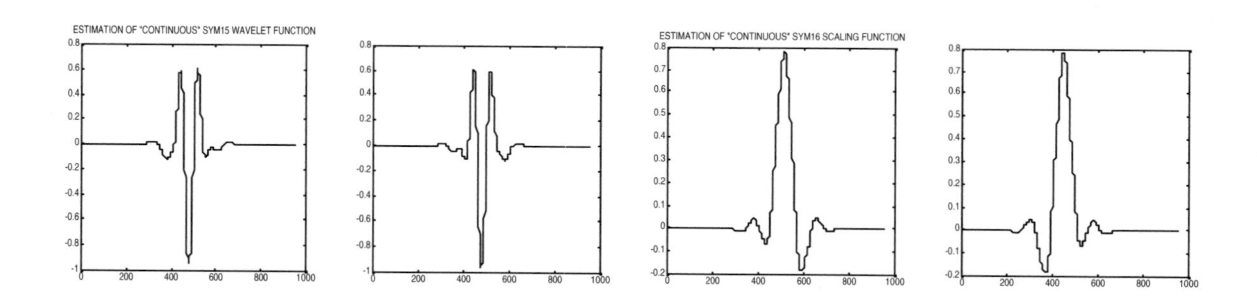

Figure 10.3–6 The 946-point estimation of the *"continuous" wavelet function* is obtained by repeatedly upsampling the 16-point Symlet highpass reconstruction filter, *hirsym*, and lowpass filtering at each step by *lorsym*. This is shown in the first graph. The 2nd graph is the same interpolation process starting with *hidsym* and lowpass filtering by *lodsym*. Note that graphs 1 and 2 are flipped similar to **H'** and **H** for the Daubechies wavelets. Graph 3 is the estimation of the *"continuous" scaling function* obtained by interpolation of *lorsym* (using *lorsym* also as the lowpass filter). The last graph is flipped similar to **L'** and **L** for the Db wavelets and is obtained with repeated dyadic upsampling of *lodsym* with filtering by *lodsym* at each step.

* Strictly speaking, *de-convolved* in the time domain.

Being nearly symmetrical, the larger Symlets (Sym12, Sym16, etc.) have also nearly linear phase. Other than the symmetry and phase, Symlets share the same properties as the Daubechies wavelets. They become more regular with larger N ("SymN"), They have the same compact support as the Daubechies for a given N, they have the same number of vanishing moments as the DbN family, and they have the perfect reconstruction and alias cancellation capability that allows them to be used in both the CWT and the DWT.

We have discussed the perfect symmetry and strict linear phase of the *biorthogonal* wavelets earlier. At first glance, the Symlets may seem obsolete. However, Symlets have the same orthogonality relationships as the Daubechies family and thus they are still strong contenders.

In the natural images all around us we seldom see things with perfect symmetry (or perfect *asymmetry*). This "match" of wavelet to image may be one of the reasons, along with the orthogonality and vanishing moments, that the Symlet is used in image processing.

Symlets are also used in applications as diverse as *power load consumption signals* and *composite structures*.

COIFLETS—The Coifman Wavelets or "Coiflets" were developed by Ingrid Daubechies at the request of wavelets pioneer Ronald Coifman to invent an orthogonal wavelet (filter set) that had vanishing moment capabilities for both the highpass and lowpass filters. Figure 10.3–7 shows the four basic 6 point filters interpolated (upsampled and lowpass filtered) to look like the Coif6 waveforms found in most texts. Note: As we have seen, different texts use different nomenclature. Wolfram calls this a C6 while Mathworks calls it a Coif1.

These Coiflets have the same orthogonality relationships as the Daubechies and the Symlet filters. As can be seen, they also have a high degree of symmetry and thus almost linear phase. They are suitable for use in any of the DWTs because they have alias cancellation and perfect reconstruction capabilities.

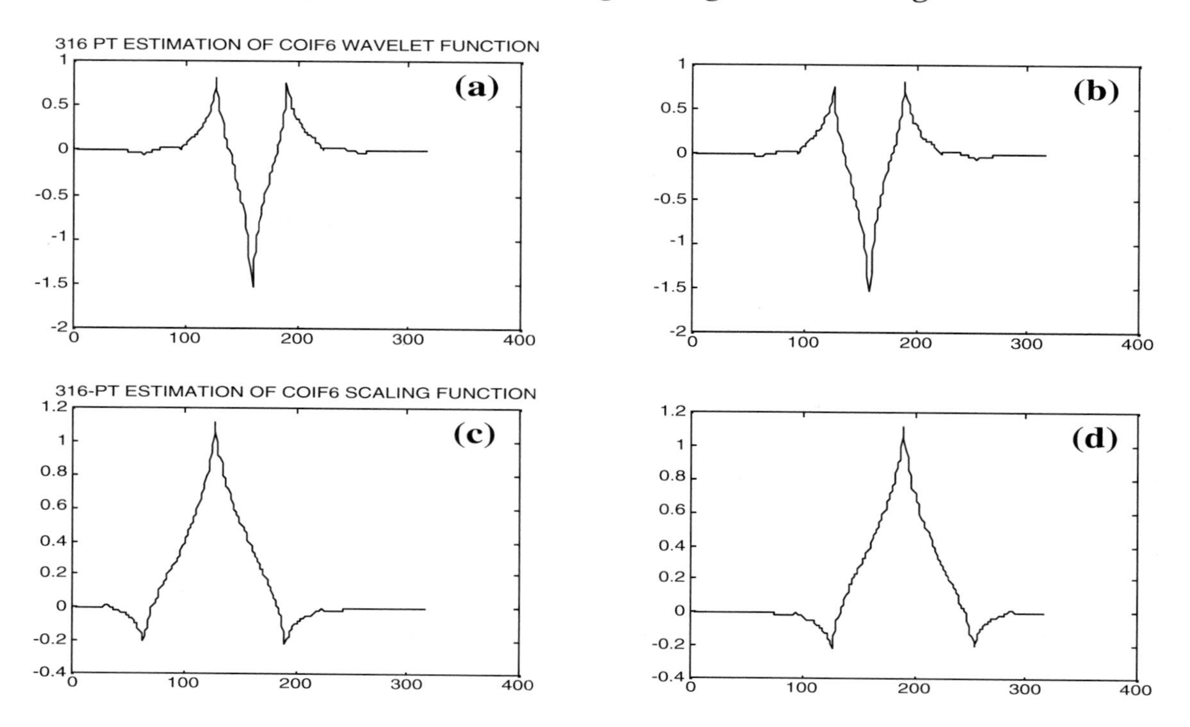

Figure 10.3–7 The 316-point estimation of the *"continuous" Coiflet wavelet function* **(a)** is obtained by repeatedly upsampling the 6-point Coiflet highpass reconstruction filter and lowpass filtering at each step by the lowpass reconstruction filter. The interpolated highpass decomposition filter **(b)** is shown next. Note the it is flipped horizontally from the decomposition filter in the first graph. The lowpass reconstruction filter interpolated to an estimation of the *"continuous" scaling function* is shown next **(c)** followed by the interpolated lowpass decomposition filter **(d)** repeatedly upsampled and lowpass filtered by the lowpass *reconstruction* filter.

The "cost" of using Coiflets is that they have one less vanishing moment than a comparable Daubechies or Symlet wavelet. This may or may not be a problem. Like the author's "fake wavelet" made by rounding off the Db4 filter coefficients (ref. Section 9.3) the Coiflets also produce a slightly different halfband filter than their Daubechies or Symlet counterparts.*

The Coiflets are used in many of the same applications as the Daubechies and Symlets. One interesting use is in the detection of self-similarities. Figure 10.3–8 shows a Von Koch curve. A CWT of this signal using an 18-point Coiflet as the basic filter is shown at right.

* The "fake" rounded-off wavelet filters also had the "cost" of one less vanishing moment and a slightly different set of halfband filters but did not produce the symmetry found in Coiflets or the promise of at least one vanishing moment on the lowpass or Approximation path in the DWT.

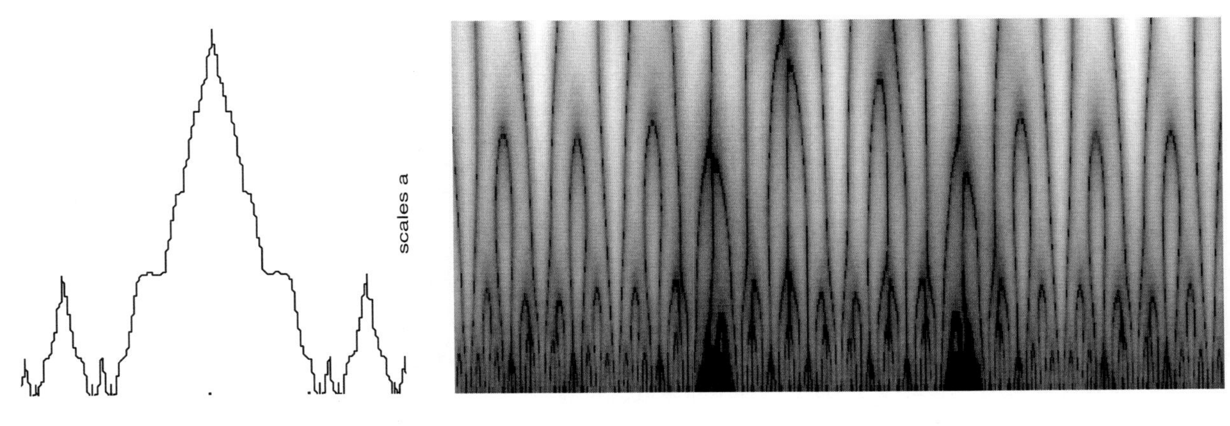

Figure 10.3–8 A Von Koch curve with self-similarities is shown at left. A 24-point Coiflet Coif24 is used in a Continuous Wavelet Transform. The display, at right, shows these self-similarities.

DISCRETE MEYER WAVELETS—Originating in the frequency domain, the "continuous" Meyer wavelets produce excellent frequency characteristics (ref. Fig. 10.1–6). Having a discrete version with Finite Impulse Response (FIR) filters that could be used in the various DWTs would be very desirable, thus the discrete Meyer wavelets (filters) were crafted from the crude version.

Recall that both an estimation of the "continuous" wavelet function and the "continuous" scaling function were created in the crude version. Using equispaced points on both functions would give us the highpass reconstruction filter **H'** *and* the lowpass reconstruction filter **L'**. Also recall that with orthogonal wavelets such as the Daubechies that **L** is the flipped version of **L'** and **H** is the flipped version of **H'** (ref. Fig. 8.1–2). With these nearly-symmetrical Discrete Meyer wavelets we can thus determine **H** and **L** from **H'** and **L'**. Thus all we need to do is check that they pass a few "tests" such as orthogonality and perfect reconstruction (they do!). The basic Four filters are each 62-points long and are shown in Figure 10.3–9.*

* Looking at the long "tails" in Fig. 13.3–9, it seems like we could "trim" these to shorter wavelets, perhaps a 31 point. We learned earlier working with the "continuous" Meyer wavelets that 32 points did not produce a good halfband filter (ref. Fig. 10.1–5) so downsampling by 2 is not an option. We could instead keep the only the center 31 points. However, like the Sinc Wavelet we have studied, truncation in the time domain can lead to longer transition bands in the frequency domain.

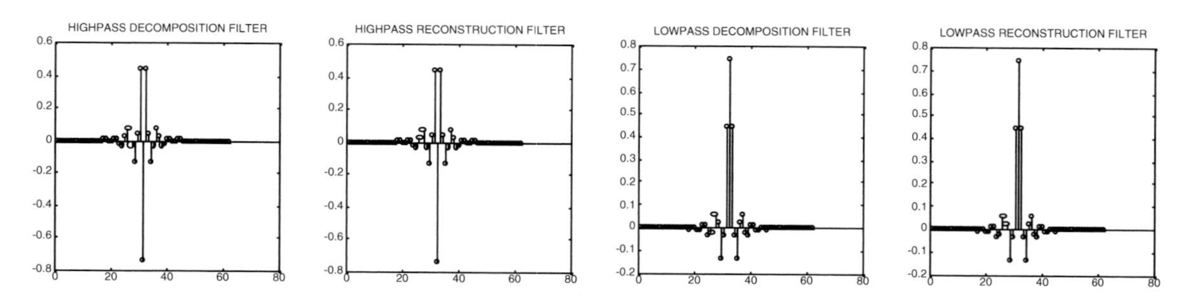

Figure 10.3–9 The four Discrete Meyer basic wavelet filters (**H, H', L, L'**). Note the high degree of symmetry.

The frequency response of the 62-point highpass reconstruction filter **H'** (2nd graph in the above figure) is shown below in Figure 10.3–10 . Stretching the filter by interpolating (upsampling by 2 and lowpass filtering by **L'**) shows the constant Q behavior.

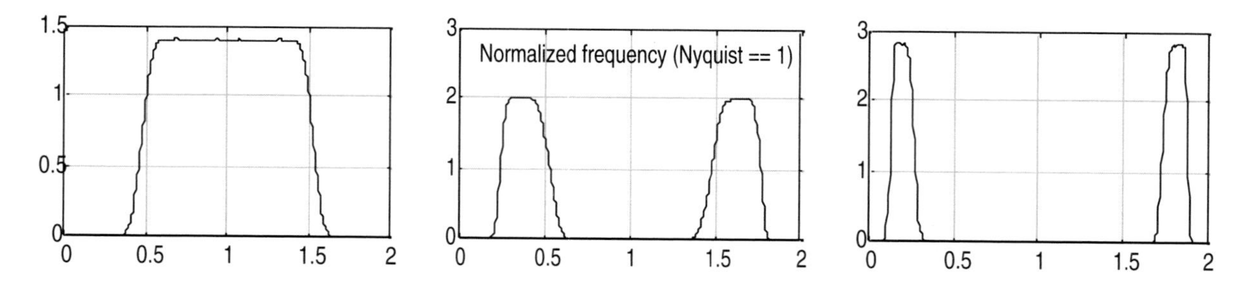

Figure 10.3–10 At 62 points, the discrete Meyer wavelet filter (**H'**) is bandpass in that it passes frequencies from about 0.5 to 1.5 Nyquist (first graph). As the filter is stretched to 184 points (about a factor of 3 with the upsampling of **H'** to 123 points then lowpass filtering by the 62-point filter **L'**) we see the constant Q behavior in the middle graph. Further stretching to 428 points produces the frequency response and further constant Q behavior shown in the right graph.

These 62-point filters are also orthogonal (and orthonormal to themselves). Figure 10.3–11 shows this for **H'** and **L'** and then for **H'** and **H'** as examples. Recall that a cross-correlation is a series of dot (inner) products. We can see in the left graph that the dot products are zero (indicating orthogonality) in every other point. The right graph indicates, like earlier orthogonal filters we have studied, that every other point is zero except where they align perfectly. This is orthonormal behavior. Thus the Discrete Meyer wavelets satisfy the orthogonality conditions.

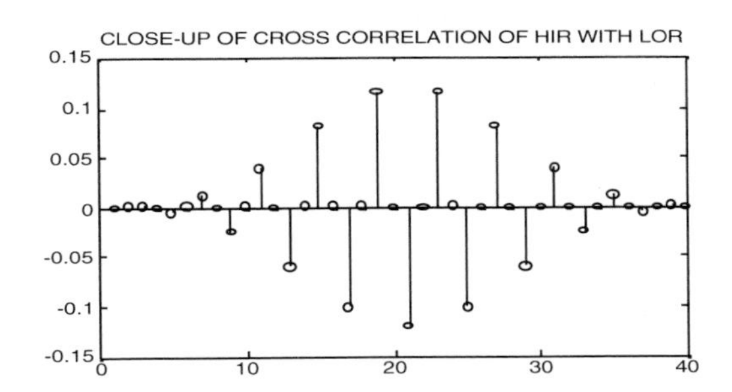

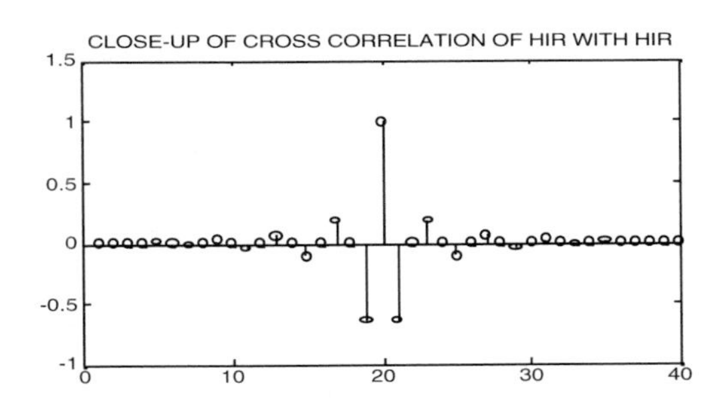

Figure 10.3–11 Cross correlation of the *discrete Meyer* **H'** with **L'** (at left) shows every other point being zero and thus orthogonality. Cross correlation of **H'** with itself (right graph) shows orthonormality.

Another required condition for use in the DWTs is that they can achieve perfect reconstruction. The halfband highpass filter is obtained by the convolution of **H** and **H'**. Figure 10.3–12 shows this in the time and the frequency domain. The lowpass halfband filters also show the desired behavior. If we add the 2 halfband filters in the time domain we have a Kronecker delta (spike) with a time delay. Discrete Meyer wavelet filters have the perfect reconstruction capability (to within a delay and a constant of multiplication) we have seen in the other orthogonal wavelets.

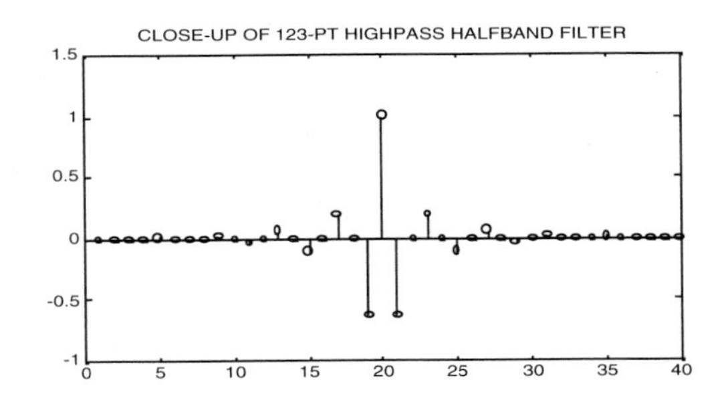

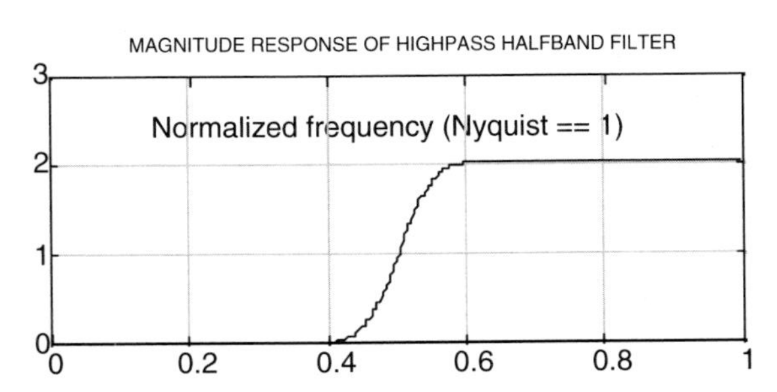

Figure 10.3–12 Discrete Meyer wavelet highpass halfband filter in time (left) and frequency. The right graph shows the exact symmetry we see in halfband filters around 1/2 Nyquist and around 1/2 magnitude we have seen in other halfband filters.

Notice that the first graph in figure 10.3–12 looks like the 2nd graph in Fig. 10.3–11. This is because the filters are nearly symmetrical and the cross correlation of **H'** and **H'** is the same as the convolution of **H'** and **H**.

Thus the Discrete Meyer Wavelet meets all the conditions to allow it to be used in the various DWTs. It does not have vanishing moments, however. At length 62 it is one of the longer basic wavelet filters, but it outperforms a 64-point Db64 filter in the frequency domain. Add to this the nearly perfect symmetry and accompanying nearly linear phase and we have a valuable addition to the orthogonal wavelets.

10.4 Biorthogonal and Reverse Biorthogonal Wavelets

In this chapter so far we have discussed crude wavelets (both real and complex) with symmetry and linear phase and how the bandpass *wavelet function filter* (**H'**) from the defining equations can be stretched and therefore used in a CWT.

We also studied orthogonal wavelet filters that are close to symmetric and that have the *orthogonality* and *perfect reconstruction* capabilities that allow them to be used in a DWT as well as a CWT. The Discrete Meyer wavelet just discussed, for example has these qualities. With long filters, no vanishing moments and imperfect symmetry, however, the discrete Meyer wavelet may be limited in its applications.

If we are willing to work with filters of differing lengths* (ref. Fig. 8.6–2) and differing orthogonality relationships (ref. Figs. 8.5–6 and 8.6–5) we can achieve perfect symmetry, vanishing moments, and the perfect reconstruction with alias cancellation required for use in a DWT. The FIR filters can also be as short as 3 points (in the 3/5 biorthogonal set).

BIORTHOGONAL WAVELETS—As previously discussed, Biorthogonal Wavelets are in wide use in image processing because of their perfect symmetry. We mentioned that the human eye is more tolerant of symmetrical imperfections. Also, extending the image symmetrically to avoid edge effects leads to better image processing. Image compression and denoising can be accomplished efficiently using the biorthogonal filters.

These filters deserve a closer look. Here are the four biorthogonal basic filters (**H**, **H'**, **L** and **L'**) that comprise the 7/9 biorthogonal wavelet set as

* A 4/4 biorthogonal wavelet has 4 points for all the filters, but **H'** is anti-symmetrical

shown in Figure 10.4–1. The original points that generated them are superimposed on the estimation of the "continuous" waveforms.

The 10 original points were upsampled and lowpass filtered to produce these 1144-point estimations of continuous functions (**H** and **L** are lowpass filtered by **L** while **H'** and **L'** are lowpass filtered by **L'**).

Some texts and software refer to these "continuous" estimations (approximations) as the *decomposition wavelet function, reconstruction wavelet function, decomposition scaling function* and the *reconstruction scaling function* respectively. In this book we are already familiar with "continuous" estimations of all four filters (e.g. ref. Fig. 8.5–6).

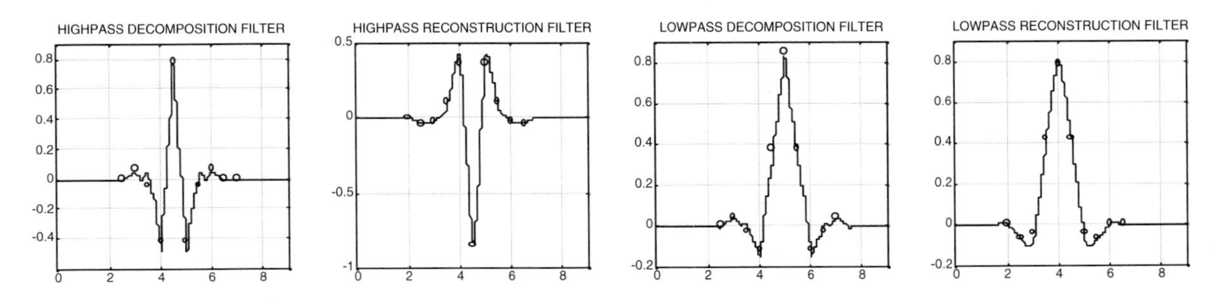

Figure 10.4–1 H, H', L, and L' for the *biorthogonal 7/9* wavelet filter set. The original 7 or 9 points have leading and/or trailing zeros to make all four filters 10 points long to aid with orthogonality (see 3/5 example in fig. 8.6–3). Note the exact symmetry.

We discussed *frequency B-splines* earlier in this chapter (splines in the frequency domain). Biorthogonal wavelets are sometimes constructed from splines in the *time* domain. For example, the 3/5 biorthogonal wavelets can be expressed as (ref. Fig. 8.6–2)

[1 -2 1]/4 and [1 2 -6 2 1]/4

Ignoring the divisors these are the coefficients of the polynomials

$$x^2 - 2x + 1 \text{ and } x^4 + 2x^3 - 6x^2 + 2x + 1^*$$

We have mentioned the wide use of biorthogonal wavelets, particularly the 7/9 filter set for image compression. In fact, this was the first "Killer App" (extremely powerful application) of wavelets. JPEG with the 7/9 filters is so common that it is hard to find images without their JPEG-compressed coun-

* These two polynomials can be factored further into **(x–1)(x–1)** and **(x –1)(x–1)(x2+4x+1) = (x –1)(x–1)(x+(2+sqrt(3))(x+(2–sqrt3)**.

terparts. The short length and perfect symmetry of these biorthogonal filters are still being explored in new applications.

REVERSE BIORTHOGONAL WAVELETS—In image processing we are concerned with visual effects such as blocking, checkerboarding, and ringing. Although a full discussion is beyond the mission of this book, it would be handy to have some control over where the longer and the shorter of the filters should go. For example, in a 3/5 biorthogonal set of filters the 3-point filters (not counting the zero padding to 6 points) are **H** and **L'**. Using a reverse biorthogonal set of filters we have **H'** and **L** as the 3-point filters (with **H** and **L'** as the 5-point filters).

For comparison purposes with the above figure, here are the 7/9 *reverse biorthogonal* filters and the 1144 point estimations of the *decomposition wavelet function* (**H**), *reconstruction wavelet function* (**H'**), *decomposition scaling function* (**L**) and the *reconstruction scaling function* (**L'**) as shown in Figure 10.4–2.

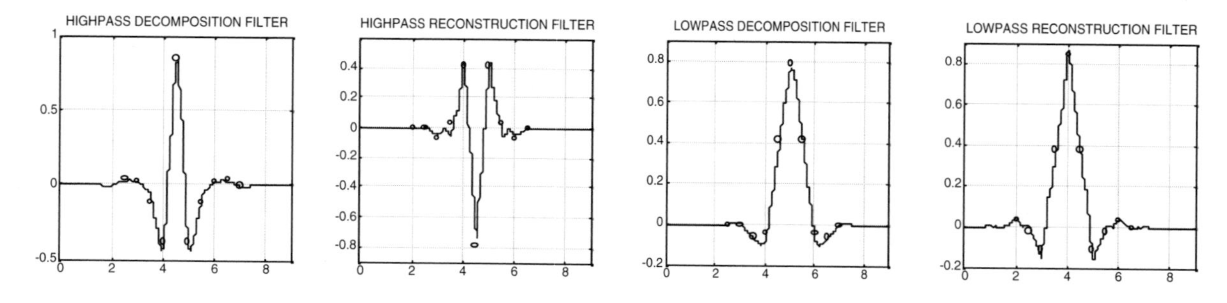

Figure 10.4–2 H, H', L, and L' for the *reverse biorthogonal 7/9* wavelet filter set. Again, the original 7 or 9 points have leading and/or trailing zeros to make all four filters 10 points long to aid with orthogonality .

Notice that we still have the same filters as the regular biorthogonal on the highpass path but that **H** and **H'** are swapped. Also **L** and **L'** are swapped on the lowpass path. The *halfband highpass filter* will be unchanged because

```
conv(H, H') = conv(H', H)
```

Similarly, the *lowpass halfband filter* will remain the same. This means that we have perfect reconstruction for both the *biorthogonal* and the *reverse biorthogonal* filters.

With the "wavelet filter", the *highpass reconstruction filter* or **H'**, being the shorter of the highpass filters, it is interesting to see the vanishing moment capabilities on a smaller 3/5 *reverse biorthogonal* set of filters. Removing the leading and trailing zeros we have **H'** = [**1** –**2** **1**] (or a pre-divided version). It turns out we still have 2 vanishing moments and can still obtain a dot product of zero for a constant or a slope. As a demonstration, consider an arbitrary linear slope [–**3** –**4** –**5**]. The dot product is then

```
(-3 x 1) + (-4 x -2) + (-5 x 1) = -3 + 8 -5 = 0
```

The biorthogonal and reverse biorthogonal wavelets are available as short filters like the 3/5 configuration* and longer filters such as the 11/17. The filters can be made purposely varied in length like the 4/20 versions. Couple this flexibility with the innate ability of the wavelets to act as a "Mathematical Microscope" by stretching and shifting and you have a very versatile and powerful tool.

10.5 Summary and Table of Wavelets and their Properties

Having gained a conceptual understanding of the major properties of wavelets we have discussed the various types of wavelets and some of the better known wavelets within these types.

We first reviewed the crude wavelets with real coefficients. These wavelet filters are bandpass with *constant Q* behavior. This means that as the basic "wavelet filter" is stretched by interpolation using the explicit equation to generate more points that the passband becomes narrower, the center frequency becomes less, and the amplitude becomes larger. Because crude wavelets are not orthogonal and don't have the other 3 basic filters to be used in a DWT, they are for CWT-use only and thus are usually designed to be symmetrical and smooth. They are theoretically infinitely long, but go to zero outside a narrow time range ("effective support"). Wavelets of this type include the Mexican Hat, Morlet, Gaussian, and the Meyer.

We next examined Complex Crude Wavelets. These are very similar to the (real) Crude Wavelets but, because they are complex, we can *bandshift*

* The Haar wavelet is also considered "a 2/2 biorthogonal wavelet" and is thus technically the shortest of this family.

them. The Shannon (Sinc) wavelet is an example of the familiar Sinc Function ($\sin(\pi x)/(\pi x)$) *lowpass filter* that is bandshifted to become a *bandpass filter* with constant Q characteristics. The Frequency B-Spline wavelet is actually a family of wavelets that appear as connected polynomials (Basic Splines) in the frequency domain and thus provide additional flexibility of use. The Complex Morlet wavelet combines the sinusoidal real part with a sinusoidal imaginary part to produce a Gaussian-looking modulus. Complex Gaussian wavelets are made from both the real and the imaginary parts of a Gaussian. They thus provide symmetry and anti-symmetry. Higher order Complex Gaussian wavelets are longer but give better frequency resolution.

We revisited the Orthogonal Wavelets. These have the four basic orthogonal filters required for use in a DWT (the **H'** filter can also be used in a CWT of course). They combine to form highpass and lowpass halfband filters and can be used for alias cancellation and perfect reconstruction. These filters are clearly finite (compact support).

The Daubechies family of orthogonal wavelets is referred to as DbN in this book where N is the number of points in the basic filter. DbN wavelets have N/2 vanishing moments. Beginning with the Haar wavelet—which can also be called a "Db2"—the smoothness increases with N. The Daubechies family does not have symmetry or linear phase (the Haar has anti-symmetry).

Symlets and Coiflets, also designed by Ingrid Daubechies, share most of the attributes of the DbN but are more symmetrical. The Discrete Meyer Wavelet is a version of the Crude Meyer wavelet but with four basic 62-point FIR filters. Like the Symlets and Coiflets, they are orthogonal and can be used in the various DWTs but they do not have vanishing moments.

We looked again at the Biorthogonal wavelets. The basic four filters in this family of wavelets have perfect symmetry with linear phase and can still be as short as 3-points. They are usually of unequal length as indicated by their names (3/5, 7/9, 4/20 etc.). This inequality can be used to our advantage and we can trade the decomposition and the reconstruction filters for better processing. This is accomplished using the Reverse Biorthogonal wavelets.

Although the Biorthogonal and Reverse Biorthogonal filters have different orthogonality relationships than their *orthogonal wavelet* cousins, they still have the properties of alias cancellation, perfect reconstruction, and vanishing moments and thus are a popular choice for use in the various DWTs—especially in image processing.

Table 10.5–1 summarizes the various properties for the various wavelets we have discussed (in the order of discussion).

Table 10.5–1 - Attributes of the various Wavelets (filters)

	Crude/ Ortho/ Biorth	***CWT or DWT***	***Symmet and Lin Phs***	***Vanish Momnts***	***Real or Cmplex***	***Smooth and Regular***
Mexican Hat	**Crude**	**CWT**	**Exact**	**None**	**Real**	**Infinitely**
Morlet	**Crude**	**CWT**	**Exact**	**None**	**Real**	**Infinitely**
Gaussian	**Crude**	**CWT**	**Sym/Anti**	**None**	**Real**	**Infinitely**
Meyer*	**Crude**	**CWT**	**Exact**	**None**	**Real**	**Infinitely**
Shannon (Sinc)	**Crude**	**CWT**	**Exact**	**None**	**Complex**	**Infinitely**
Complex Frequency B-Spline	**Crude**	**CWT**	**Exact**	**None**	**Complex**	**Infinitely**
Complex Morlet	**Crude**	**CWT**	**Exact**	**None**	**Complex**	**Infinitely**
Complex Gaussian	**Crude**	**CWT**	**Sym/Anti**	**None**	**Complex**	**Infinitely**
Haar	**Orthog**	**Both**	**AntiSym**	**1**	**Real**	**No**
Daubechies (DbN)	**Orthog**	**Both**	**Not Sym**	**N/2**	**Real**	**If N large**
Symlets	**Orthog**	**Both**	**Close**	**N/2**	**Real**	**If N large**
Coiflets	**Orthog**	**Both**	**Close**	**N/2 – 1**	**Real**	**If N large**
Discrete Meyer	**Orthog**	**Both**	**Yes**	**None**	**Real**	**Yes**
Biorthogonal	**Biorthog**	**Both**	**Exact**	**	**Real**	**If N large**
Reverse Biorthogonal	**Biorthog**	**Both**	**Exact**	**	**Real**	**If N large**

* Meyer wavelets filters are from the IFFT of an explicit equation in the frequency domain
**Number of vanishing moments depends on filter length of highpass reconstruction filter (H')

This table (and this chapter) is by no means a complete list of the existing wavelets. Additional wavelets and wavelet families include B*inlets, Brushlets, Beamlets, Bathlets,[*] Chirplets, Contourlets, Grouplets, SURE-lets, Lemierre, and Malvar wavelets* and many more. More new wavelets are also being created as time progresses—although some of these are highly customized and not intended for general use.

Understanding and gaining intuition for the wavelets presented in this chapter will allow you to "hit the ground running" in understanding and using any of these other wavelets. Remember not to indulge in the "Sport of Basis Hunting" too much when an existing, proven, wavelet might work very well for you. Usually changing the wavelet name in a line of software code is all that is required to compare the performance of the various wavelets.

[*] Named after a Klingon battle sword (for Star Trek fans).

We will next look at a few new examples of the use of the wavelets presented in this chapter and revisit in more detail some earlier examples of the use of these amazing tools.

CHAPTER 11

Case Studies of Wavelet Applications

Having seen the properties and some general applications of the various types of wavelets, we are now ready to gain a conceptual understanding of some applications to case studies. While we will not be demonstrating all the wavelets or types discussed, we should be able to gain some intuitive insights into wavelet use.

In addition to some new examples, we will re-visit some that have been introduced earlier in the book and can now be better understood with the further knowledge and insights we have since acquired.

11.1 White Noise in a Chirp Signal

We begin by adding white noise to a chirp signal. The result is shown in Figure 11.1–1 below at left. The FFT at right shows that the white noise appears at all frequencies (hence the name "white" as in *all colors*) and will be difficult to remove using conventional FFT methods.

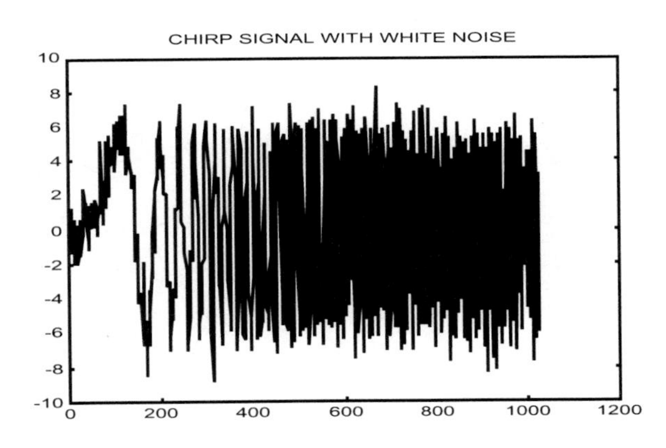

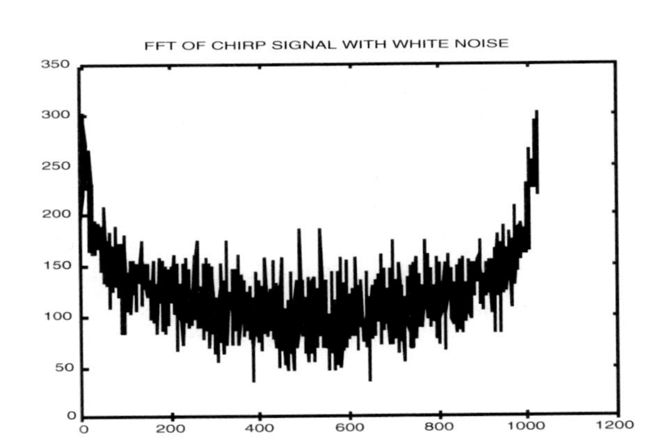

Figure 11.1–1 A chirp signal with white noise is shown in the time and frequency domains.

Using a conventional (non-wavelet) lowpass filter to keep only the low frequencies produces the time and frequency results shown in Figure 11.1–2. The low-frequency portion is fairly well denoised, but the high-frequency portion (most of this chirp signal) has been severely attenuated.

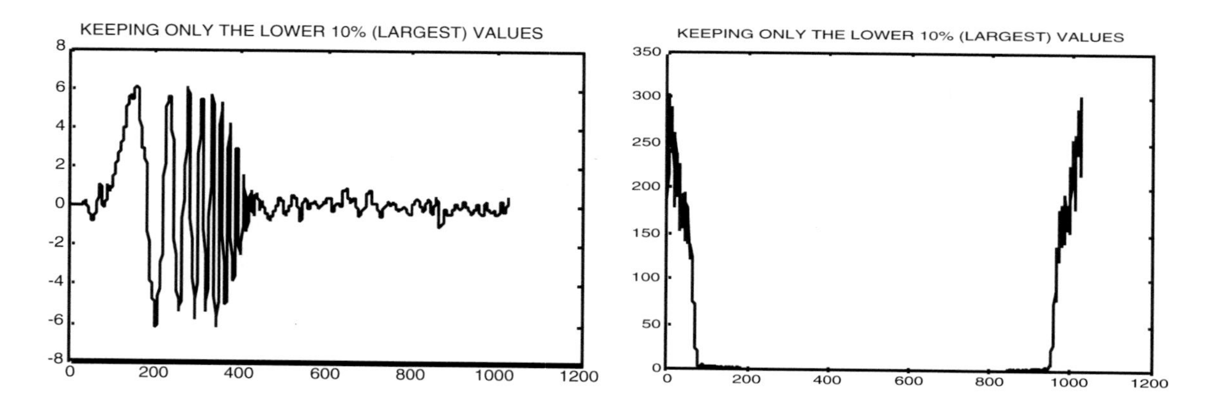

Figure 11.1–2 Conventional lowpass filtering accomplishes some denoising at the start of the signal (left graph) but destroys most of the signal as well as the noise. This is also seen in the FFT shown at right.

If we adjust the lowpass filter to be less severe we have the denoised signal as shown in Figure 11.1–3. More of the signal is preserved but the highest frequencies are still lost and the beginning of the signal is not de-noised very well.

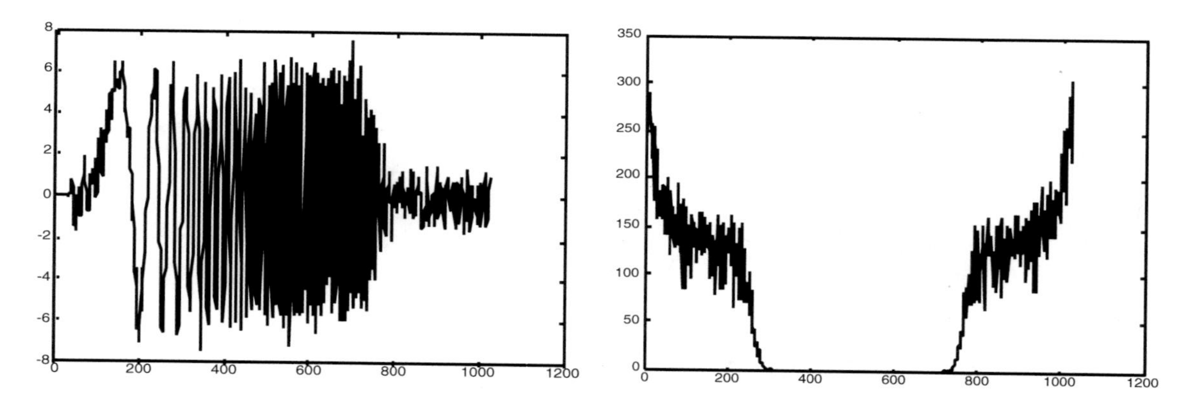

Figure 11.1–3 . Less severe conventional lowpass filtering keeps more of the signal (but not all) and the high frequency noise is still present, especially at the beginning.

We use a conventional DWT to attempt to denoise this signal. The decomposition is shown if Figure 11.1–4. We try first a Db6 filter because it looks like it might "match" portions of this asymmetric signal (ref. Fig. 6.4–2) and because it is short (6 points).

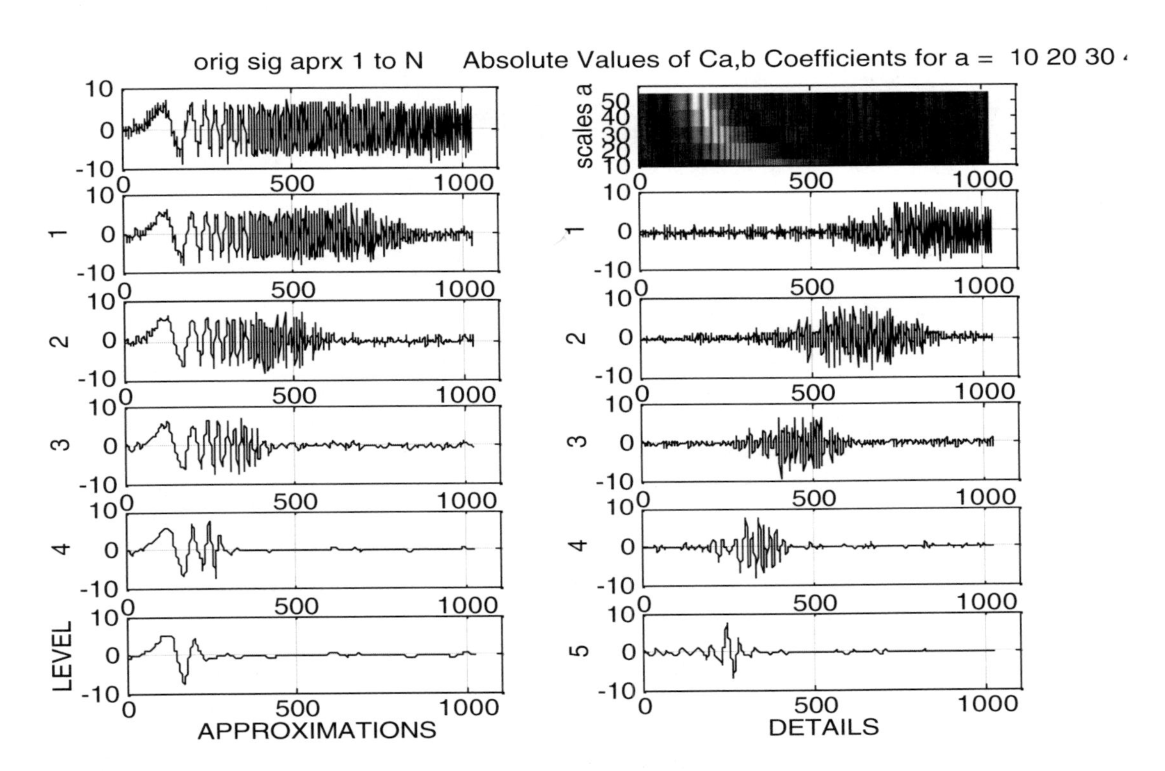

Figure 11.1–4 Display of a conventional DWT of signal using a Db6 wavelet.

The signal is 1024 points (2^10) long so we could downsample 10 times, however a 5-level DWT, as shown here, is sufficient. The original noisy chirp signal is shown in the upper left graph while a mini-CWT of the signal is shown at upper right (every 10th scale is sufficient in this case). The 5 levels of *Approximations* are at left while the 5 levels of *Details* are shown at right.

A frequency allocation diagram for a 5-level DWT (or UDWT) is shown in Figure 11.1–5. We now demonstrate the time/frequency manipulation capabilities of *wavelet technology*. We will keep only *some* of the data within a *certain time* and within a *certain frequency range*.

We notice in the above DWT display that different parts of the noisy chirp signal appear in levels **D1** through **D5**. For example, in **D1** (graphic directly beneath the mini-CWT) it appears that the "signal" portion begins at about **time** = **650** and that everything before that can be attributed to noise. We thus set the first 650 **D1** values to zero to attempt to de-noise the signal.

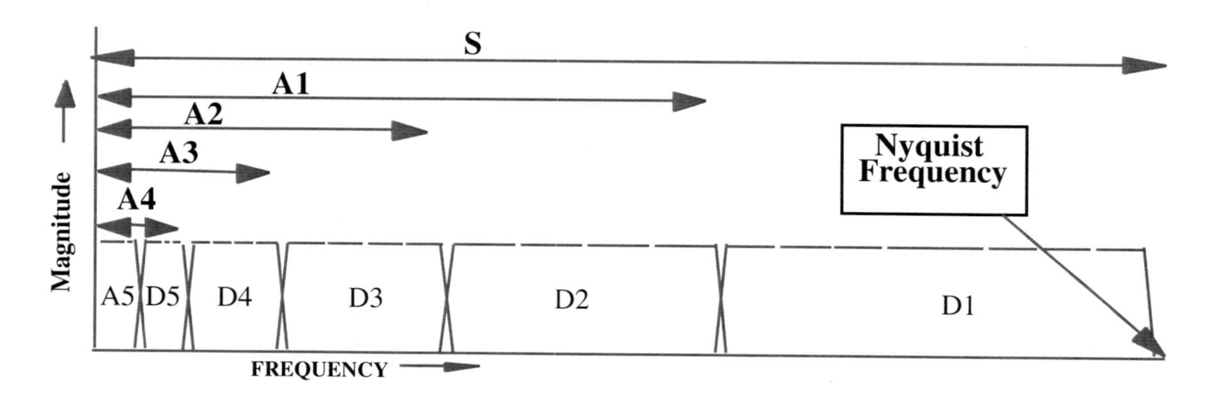

Figure 11.1–5 Frequency allocation for a 5-level DWT. Notice that **S** = **D1+A1** = **D1+D2+A2** = • • • = **D1+D2+D3+D4+D5+A5**. Note also that the filters are imperfect.

For **D2** it appears that the signal portion is located between about 400 to 800. Thus we place zeros everywhere else.* This process is shown below in Figure 11.1–6 for **D1** and **D2**.

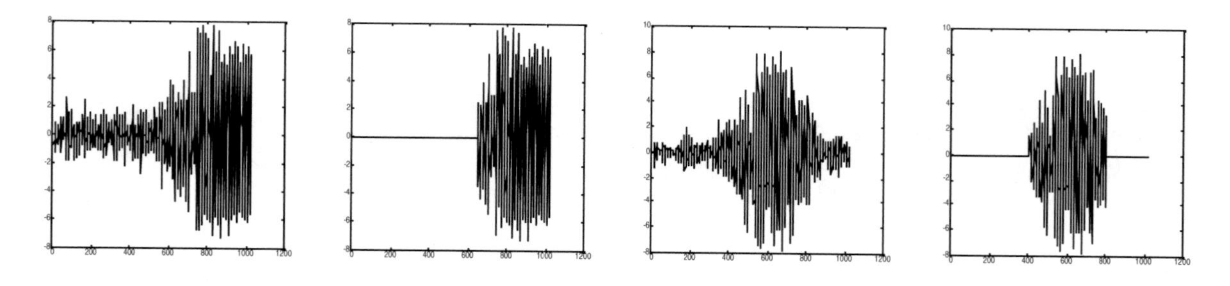

Figure 11.1–6 **D1** (1st graph) and **D2** (third graph) are adjusted to be zero except for the specified times (650 to 1024 for **D1** and 400 to 800 for **D2**).

We continue this process through **D5**, keeping only the signal portions. We are now ready to add the denoised details **D1** through **D5** together with **A5** (ref. Fig. 11.1–5) to reconstruct our signal. We are not seeking perfect reconstruction here but instead we wish to denoise the signal. Figure 11.1–7 shows the final result with this method of denoising.

* Although this process is being done somewhat "by hand" here, the MATLAB Wavelet Toolbox has interactive graphics called "*interval-dependent thresholding*" that allows the user to quickly discard unwanted data in specified periods of time.

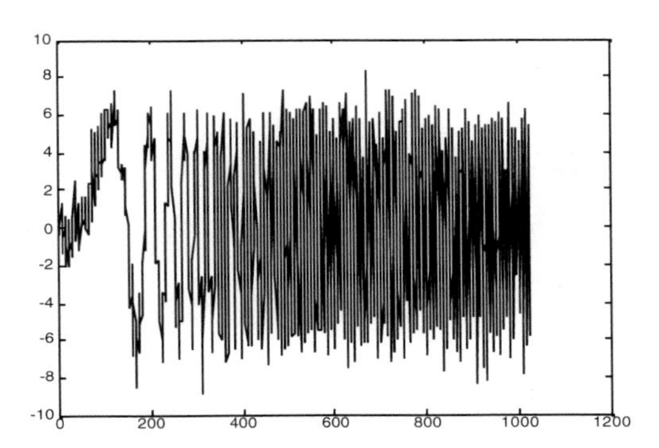

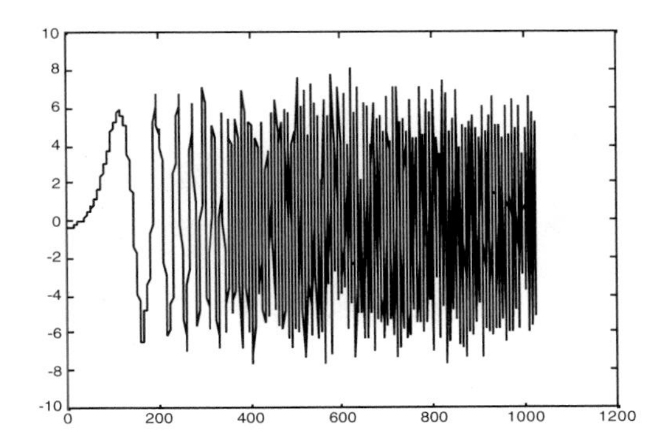

Figure 11.1–7 The original noisy signal is shown at left. Denoising using a 5-level DWT with Db6 wavelet filters gives the result at right. Compare with Figs. 11.1–2 and 11.1–3.

The results are impressive, especially when compared with conventional DSP methods. They can be possibly improved by trying other wavelets or by performing the time/frequency manipulations on additional levels.

We will talk further in an upcoming chapter about "throwing out the baby with the bathwater"—throwing away some alias cancellation capability as we discard the noisy parts of **D1** through **D6**. For now, we can state that in each of the levels the noise was far less than the signal and the effect of aliasing was minimal. We will compare this later with the results obtained using the Undecimated DWT which has no such aliasing problems.

11.2 Binary Signal Buried in Chirp Noise

This next example is similar to the first except we have a binary signal and the *noise* is in the form of a chirp.* The process is similar to that of the last section. Figure 11.2–1 shows a Binary Phase Shift Keying, Polar Non-Return to Zero (BPSK PNRZ) signal in both the time and frequency domains. This example was mentioned in the overview Chapter One. We now provide more details.

* A constant frequency jammer can be easily removed from data by conventional *notch filtering* using FFT methods. A *chirp jammer* is more difficult because the frequency keeps changing. Wavelets work well here.

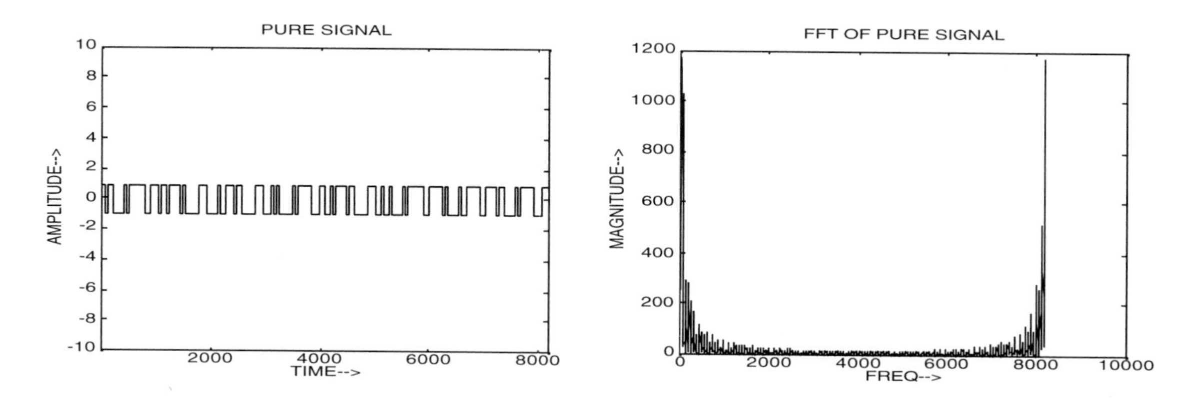

Figure 11.2-1 The original binary signal is shown in the time domain. Note the values alternate between −1 and +1. The FFT of this signal is shown at right.

The signal is next buried in 80 dB* of chirp noise as shown in Figure 11.2–2. The left plot shows the chirp noise values from −10,000 to +10,000. Although we can't see the small binary signal with this much noise added, we show a close-up (right graphic) of the noise from −100 to +100 with the original binary signal over-plotted for comparison (the signal overplotted on the full noise would look like a straight line). Looking at the signal in the frequency domain does not offer much hope of finding it either (ref. Fig. 1.9–2).

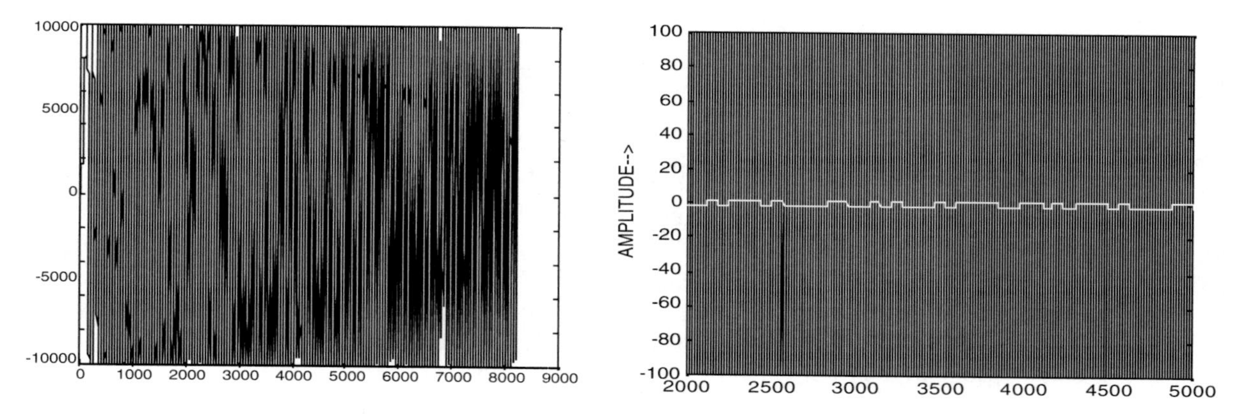

Figure 11.2-2 Binary signal from −1 to +1 with noise from −10,000 to +10,000 added is shown at left. A close-up is shown at right with the original binary signal overplotted (the signal would not be visible in 80 dB of noise). Note the dimensions.

* *Decibels*, not *Daubechies*—the decibel is named for Alexander Graham **B**ell and is abbreviated "dB" while a Daubechies wavelet is named for Ingrid **Dau**b**echies and is abbreviated "Db".

This is another instance where we use wavelets. We talked earlier about matching the wavelet to the signal. In this case we might consider a Haar wavelet because it looks like the binary signal. However, with this much noise we wouldn't be able to find it! Instead, we will match the wavelet to the noise. Because the signal is 8192 points long we will use a longer Db40 wavelet as shown in Figure 11.2–3. This looks a lot like the chirp signal and, as we mentioned for large Daubechies filters, is fairly smooth. Also, this longer wavelet should provide for fairly good frequency discrimination.

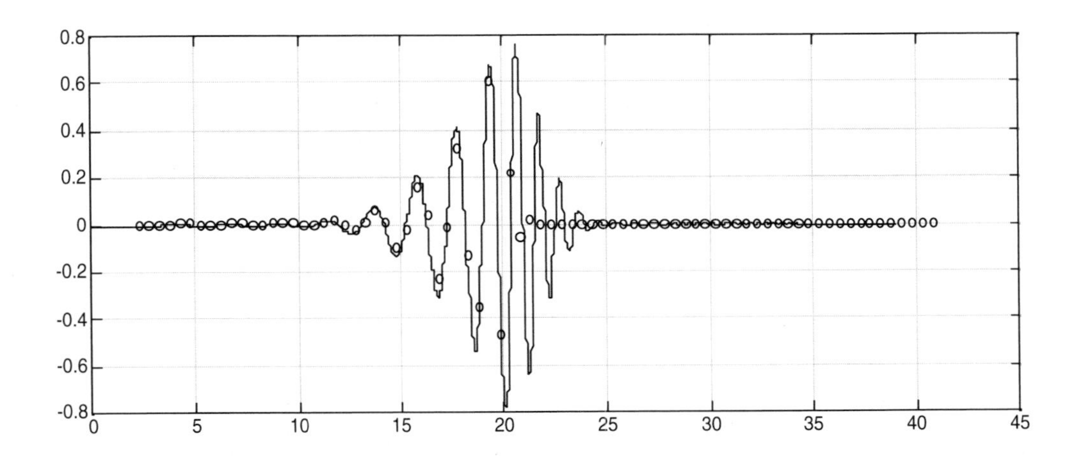

Figure 11.2–3 4954 point estimation of the "continuous" Db40 wavelet with the 40 points of the **H'** filter that created it superimposed (along with trailing zeros).

We will use a 7-level DWT here. The results are shown in Figure 11.2–4 below. We notice that the match of the wavelet to the noisy signal is excellent and that the highest frequency portion of the chirp noise is found in right-hand part of **D1** (ref. Fig. 11.1–5). Mid-frequency portions are found in the center of the **D2** through **D5** plots and the lowest frequencies are found at the left of the **D6** and **D7** plots.

We can also literally see how **A1** and **D2** combine to make the signal (left top graph), how **A2** and **D2** combine to make **A1** and so on.

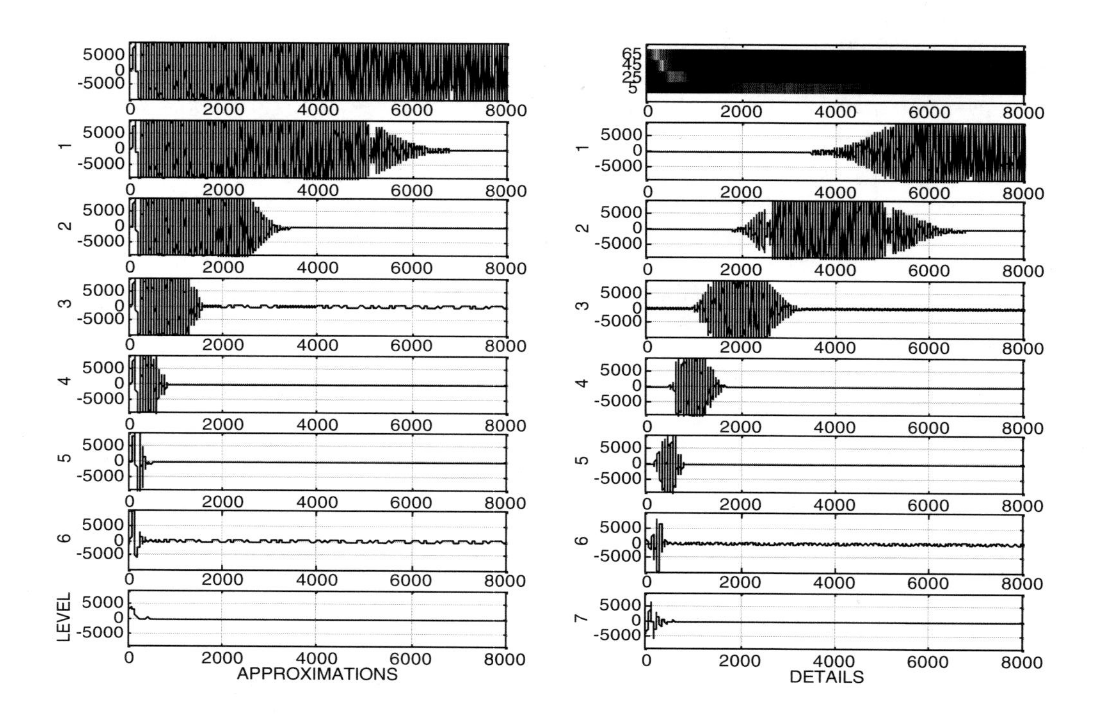

Figure 11.2–4 7-level DWT of binary signal with 80 dB of noise added. Wavelet filters used are Db40.

Looking at **D2** (Details, #2) on the above display we might assume we would be "safe" to delete the values from about 1500 to 7500 because the chirp portion seems to be isolated in this area. A close-up look at **D2** limiting the values to the range –2 to +2 (rather than –10000 to + 10000) tells a different story as shown in Figure 11.2–5 (left). We see a binary pattern in the first 1000 points and perhaps a few more points at the end. Rather than plotting close-ups and setting the chirp jammer portions to zero by hand, we can automate this process. We use a "reverse threshold' in which any values of the signal greater than, say, 2 (or less than –2) are set to zero.*

* We can set up this "reverse" threshold in software or we can use *median filtering* to keep the binary parts.

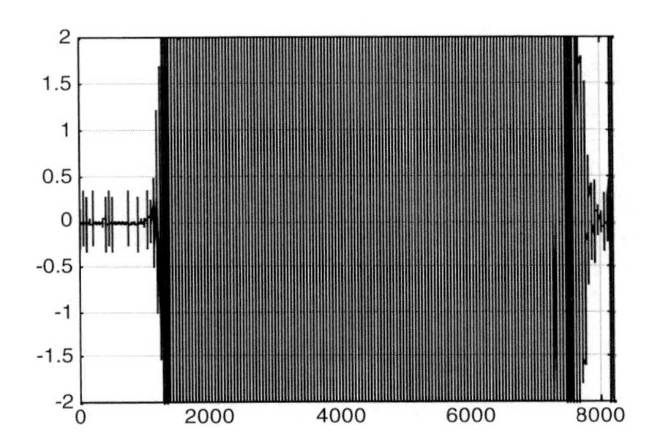

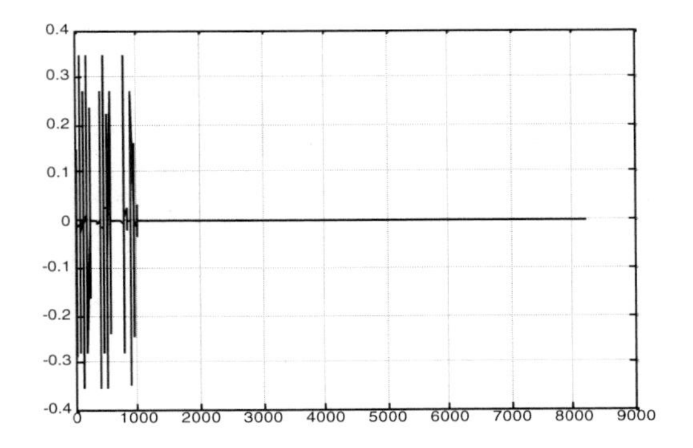

Figure 11.2–5 Close-up of **D2** showing remnants of the signal at left. After reverse thresholding using a median filter we keep only the values *less than* 2 and set to zero all the large values as shown in the right graph.

The right graph of Figure 11.2–5 above shows the result. We have only the "scraps" left over (in this frequency bandwidth of **D2**). However, looking again at the 7 Details on the above DWT display we can see that these "scraps" will be located at different times. **D1** with the chirp noise removed gives us remnants at the beginning while **D3** and **D4** give us remnants in the middle and **D6** and **D7** give us remnants at the end.

With the chirp portions thresholded to zero we now combine these de-noised Details as

```
Sig' = A7 + D7' + D6'+ D5' + D4' + D3' + D2' + D1'
```

where the prime indicates de-noised. A close-up of the denoised signal is shown in Figure 11.2–6 along with the original binary signal for comparison. The denoising is not perfect, but allows us to reconstruct a recognizable binary signal. This would not have been possible using conventional DSP methods.

This is a good example of how wavelets are useful to match either the signal or the noise and how the time/frequency nature of wavelet processing allows us flexibility we would not find in either the time or the frequency domains by themselves.

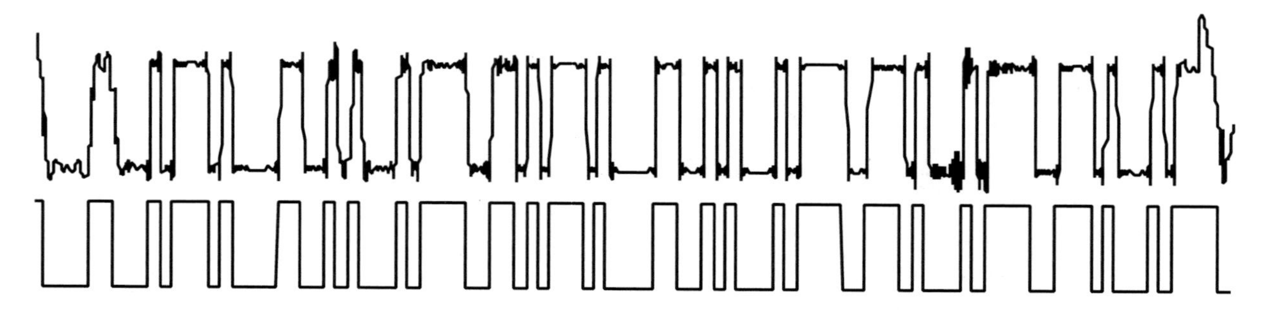

Figure 11.2–6 Portion of signal pulled from 80 dB of noise using time-specific thresholding with a 7-level conventional DWT is shown at top. Original binary signal. is redrawn at bottom for comparison.

11.3 Binary Signal with White Noise

This time a binary signal has intermittent white noise added as shown in Figure 11.3–1. The signal has 16 "chips" (binary value +1 or –1) per bit. For example, the binary sequence **[1 –1]** is represented by 16 values of +1 followed by sixteen values of –1. The "pure" signal here is 1024 chips (points) long and represents 64 bits. The first 8 bits are **[1 –1 1 –1 1 1 –1 –1]**.

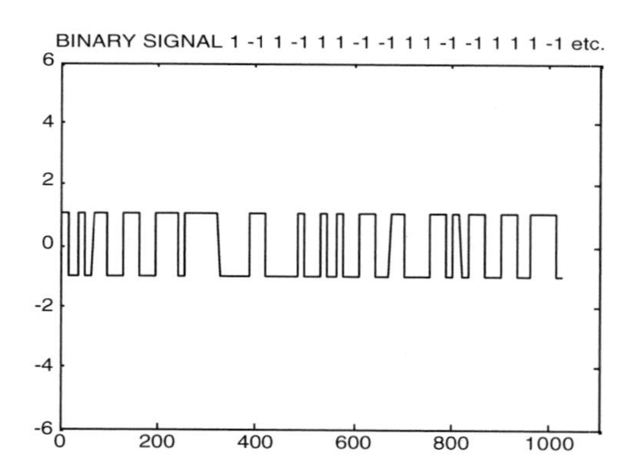

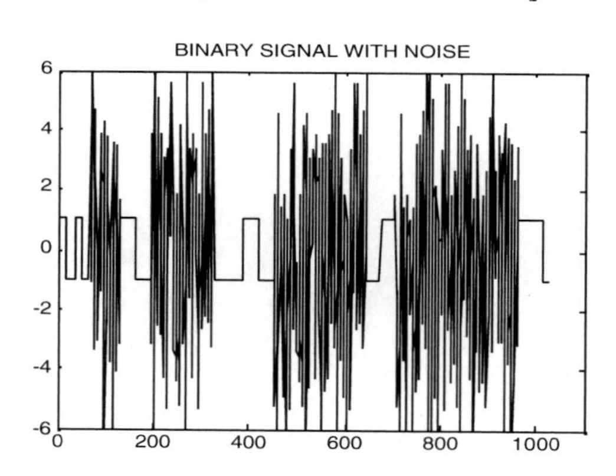

Figure 11.3–1 Binary signal with 16 "chips" per bit is shown at left. Intermittent pseudo-random noise is added as shown at right.

We can actually see part of the signal in the time domain, but not enough to decode it. Figure 11.3–2 shows the pure binary signal and the noisy signal in the frequency domain.

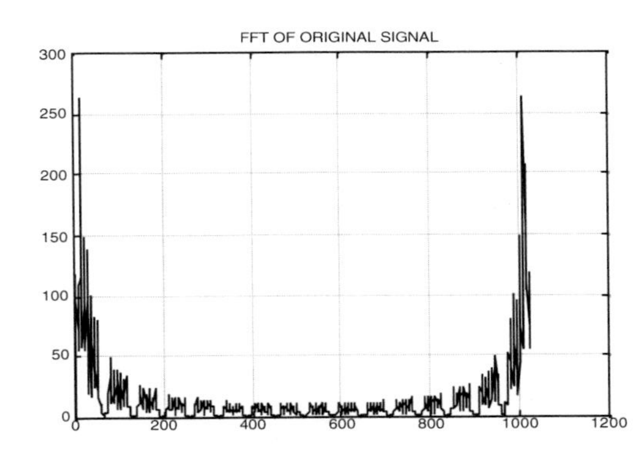

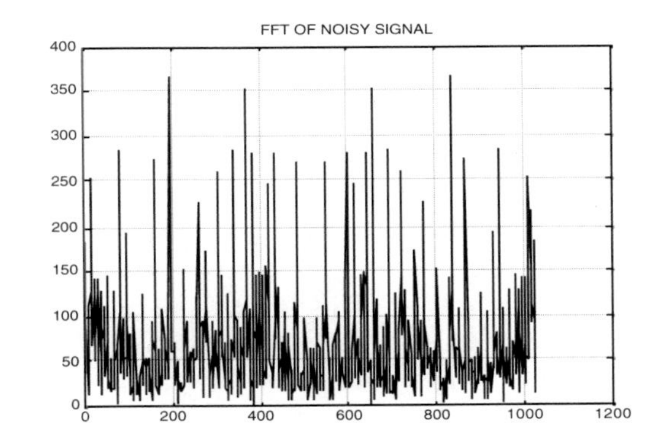

Figure 11.3–2 FFT of the original binary signal at left and the FFT of the signal with intermittent noise at right.

We will encounter problems using conventional DSP lowpass filtering techniques because of the amount of high frequencies that come from the "untouched" portions of the pure binary signal (ref. Fig. 11.3–2, left). In other words, both the signal and the noise have high frequency components.

A square wave, as most audio enthusiasts know, is made up of many frequencies.* Also, we have tried to make the binary signal realistic by including "wide" areas (e.g. sections where the bit pattern is "**–1 –1 –1 –1**" etc. instead of just "**1 –1 1 1 –1 1 –1**" etc.). Thus having sections in the signal with the (variable-length) square bits may actually make it harder to separate from the sporadic noise using conventional filtering techniques.

One conventional method might be to use a series of Short Time Fourier Transforms (STFTs) on the "clean" and noisy sections separately. However, wavelet technology incorporates this same time/frequency capability with a better *match* to the signal than the FFT sinusoids.

We have seen in the previous 2 examples methods of using the simultaneous time/frequency capabilities of wavelets by selectively denoising the Details at

* Early Rock musicians would intentionally crank up their tube-type amplifiers to distortion. Instead of "clean" sinusoids the tops and bottoms of the sine waves would be *clipped* flat and would look more like square waves and sound to the human ear like a combination of many high-frequency harmonics and overtones. The next generation of amplifiers provided a smoother attenuation and less harmonics—which caused the old tube amplifiers to be highly sought after by the Rock musicians! Amplifier manufacturers finally caught on and developed solid-state units that provide flat clipping and the harmonic distortions so loved by some young musicians (and tolerated at best by most senior DSP professors).

specified times and we could do that again here. However, we will show instead the power of a good match of wavelet to signal.

This time we will begin by choosing a wavelet that matches the binary signal—the Haar would be a good choice. Using a conventional DWT we could decompose up to 10 levels (**1024** = **2^10**) but 5 levels will adequately demonstrate the process. First we will perform the DWT on the pure noiseless binary signal using the Haar wavelet. The display is shown in Figure 11.3–3.

We notice that the Details in levels 1 through 4 are zero. This is true for *any time interval* of the binary signal. For example the "skinny" square waves at the beginning of the pure binary signal (ref. Fig. 11.3–1 at left) as well as the "fat" square waves closer to the middle (times roughly 200 to 500) all produce zero values for the Details in levels 1 through 4. This means that for *any* binary signal similar to this test case the information is captured in the higher levels (**D5**, **D6**, etc.) that we have learned represent the lower frequencies

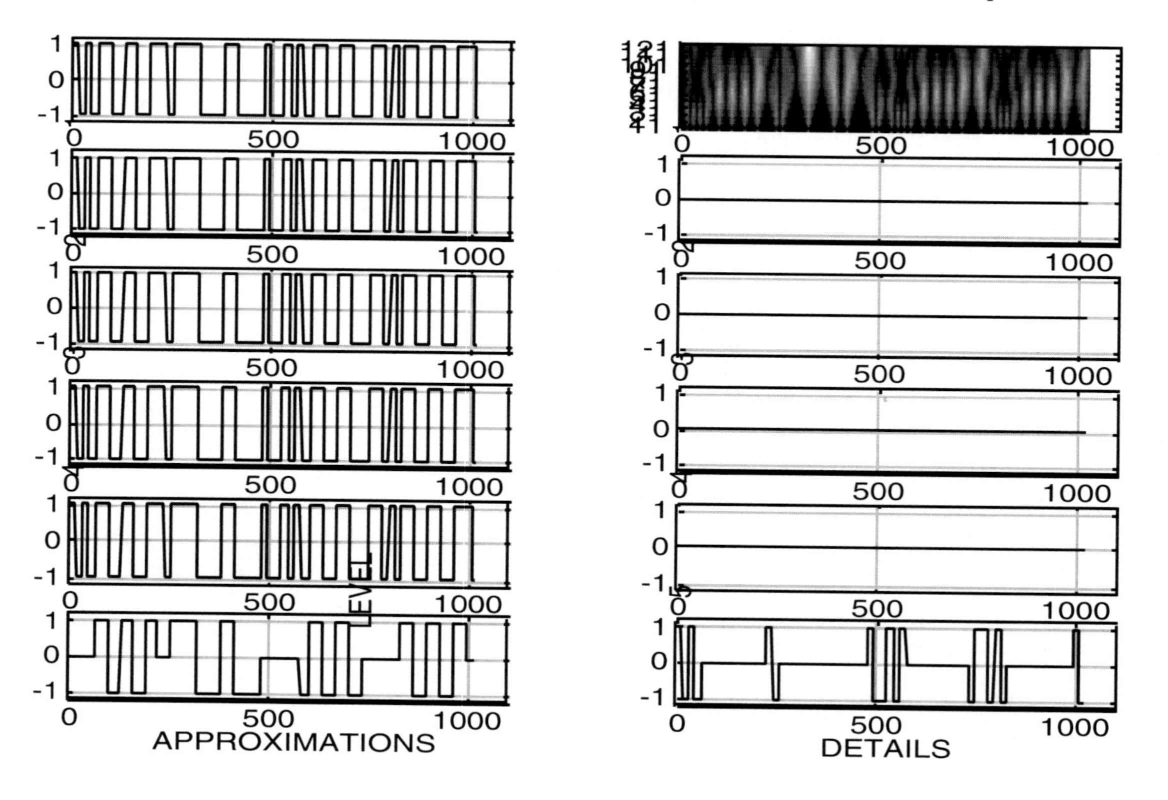

Figure 11.3–3 DWT of the original binary signal (top left) with CWT (top right).

Specifically, we know that the pure binary signal, S, can be represented by

```
S = D1+A1 = D2+D1+A2 = . . . = D4+D3+D2+D1+A4
```

But with the Details being zero on the first 4 levels we have

```
S = 0 + 0 + 0 + 0 + A4 = A4
```

We can see this in the above DWT display as we compare the signal (top left) to the Approximations in levels 1 through 4.

The DWT display of the binary signal with intermittent noise is shown in Figure 11.3–4.

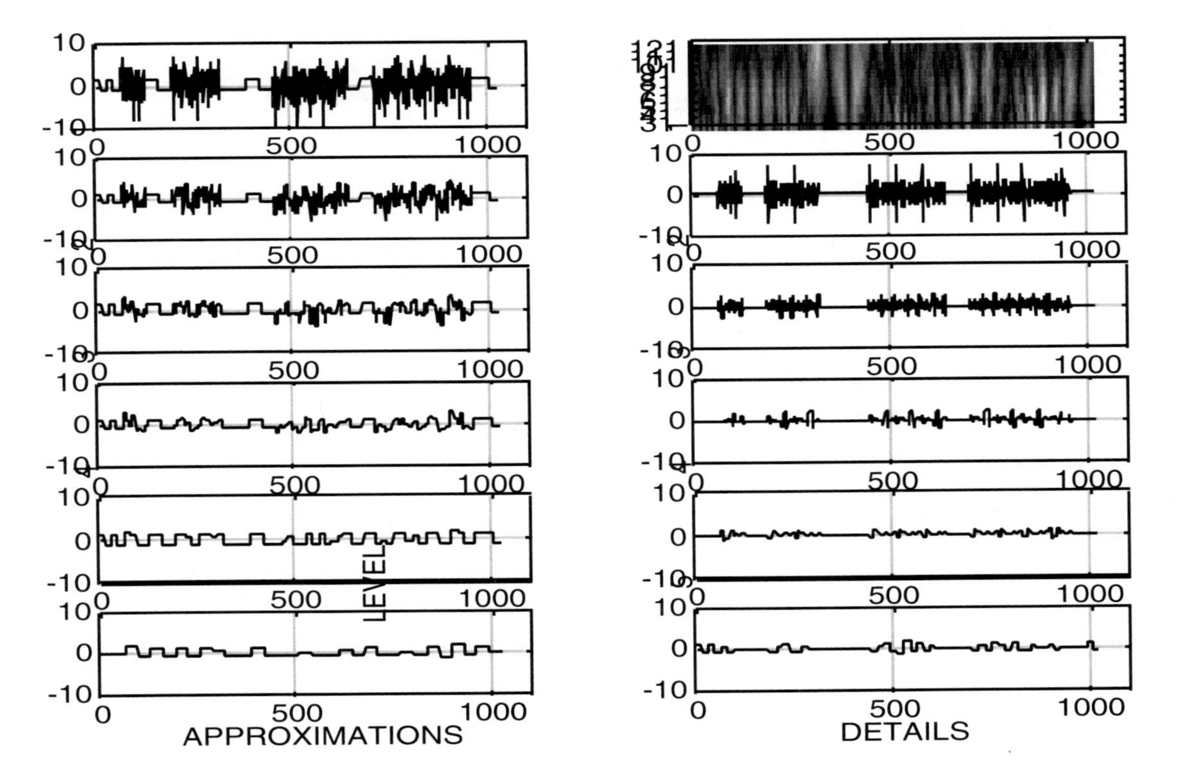

Figure 11.3–4 DWT of the original binary with intermittent noise added.

We notice that there is information in all 5 levels of Approximations and Details. However, we now know that for a binary signal similar to our test case that the information in levels 1 through 4 of the Details is all noise. This means that the level 4 Approximations (**A4**) contains the signal with much of the noise removed.

Figure 11.3–5 shows at left the partially de-noised signal found in **A4**. The right graph shows the original signal for comparison. The original signal is also superimposed on the denoised signal (dotted lines).

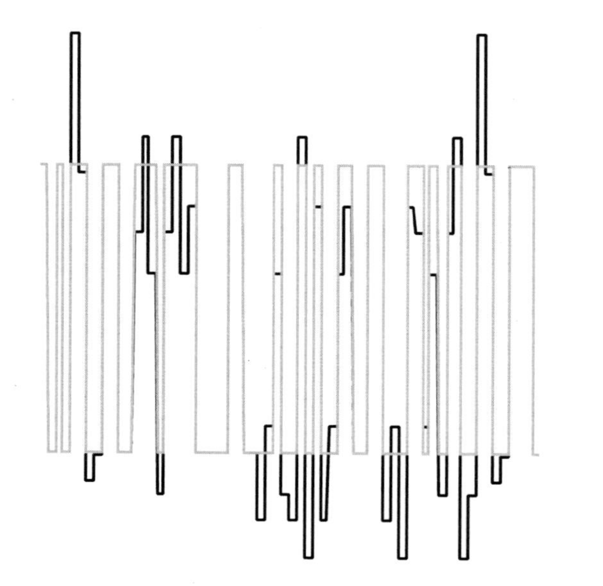

Figure 11.3–5 Denoised binary signal at left. For comparison, the original pure binary signal is superimposed on the left graph and presented by itself at right.

Although not perfect, we can tell which bits are positive and which are negative and thus +1 or –1. The bit pattern is thus preserved after denoising.

The CWTs we saw in the upper right corner of the DWT displays deserve a closer look. Figure 11.3–6 shows the CWTs for the pure binary signal and the signal with the intermittent noise added.

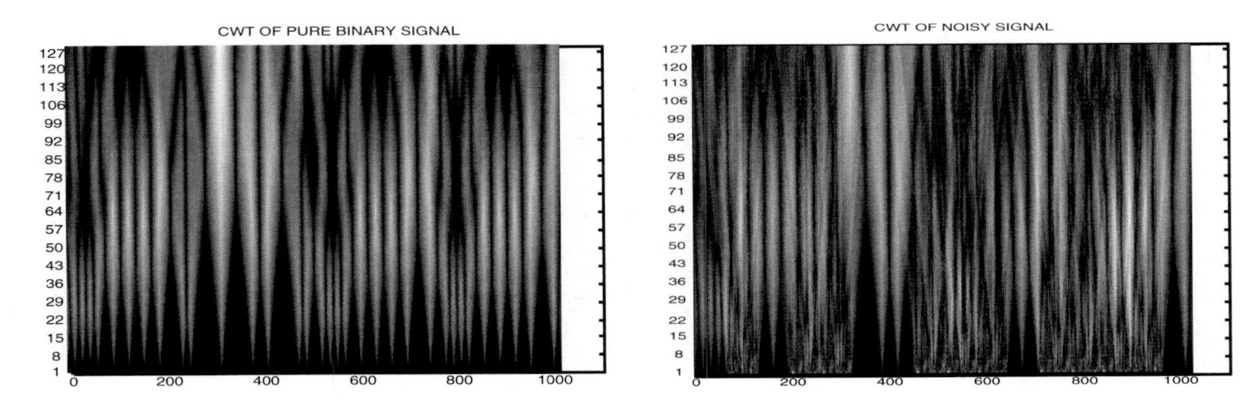

Figure 11.3–6 CWT displays using Haar wavelet of pure binary signal and of noisy signal.

The displays look somewhat like a row of pipes on a pipe organ. The smaller "pipes" indicate the "skinny" square waves as found at the left side of each of the CWT displays (times from about 1 to 40). The larger "pipes" indicate the "fatter" waveforms just to the left of the middle of each display (about **time = 400**). These displays show the magnitude of the values so **–1** will appear as bright as **+1**. We can, however, adjust the display so that negatives are dark and positives are bright so we can discern the bit patterns. Notice that we can "see past the noise" enough to discern bits in the noisy signal (right graph).

Figure 11.3–7 (left) shows the CWT of the Haar-denoised signal. Notice how well it agrees with the original binary signal (ref. Fig 11.3–6). We chose the Haar wavelet because it looked like the binary signal. It is interesting to see what would happen if we chose a wavelet that looked nothing at all like the signal (or the noise). The right side of Fig. 11.3–7 shows the results of using a Db20 "chirp" wavelet. The DWT (not shown) for the Db20 is also of no use.*

This is why conventional DSP falls short—especially with signals that do not look sinusoidal and do not match the sinusoids of the Fourier Transform.

* In practice, if we are unsure about the shape of the signal we would start with a more general-purpose wavelet such as the Db4. In this case, however, we can see portions of the binary signal and can tell up front that a Haar wavelet would be an excellent choice but that a 20-point chirp wavelet (as used in the previous example) would be a poor match to either the signal or the noise.

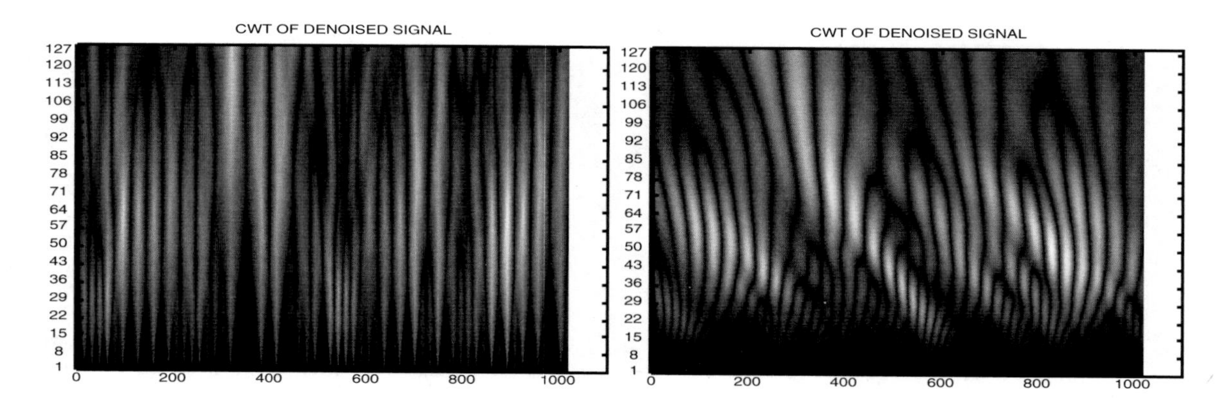

Figure 11.3–7 CWT of Haar-denoised signal using the Haar wavelet at left clearly shows the bits. At right, the CWT of the denoised signal using a Db20 wavelet is interesting-looking, but gives no practical visual information.

11.4 Image Compression/De-noising

Compression and/or de-noising using wavelets are in wide use in image processing. We saw an example of JPEG image compression earlier (ref. Fig. 1.9–4). While a full discussion of the use of wavelets in image processing is beyond the scope of this book, we can provide a brief overview.

Image processing means two-dimensional processing. Instead of the simultaneous *time/frequency* capabilities of wavelets we are usually talking about *space/frequency or distance/*frequency. In other words, a signal would have various amplitudes as we vary the time while a monochromatic image would have various brightness as we vary the position (space) on the image (e.g. "3 centimeters to the right and 4 centimeters down on the photo").

To get our bearings we will look first at a simple 256-point split sine signal and a single-level one-dimensional conventional DWT display as shown in Figure 11.4–1 (top left graph). We will use the set of 4 Haar wavelet filters (the **H'** filter is used by itself to produce the CWT at top right). Notice that the Details and Approximations at level 1 (**D1** and **A1**) combine to produce the original signal. Note also that both halves of **A1** have high amplitudes (like the original signal) while the left half of **D1** has much smaller values than the right half). This is of course because **D1** represents the higher frequencies while **A1** represents the lower frequencies as we have discussed (ref. Fig. 4.5–1).

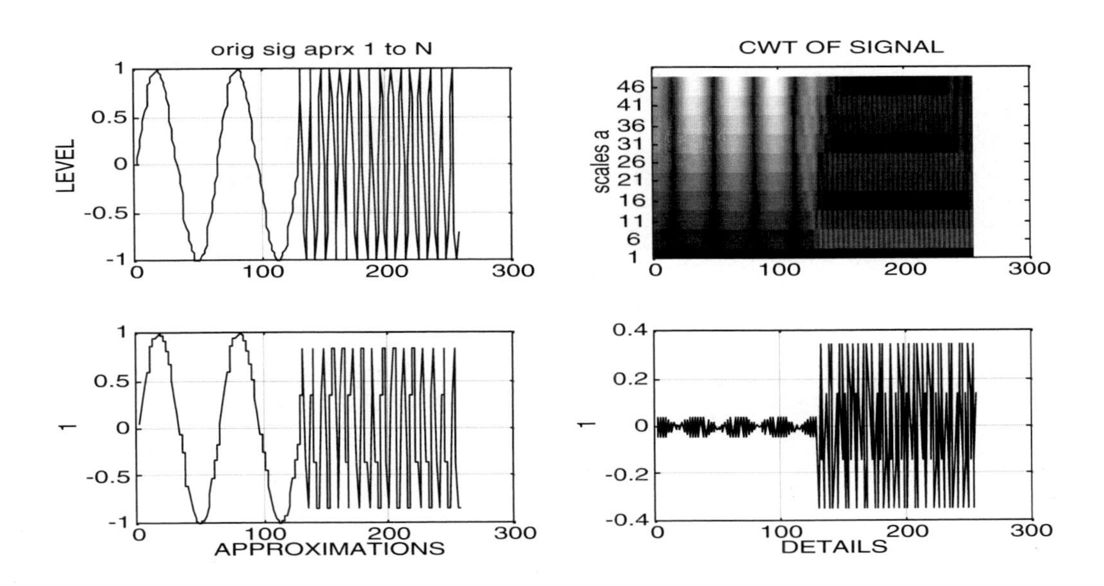

Figure 11.4–1 Single-level DWT of a split-sine signal using the Haar wavelet filters.

We next construct a 2-dimensional 256 by 256 image "test pattern" The first 128 *rows* will be identical copies of the above split sine signal. The image is shown below in Figure 11.4–2 at left. Comparing with the 1-D signal we can discern the tops of the 2 low-frequency sine waves (cycles) and the tops of the 16 high-frequency sine waves as bright spots. We use a shorter 128 point split sine signal with one low-frequency sine wave and 8 high-frequency sine waves (not shown) to fill the lower half of the test pattern (right graph) by constructing 256 identical *columns* each 128 points long. As with the longer split sine test signal we can see the bright spots corresponding to the top of the low-frequency sine wave (single cycle) and the 8 high-frequency sine waves.

We now proceed with a single level *two-dimensional* conventional DWT of this test-pattern image. Whereas the 1 level 1-D DWT would decompose the signal into **A1** and **D1**, the 2-D DWT converts the image into **A1** (the lower-frequency Approximation), **H1** (a vertical scan yielding Horizontal components), and **V1** (a horizontal scan yielding Vertical components).*

* Some software also uses a diagonal scan and/or additional methods to further decompose the data at each level.

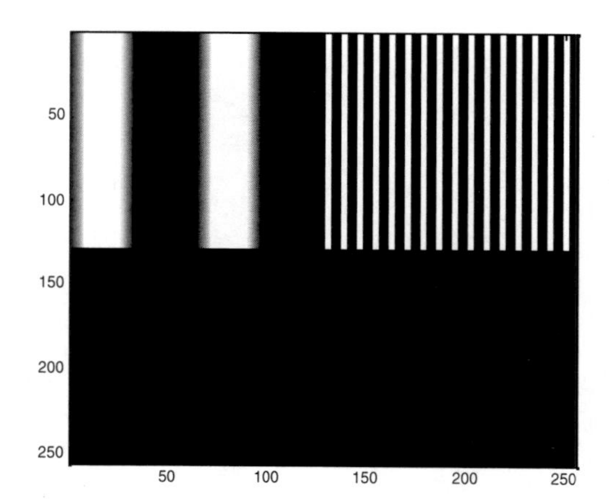

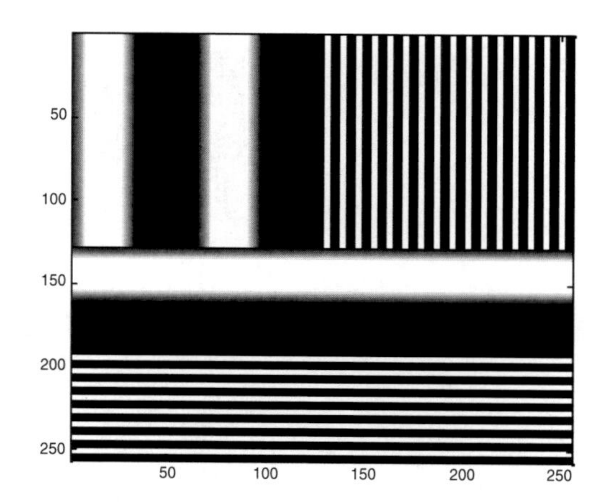

Figure 11.4–2 Test Pattern produced by 128 identical rows of the 256-point split-sine signal (left image) followed by the addition of 256 columns of a shorter 128 point split-sine signal (bottom half of complete test pattern image as shown at right).

Figure 11.4–3 shows the **A1**, **V1** and **H1** portions of the decomposed image. The Approximation, **A1**, looks much like the original image (as did the 1-D **A1** from Fig. 11.4–1 look much like the original signal). Just as the one-dimensional **D1** (right lower plot from Fig. 11.4–1) had very low values until the signal changed to high frequency, the upper half of **V1** (the center image of Fig. 11.4–3 below) is dark for the first half then we can see the higher frequency sine waves as vertical components from the horizontal scans.

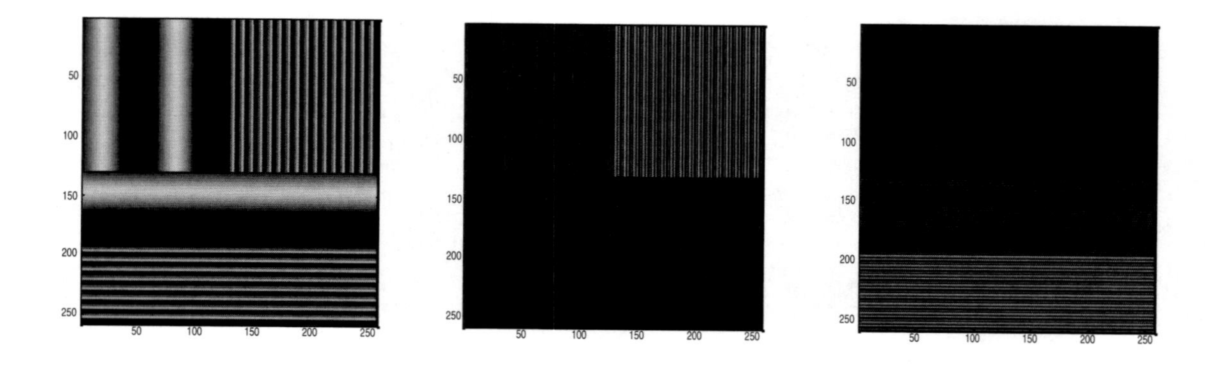

Figure 11.4–3 Single-level 2-D DWT of test pattern produces the lower frequency Approximation (**A1** at left), the **V**ertical components from the horizontal scan (**V1** in center) and the **H**orizontal components from the vertical scan (**H1** at right).

The lower half of **V1** is all dark because horizontal scans of the test pattern in that area will produce constant values (whether bright or dark, the values don't change horizontally) and thus zero frequency. The vertical scans in the rightmost graphic will also produce zero values (dark portions) until they encounter the high frequency portions of the shorter split-sine signal.

Similar to the 1-D cases, the image can be reconstructed by combining **A1**, **V1**, and **H1**. If we do this without changing the components we will have perfect reconstruction. However, our goal here is to remove some noise, compress the signal or both.

Figure 11.4–4 shows the familiar "*Barbara*" image (left). We have taken a 200 x 200 pixel close-up to show the facial quality. Then we have added some noise as shown at right. Notice particularly the "freckles" we have added around the forehead, cheeks, and nose areas.

Figure 11.4–4 Classic "Barbara" image 200 x 200 close-up is shown at left. Noise is added along with facial skin imperfections ("freckles") to the "pure" image giving us the image at right.

We will now compress this image. Since compression often involves removing high frequency components we might expect a possible improvement in skin quality (i.e. freckles and other skin imperfections less pronounced).

Since this image is small (200 x 200) we will want to use a small wavelet (filter). The 2-point Haar comes to mind. The result of a 9 to 1 compression is shown in Figure 11.4–5 at left. Even after fine-tuning, the facial quality will still be very poor. However, notice how the fabric of her scarf is very pronounced. The Haar wavelet provides for good edge detection.

A better choice for her complexion would be a biorthogonal wavelet. We choose the 7/9 wavelet because it is perfectly symmetrical and still short (9 points maximum filter length). The results of a 9 to 1 compression using this biorthogonal wavelet is shown below at right. Notice her complexion has cleared up considerably and that she now has a "softer" look.*

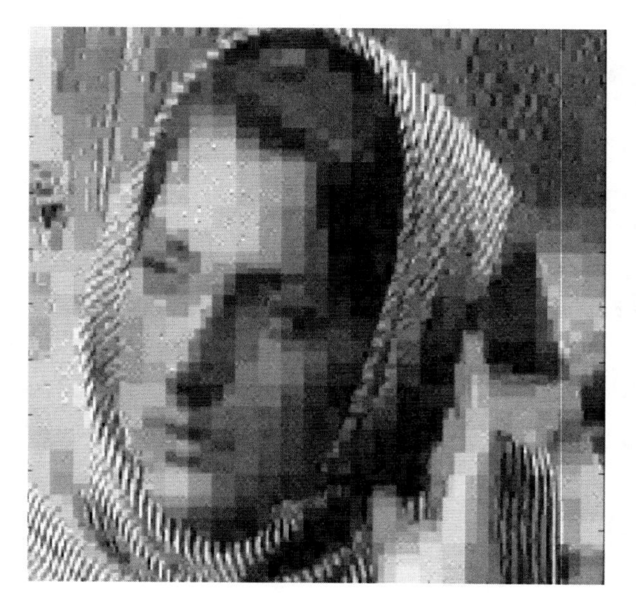

Figure 11.4–5 9 to 1 compression using a set of 2-point Haar filters is shown at left. The same compression using a set of Biorthogonal 7/9 filters is shown at right.

We saw earlier how wavelets can be used to denoise a signal at specific time intervals. Similarly, with wavelet image processing we can denoise specific areas of an image. In the above example we could use heavier filtering on the freckles areas and lighter filtering on the rest of the image.

* In the early days of Hollywood, long before Digital Image Processing, some older screen actresses would insist upon a gauze "filter" stretched across the movie camera lens. This "soft" effect would hide wrinkles and age spots. One young actress also insisted on this "soft" effect—not to hide wrinkles but to hide her freckles!

11.5 Improved Performance using the UDWT

So far in this chapter we have successfully used the conventional DWT for de-noising and compression. As mentioned earlier, when we remove noise we also remove part of the alias cancellation capability of the conventional DWT. In the examples so far this has not been a problem but in this last example we highlight a pathological case where aliasing *is* problematic and show why the Undecimated DWT (UDWT) can provide better results.

Figure 11.5–1 (left graph) shows a generalized test signal that begins with Frequency Shift Keying (FSK) and ends with Frequency Modulation (FM). The FFT of this signal (right graph) shows the low and high frequency portions (peaks) from the FSK modulation and from the linear FM modulation (wide portions at low frequencies).

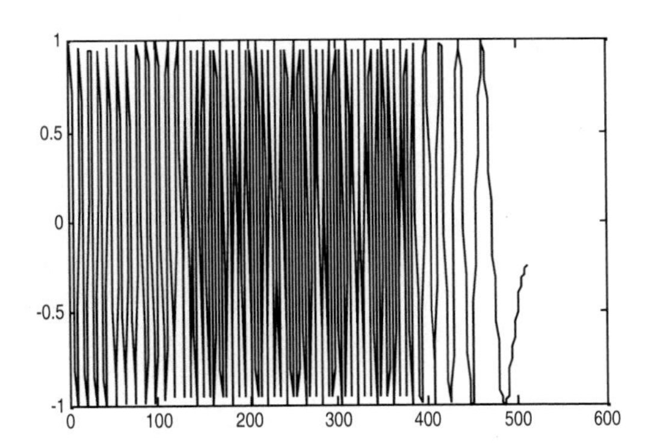

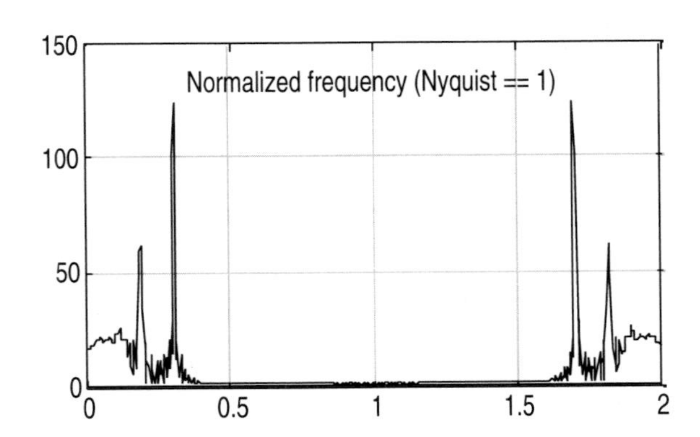

Figure 11.5–1 Composite demonstration signal with FSK modulation followed by FM modulation shown in the time domain at left and in the frequency domain at right.

We next add some high-frequency noise in the form of harmonics/intermodulation effects. In other words, as the modulated frequency of the original signal changes, so does the frequency of the noise. This is shown in Figure 11.5–2

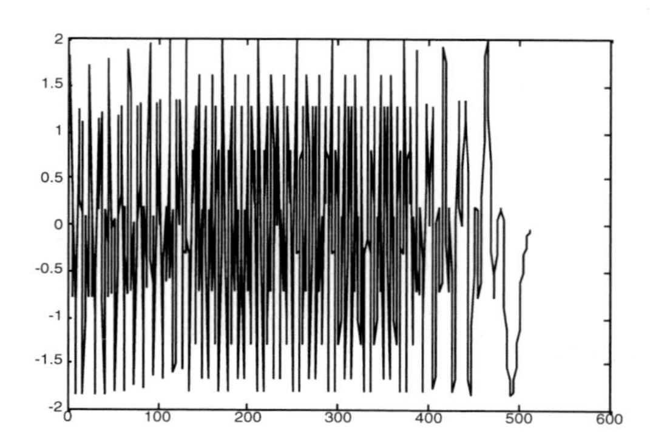

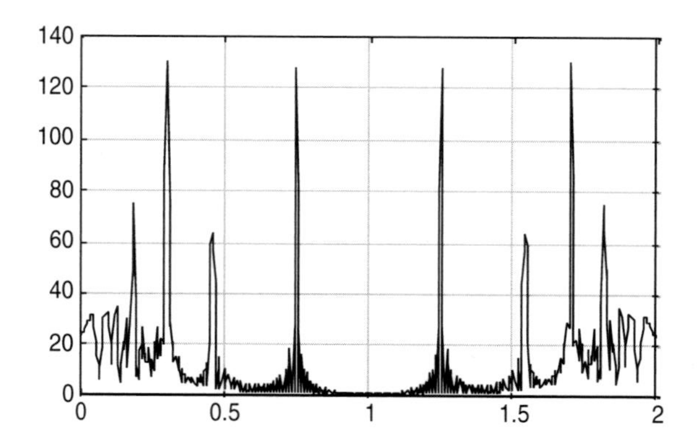

Figure 11.5–2 Noisy composite demonstration signal shown in the time domain at left and in the frequency domain at right.

As can be seen by comparing the FFT of the original signal to the FFT of the noisy signal (right graphs of Figs. 11.5–1 and 11.5–2) we cannot isolate the noise from the signal using conventional DSP filtering techniques. We turn again to wavelet processing.

We use a "general purpose" set of Db4 wavelet filters and perform a conventional DWT and a UDWT on the noisy signal as shown below in Figure 11.5–3. As we compare the DWT and the UDWT displays we notice differences, especially in the higher frequency sub-bands of **D1**, **D2**, and **D3**. Furthermore, the UDWT looks "cleaner" in isolating the sections of the signal. The frequency allocation is the same for both the DWT and UDWT (ref. Fig. 11.1–5) and the wavelet is the same (Db4) causing us to ask "Why the differences in results from using the 2 methods?"

The answer, as we will demonstrate, is non-canceled aliasing in the conventional DWT while the UDWT (which uses stretched filters instead of downsampling) has no such problems. We note in passing that the CWT (which also uses the stretched "H'" filter) is the same for the DWT and the UDWT customized displays (upper right graphs in both displays).

To better see what's going on, we look at the center part of the signal by itself. We would do something similar to this "isolation in time" as we exploit the time/frequency capabilities of wavelet processing to order to impose different thresholds on the various time segments of the signal (interval dependent thresholds) as shown earlier in this chapter (ref. Figs. 11.1–6 and 11.2–5).

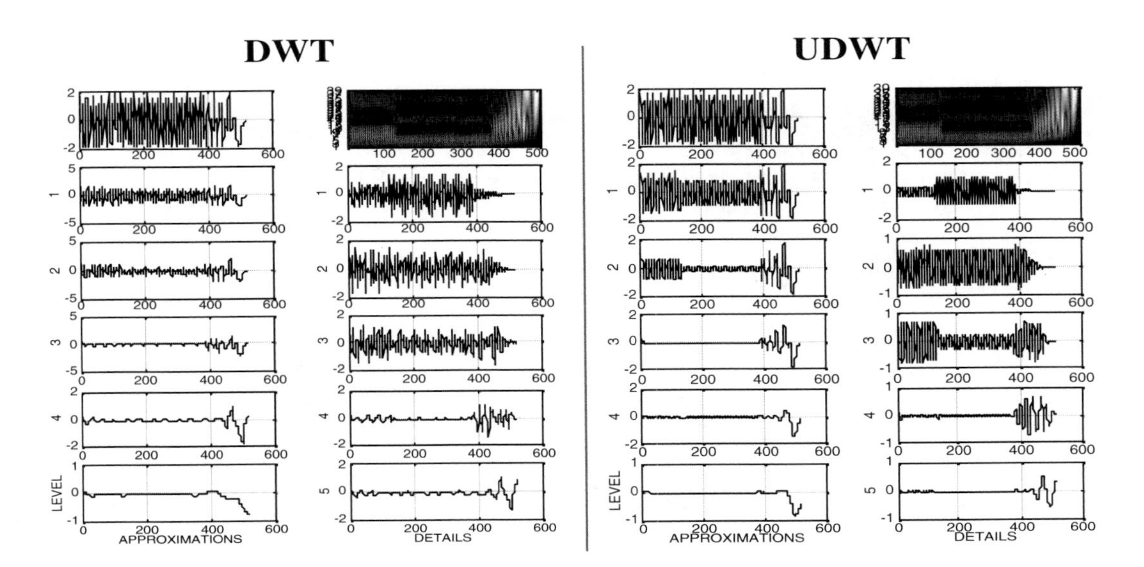

Figure 11.5–3 Db4 wavelet decomposition of the noisy composite FSK/FM signal using a convention al DWT at left and an Undecimated DWT at right.

Figure 11.5–4 shows the center 256 points of this 512 point noisy signal. It is composed of a 0.3 Nyquist sinusoid with a 0.75 Nyquist sinusoid as the noise. Before proceeding, we note that this center portion of this FSK/FM noisy signal is a *stationary signal* in that it does not change frequencies or amplitude (envelope) over time. If the entire 512 points of the signal looked like this we would be better off using a sinusoid as the "wavelet"—in other words to use conventional Fourier techniques instead of wavelet processing. However, since this "sum of sines" can occur, even for a very short time,* in a large number of ways and in a variety of signals we will proceed.

Another glance at the frequency graph from Fig. 11.5–4 below (also ref. Fig. 11.1–5) shows the frequency subband **D1** to be roughly from 0.5 to 1.0 Nyquist and the subband **A1** to be from roughly 0.0 to 0.5 Nyquist (with some overlap due to the imperfect filtering of the Db4). Because the signal at 0.3 Nyquist is well-isolated from the noise at 0.75 Nyquist it appears that the signal should be found in **A1** with a minimal amount in **D1** while the noise should be found almost entirely in **D1** with minimal amounts in **A1**. A single-level DWT or UDWT should show this.

* Although a Short Time Fourier Transform (STFT) could be used for stationary portions of a signal, if the time is too short we won't achieve meaningful results. The Wavelet Transforms are a better choice here.

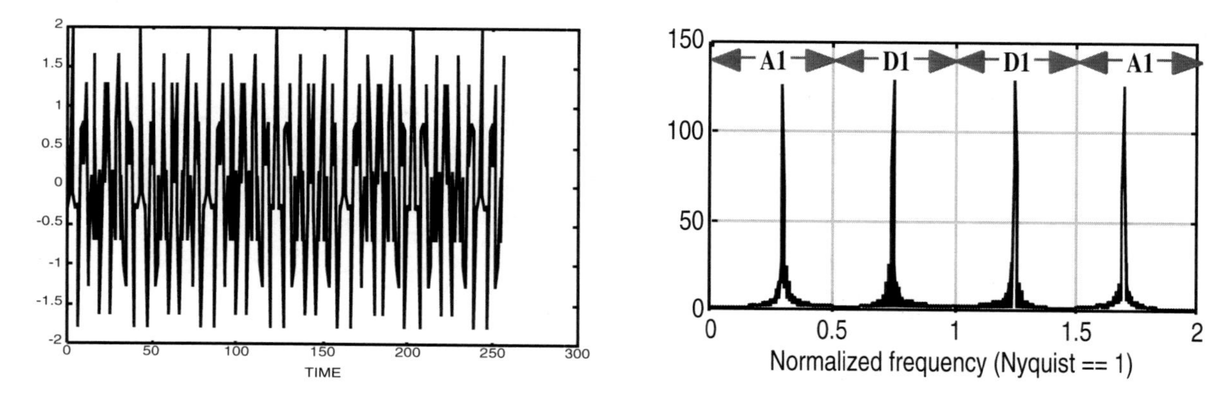

Figure 11.5–4 Center 256 points of noisy 256-point signal shown in both the time and frequency domains.

As a "sanity check",* we will first look at a single-level decomposition of the *noiseless* signal by itself using both the conventional (decimated) DWT and the Undecimated DWT (UDWT). Figure 11.5–5 shows the single-level DWT and then the UDWT displays for the *noiseless signal*. The signal is found almost entirely in **A1** for the UDWT with very little in **D1** (note different scales) as expected. But the conventional DWT has much more high-frequency components in **D1**. We will show that this is not due to noise, but instead actually due to the aliasing of the noiseless *signal*!

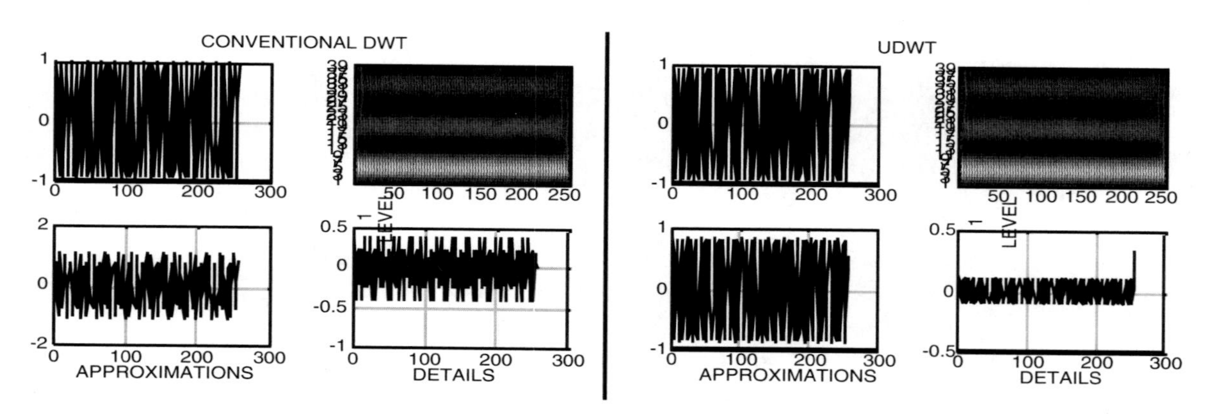

Figure 11.5–5 Results of a single-level decomposition of the pure signal with no noise using a Db4 filter set. Conventional DWT results are compared with UDWT results.

* Some would argue that it will take more than this simple check to establish the author's sanity—but it is a good idea when faced with puzzling results to look at the basics and build from there.

The block diagrams for the single-level conventional DWT and the UDWT are drawn again in figure 11.5–6. We can see the potential for aliasing from downsampling in the Conventional DWT here.

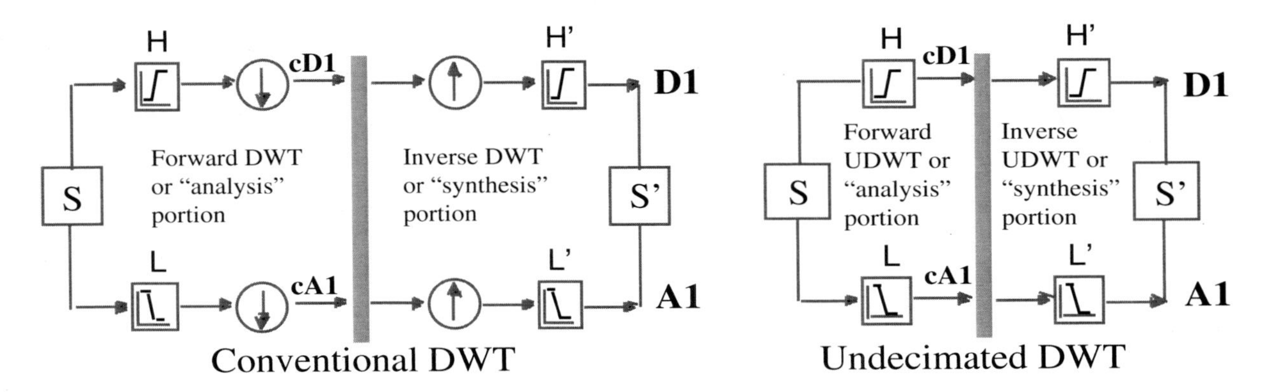

Figure 11.5–6 Single-level Conventional DWT and Undecimated DWT. The 2 are identical (for the single level case) except for the lack of downsampling (decimation) by 2 in the UDWT.

A look in the frequency domain will illustrate the problem. Figure 11.5–7 shows the *noiseless* signal after the single-level decomposition using a Db4 UDWT. As expected, most of the 0.3 Nyquist signal is found in **A1** (left graph). Because of imperfect filtering we have a small amount of the signal in **D1** (right graph). Note the difference in scales however with **A1** containing almost 9 times the signal content of **D1**.

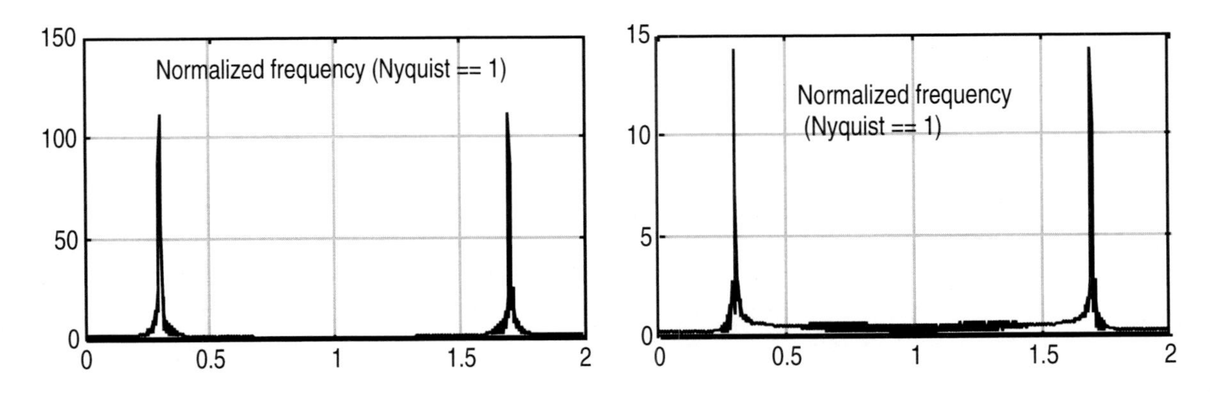

Figure 11.5–7 Frequency domain results for the Undecimated (non-downsampled) UDWT *noiseless* pure signal. **A1** (left graph) contains the signal at 0.3 Nyquist with no noise as expected. **D1**, as shown in the right graph, contains some small aliasing due to imperfect filtering.

We next look at frequency domain results using the same *noiseless signal* in the **A1** and **D1** subbands produced by the *conventional* DWT as shown in Figure 11.5–8. The results show aliasing present in both **A1** and **D1**. Recall from DSP that for a signal at 0.3 Nyquist aliasing from downsampling will "reflect" the signal across Nyquist. Thus we see the aliasing components at Nyquist minus 0.3 Nyquist or 0.7 Nyquist. This is not to be confused with our earlier noise at 0.75 Nyquist—*there is no noise* in this sanity check! In other words, this alias artifact will appear even in a noiseless signal.

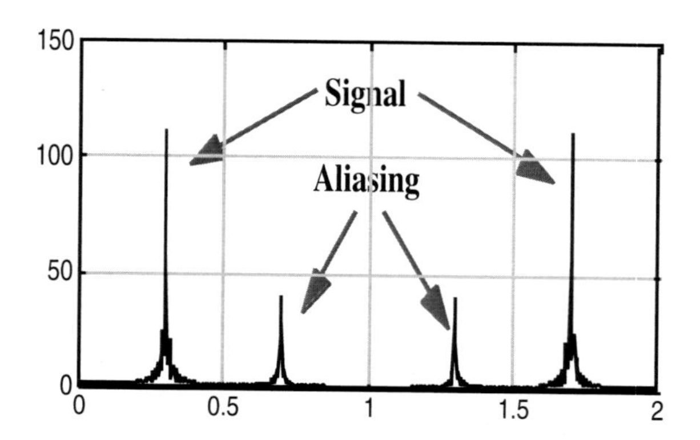

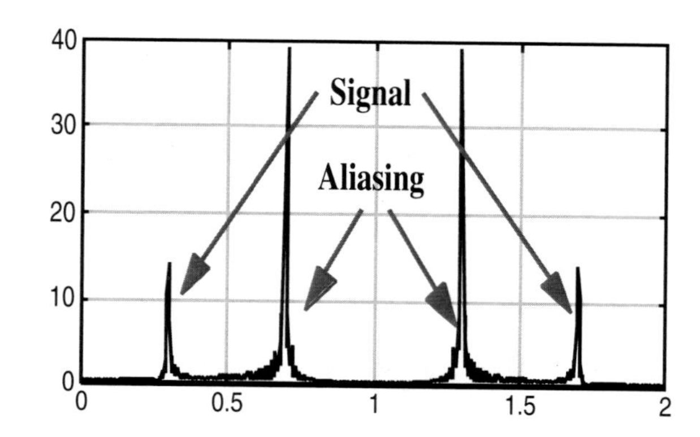

Figure 11.5–8 Frequency domain results for the *conventional* (downsampled) DWT of the *noiseless* pure signal. **A1** (left graph) has aliasing components at (**1.0** – **0.3**) = **0.7** Nyquist. **D1** (right graph) also has magnitude 40 aliasing components at 0.7 Nyquist .

Notice from the above figure 11.5–8 that the signal at 0.3 Nyquist has the same components (magnitude 120 in **A1** and 14 in **D1**) as with the UDWT case (ref. Fig 11.5–7). Notice also the *aliased components* at 0,7 Nyquist (magnitude 40) are the same size! This is important because when we add **A1** to **D1** in the conventional DWT these components cancel. But when we throw away **D1** to get rid of some high frequency noise (added later) we are also throwing away the alias cancellation components and we are left not only with the 0.3 Nyquist signal but also a substantial 0.7 Nyquist alias artifact. Note: We have shown only the magnitudes of the (complex) signal here. We will discover in the next chapter how the aliased components in **A1** and **D1** are 180 degrees (π radians) out of phase and thus cancel.

Having demonstrated the superior performance of the UDWT on the *noiseless* signal, we now compare the DWT and UDWT on the signal (center portion of the FSK/FM) with *added noise*. With the signal at 0.3 Nyquist and the noise at 0,75 Nyquist (ref. Fig. 11.5–4) we will look at the results of keeping

A1 while discarding **D1**. A look at the frequency domain of the UDWT **A1** and **D1** in Figure 11.5–9 below shows this to be a viable option for this particular signal.

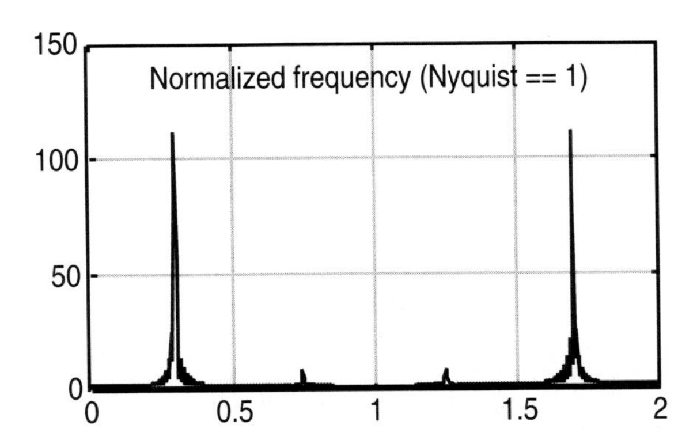

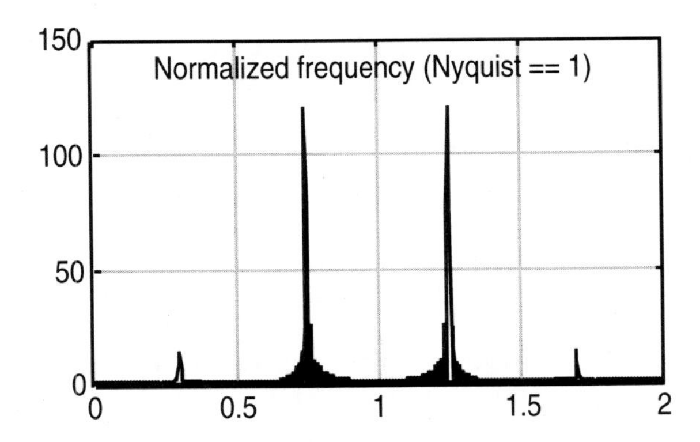

Figure 11.5–9 Frequency domain results for the UDWT noisy signal. **A1** (left graph) contains almost all the signal at 0.3 and very little of the noise at 0.75 Nyquist. **D1** (right graph) contains almost all the noise and very little of the signal.

However, a look at the *conventional* DWT **A1** and **D1** in the frequency domain indicates that this option of keeping **A1** as the "de-noised" signal is not viable because of aliasing problems caused by the downsampling. This is shown below in Figure 11.5–10

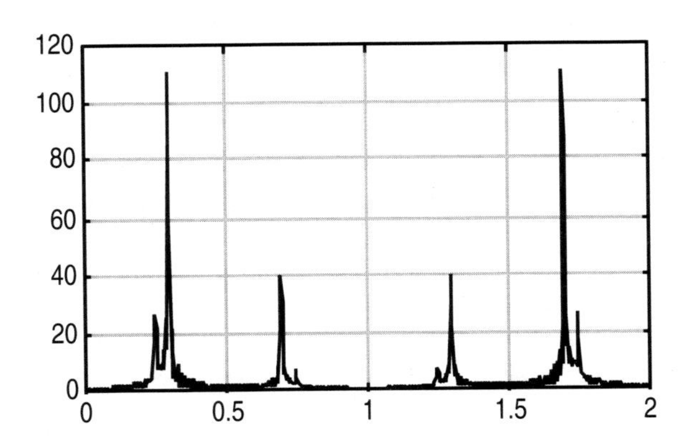

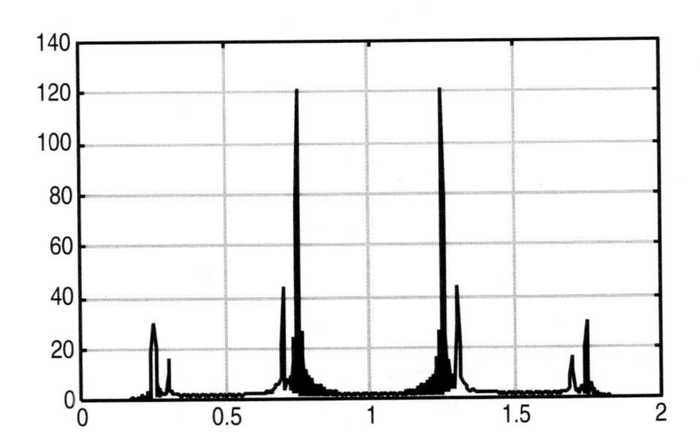

Figure 11.5–10 Frequency domain results for the conventional (downsampled) DWT noisy signal. **A1** (left graph) contains the signal at 0.3 Nyquist along with significant aliasing effects. **D1** shows the added noise at 0.75 Nyquist along with further aliasing effects.

We can now compare the results of denoising using the conventional DWT with those of the UDWT for this stationary portion of the noisy signal using a single-level Db4 wavelet transform (keeping **A1** and discarding **D1**). We saw **A1** in the frequency domain earlier in Figs. 11.5–7 and 11.5–8. We now take a look in the time domain. Figure 11.5–11 shows a close-up of the original noiseless signal and the same signal with noise added.

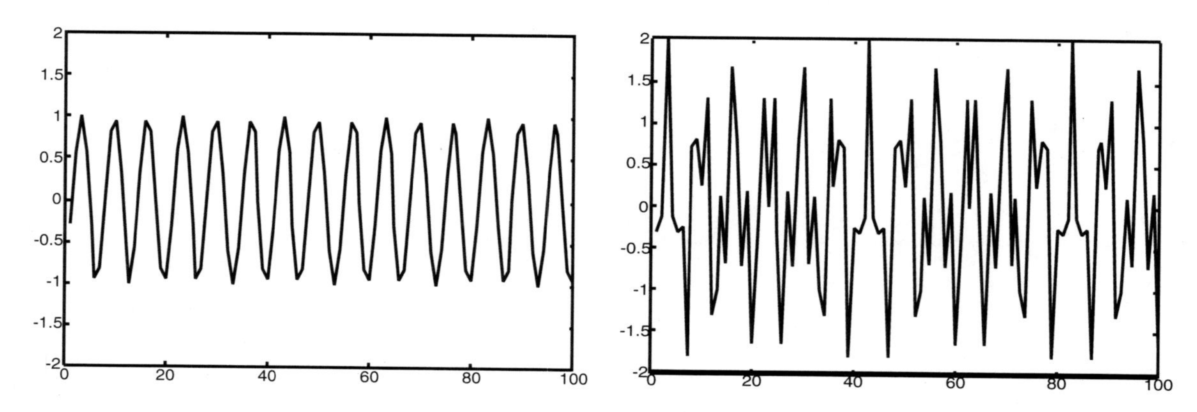

Figure 11.5–11 Close-up in time domain of original and noisy signals.

Figure 11.5–12 shows a close-up of the de-noised signal using the conventional DWT and the same signal denoised using the UDWT.* As can be seen, the UDWT does an almost perfect job of de-noising even using a single-level transform in this pathological (but very possible) case. The conventional DWT will need additional levels, a longer wavelet, and/or further processing.

We can of course use the UDWT to denoise the rest of the 512 point FSK/FM signal. Because the conventional (decimated) DWT is in such wide use, it would be a good idea to better understand and utilize it correctly! Thus, we will look in more depth at how the alias cancellation works within the conventional DWT in the next chapter.

* Another reminder that the Undecimated DWT is also referred to as Redundant, Stationary, Quasi-Continuous, Translation Invariant, Shift Invariant, and Algorithme à Trous in some texts, papers, and software.

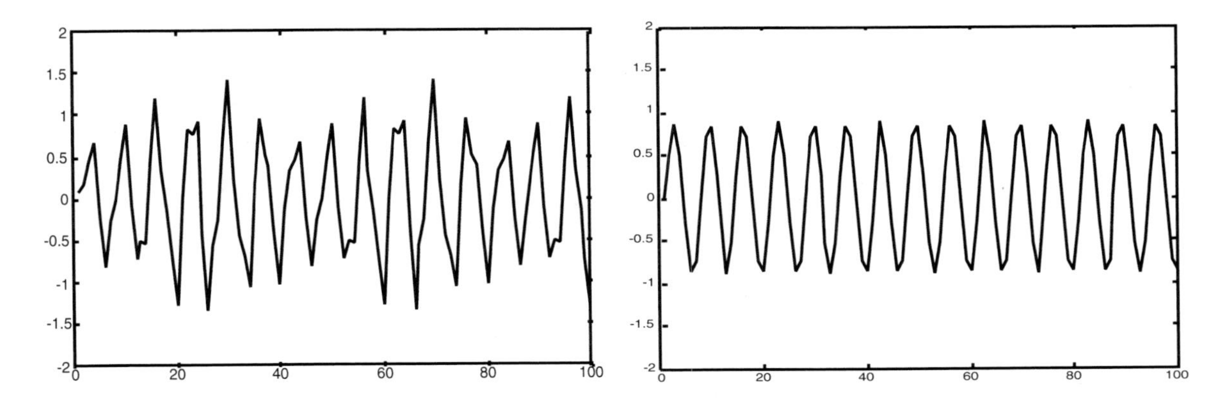

Figure 11.5–12 Close-up in time domain of conventional DWT de-noised signal (left) and UDWT denoised signal (right).

Andrew P. Bradley in his landmark paper "Shift-invariance in the Discrete Wavelet Transform" * reminds us "It should be noted that the aliasing introduced by the DWT cancels out (only) when the inverse DWT (IDWT) is performed using ***all*** of the wavelet coefficients, that is, when the original signal is reconstructed". He offers several suggestions (besides exclusive use of the UDWT) including (1) using longer filters with better frequency resolution and (2) creating a hybrid that uses the UDWT at some levels to prevent non-canceled aliasing and using the conventional DWT for improved speed and storage efficiency.

11.6 Summary

In this chapter we explored and demonstrated some of the capabilities of wavelet processing techniques using specific examples. We first added white noise to a chirp signal and then removed the (supposedly unknown) noise using a Db6 conventional DWT and exploiting the time/frequency capability of wavelet processing. We employed a somewhat similar process using a Db20 chirp wavelet to find a binary signal buried in 80 dB of noise from a chirp jammer.

Our next case was a 16 chip per bit binary signal with intermittent pseudo-random noise. The best match to the binary signal was a Haar wavelet. Performing a 7-level conventional DWT on the noiseless signal we saw there were no components of the noiseless signal in **D1** through **D4**. Thus any-

* Proc. VIIth Digit. Image Comp., Dec. 2003, pp. 29–38.

thing found in **D1** through **D4** would be noise and could be safely discarded. We were thus able to re-create the original binary signal enough to discern the binary values (+1 or −1). We also revisited the CWT for this example and demonstrated its capability.

We demonstrated image compression on first a test pattern and then using a subset of the "Barbara" image. We added some noise in the form of skin imperfections (freckles). By thresholding the levels (in two dimensions) we were able to compress the image by almost an order of magnitude and in the process selectively remove high frequency components. The final image had a "soft" look that removed the skin imperfections. The wavelet filters of choice were those of the Biorthogonal 7/9, chosen for their symmetry and short length. We used a Haar for comparison and obtained better edge detection but far worse skin quality.

In the last example, we featured a case where de-noising using a UDWT was superior to using a conventional DWT. In both the DWT and UDWT a Db4 set of filters and a single level decomposition was used. The problem was shown to be with the conventional DWT itself. In real-life Digital Signal Processing we can get aliasing when we downsample. The (conventional downsampled or decimated) DWT cancels aliasing, but when we throw away parts of the decomposition (**D1** for a single-level DWT) we also lose the alias cancellation. Alternatives include (1) using the *Undecimated* DWT exclusively; (2) using the UDWT for part of the decomposition; and (3) using a longer wavelet with better frequency resolution or (4) checking first to see if the values in the Details (for a given length of time or space) are close enough to zero that they can be safely suppressed for that interval.

In all these cases, conventional DSP methods with filtering and/or FFT methods would not work. A Short Time Fourier Transform might work on some of these examples but the dynamic nature of wavelets usually make them a better option. Also, in every case we tried to match the wavelet to either the signal or noise for best discrimination.

With these examples under our belt we can take a closer look at the alias-cancellation methods used in the conventional DWT and gain further conceptual understanding and, hopefully, increased wisdom in using the various wavelet transforms.

CHAPTER 12

Alias Cancellation in the Conventional (Decimated) DWT

Many of the basic principles found in wavelet technology are also found in other disciplines, often with different terminology (ref. Chapter Ten, introduction). For example, a "2-channel Quadrature Mirror Filter bank" as used in multirate signal processing is identical to the now-familiar single-level conventional Discrete Wavelet Transform. In both instances the objective is to be able to downsample the signal while canceling the effects of aliasing.

In the last chapter we ended our examples with a pathological case in which discarding high frequency noise from a single-level (conventional, decimated) DWT would also discard the alias cancellation capability. In this chapter we will demonstrate how perfect reconstruction, even with downsampling, is accomplished by looking first in the time domain and then in the frequency domain. This will also , hopefully, provide intuitive insights into how we can "safely" use the conventional DWT (having minimal aliasing problems) and when it is better to use the Undecimated DWT (UDWT).

We will then relate this knowledge to terminology found in traditional texts and papers as we examine the "No Distortion Equation" and the "Alias Cancellation Equation".

12.1 DWT Alias Cancellation Demonstrated in the Time Domain.

In our discussion of halfband filters (ref. Fig. 8.2–3) we reviewed the *Kronecker deltas* used in DSP (commonly called *delta functions*) and saw how the convolution of a vector with a Kronecker delta leaves the vector unchanged. For example,

```
{1 2 3 4 5]*{1] = [1 2 3 4 5]
```

where "*" in the traditional (non-MATLAB) context indicates *convolution*.

In order to demonstrate alias cancellation in the time domain we will use a Kronecker delta as the signal, S, in a single-level conventional DWT as shown in the left graph of Figure 12.1–1. If we can show alias cancellation for this single point, we can proceed to show alias cancellation for any signal.

The convolution of **S** with the highpass decomposition filter, **H**, is simply **H** again (**arrow 1**, left diagram). We now downsample the result (**H**) to obtain the result at **cA1** For the familiar Db4 wavelet the **H'** (wavelet) filter is given by **c0, c1, c2, c3** and thus the **H** filter is given by **c3, c2, c1, c0** (ref. Fig 8.1–2). Thus downsampling by 2 (we will keep the even values for now) we would obtain **c2, c0**. Upsampling by 2 gives us **0, c2, 0, c0** (**arrow 2**).

At this point we are ready to convolve with the highpass reconstruction filter H' to produce **D1**. Convolving our result so far (**0, c2, 0, c0**) with **H'** (**c0, c1, c2, c3**) is the same as *correlating* the flipped version of our result at **arrow 2** with H'.* In other words

```
D1 = xcorr([c0 c1 c2 c3],[c0 0 c2 0])
```

We show this graphically in the right graph of Figure 12.1–1 as a "sliding correlation".

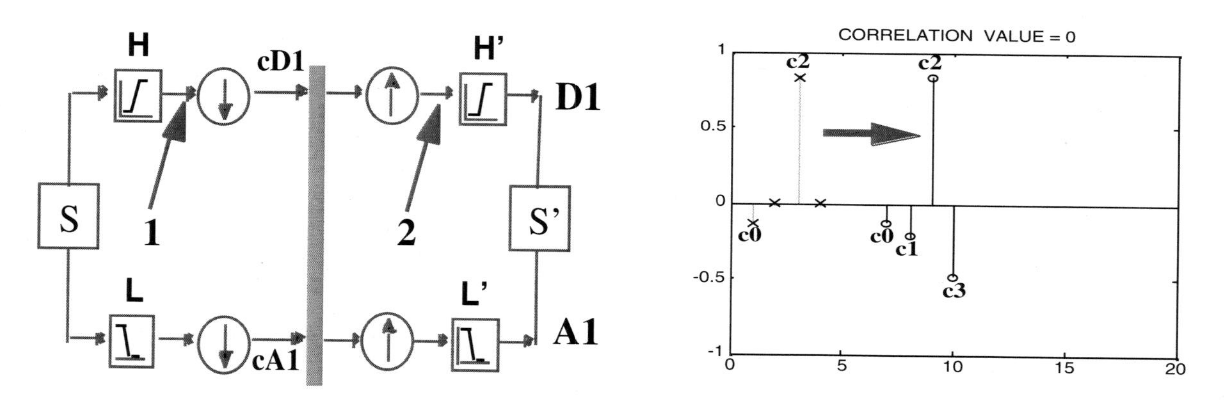

Figure 12.1–1 Conventional single-level Db4 DWT at left. We first follow the Kronecker delta signal through the upper (highpass) path to **D1**. Convolution of the "**arrow 2** results" with H' is same as *correlation* of the *flipped* "**arrow 2** results" (dotted lines with "X") with H' (solid lines with "O") as shown in right graph.

A similar process is seen on the lowpass (lower) part of the DWT that produces A1 as shown in Figure 12.1–2 at left. The Kronecker delta signal, S, is convolved with L which produces simply L at arrow 3.

* We could also have correlated the *flipped* version of **H'** directly with our "arrow 2" result (**0, c2, 0, c0**) but the former way is clearer and the coefficients (**c0, c1, c2, c3**) remain in ascending order.

We learned earlier (ref. Fig. 8.1–2) that we can produce L by alternating the signs of H'. Thus L = c0, –c1, c2, –c3. We also know that L' is the flipped version of L thus L' = –c3, c2, –c1, c0. For simplicity we will let

```
b0 = -c3, b1 = c2, b2 = -c1 and b3 = c0
```

Then we can express the coefficients of the lowpass reconstruction ("scaling function") filter, L', as simply **b0, b1, b2, b3** and L as **b3, b2, b1, b0**.

After downsampling and then upsampling by 2 and keeping the even values,*. we have the result at **arrow 4** of **0, b2, 0, b0**. The final step of convolution with L' can be expressed as a *correlation* of the flipped version of the **arrow 4** result **b0, 0, b2, 0** with L' as shown in the right graph of Figure 12.1–2. In other words

```
A1 = xcorr([b0 b1 b2 b3],[b0 0 b2 0])
```

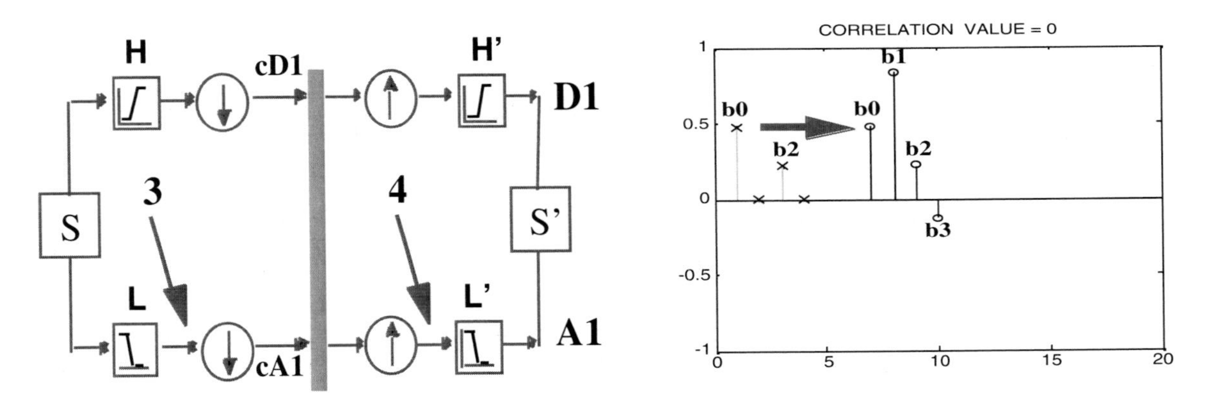

Figure 12.1–2 Conventional single-level Db4 DWT at left. We now follow the Kronecker delta signal through the lower (lowpass) path to **A1**. Convolution of **arrow 4** results with **H'** is the same as the *correlation* of flipped **arrow 4** result (dotted lines with "X") with **H'** (solid lines with "O") as shown in right graph.

As we proceed carefully along both the upper and the lower paths we will now see the alias cancellation in action (for this simple delta function signal). Because we will be summing the correlations on the upper and lower path, we present them here simultaneously point by point. In other words, **D1** and **A1** from the highpass and lowpass paths are summed point by point and we

* It turns out we could keep either the odd or even values if we do this on both the upper and lower paths. This has to do with *shift invariance* which we will discuss later.

will show how the sums cancel except when they are lined up and thus provide perfect reconstruction.

We slide the downsampled-upsampled-flipped vectors along to where they begin to overlap with the reconstruction filters **H'** and **L'** as shown in Figure 12.1–3. The correlation value of the vectors on the left and the correlation value of the vectors on the right are both zero because **0** x **c1** = **0**; **0** x **b1** = **0** and for this trivial case we have cancellation (**0** + **0** = **0**).

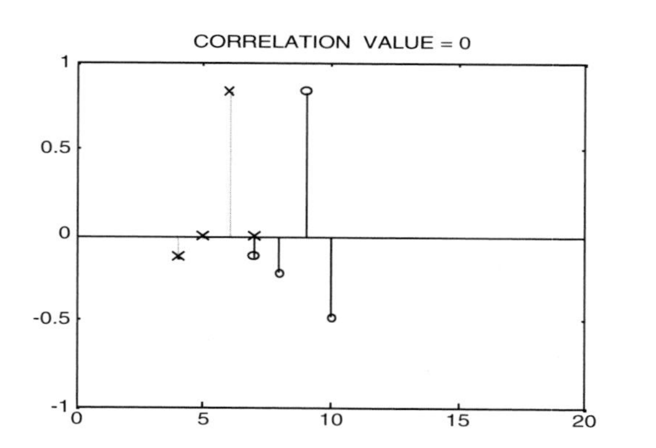

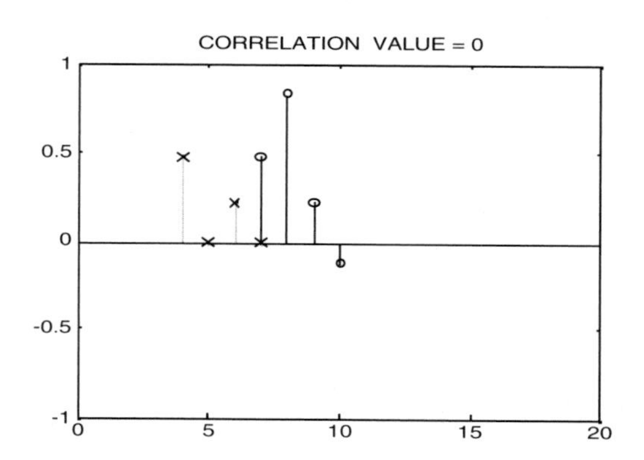

Figure 12.1–3 As we slide the vector (**c0**, **0**, **c2**, **0**) along the upper path (left graph) until they first overlap with **H'** (**c0**, **c1**, **c2**, **c3**) we have **0** times **c2** which produces **0**. The right graph shows the same overlap with **0** times **b2** which also produces **0**. The values thus cancel

As we move along another point we have our first non-trivial dot (inner) product as shown in Figure 12.1–4

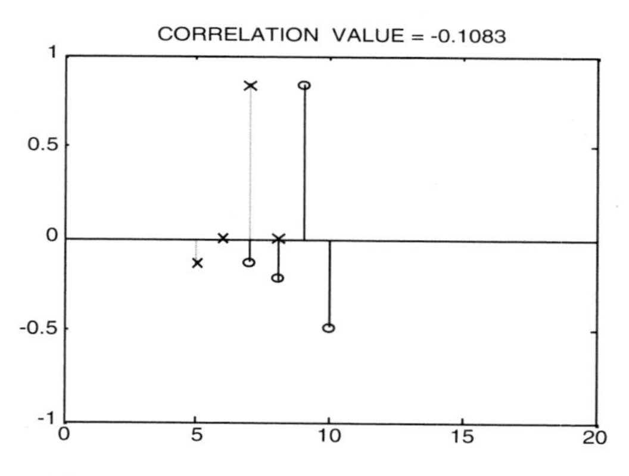

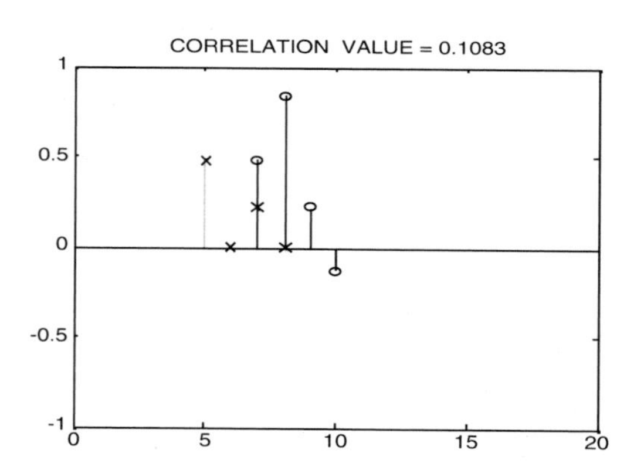

Figure 12.1–4 As we slide the vector (**c0**, **0**, **c2**, **0**) along the upper path (left graph) until the 2nd overlap with **H'** (**c0**, **c1**, **c2**, **c3**) we have **c0c2** (**c0** times **c2**) which produces –**0.1083**. The right graph shows the same overlap with **b0b2** which produces +**0.1083**. The values cancel.

As we move along the path by another point the correlation values cancel. Furthermore, we can literally *see* the cancellation as shown in Figure 12.1–5.

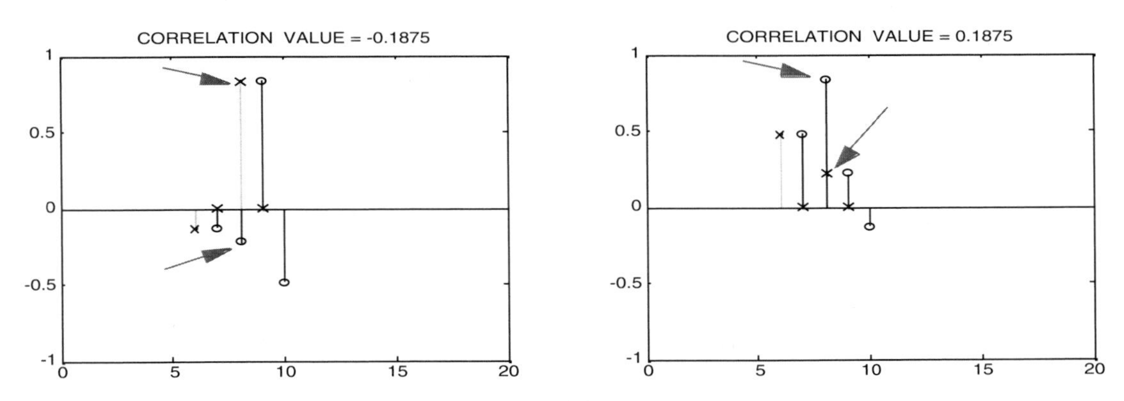

Figure 12.1–5 As we slide the vector (**c0, 0, c2, 0**) one step further we now have **c1c2** = **–0.1875**. Right graph shows the same overlap with **b1b2** = **+0.1875**. The values again cancel.

Recalling that we set **b0** = **–c3**, **b1** = **c2**, **b2** = **–c1** and **b3** = **c0** we then have **b1b2** = **c2(–c1)** = **–c2c1** which cancels **c1c2**.*

Continuing our correlation, we see that the vectors now completely overlap **H'** and **L'** as shown in Figure 12.1–6. The correlation values, however do *not* cancel but instead sum to **1.0**.

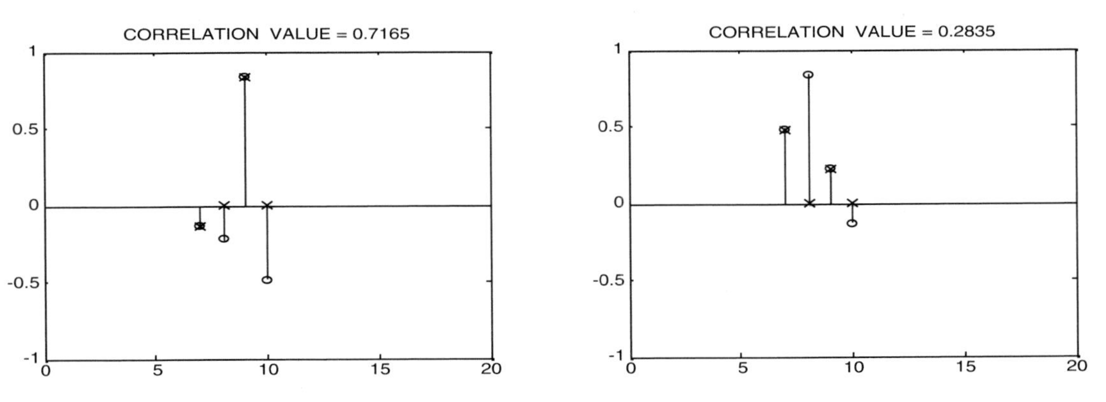

Figure 12.1–6 As we slide the vector (**c0, 0, c2, 0**) to where it completely overlaps **H'** (left graph) we have **c0c0** + **c2c2** which produces **+0.7165**. The right graph shows the same overlap with **b0b0** + **b2b2** which produces **+0.2835**. The values now sum to **1.0**.†

* We can also show that the preceding (first) overlap also cancels by *orthogonality* by substituting **(c3)(-c1)** = **c1c3** for **b0b2**. These 1/2 integer apart filters are "integer orthonormal" That is, the dot product of a filter with an integer-shifted version of itself is zero and thus **c0c2** and **c1c3** cancel.

As we continue the correlation the lower and upper paths continue to cancel twice more until the vectors and the filters no longer overlap and we have zeros again (not shown).

To review, we had cancellation at the first 3 overlaps and a summation of **1.0** at the 4th overlap when the vectors and filters were aligned. This means that as we added together the **D1** and **A1** components in the conventional DWT with **S** = [**1**] we obtained a value of

```
S' = [0 0 0 1 0 0 0].
```

Thus we have our original signal reconstructed to within a delay of 3.

We note in passing that for a signal twice as large

```
S = [2]
```

we would still have perfect reconstruction with S' = [**0 0 0 2 0 0 0**].

For a signal delayed in time by 1 (**S** = [**0 1**] we must now look a little further into the process. Looking again at Figs. 12.1–1 and 12.1–2 we would have for this new "delayed delta" signal at **cD1** the convolution with the highpass decomposition filter **H**. In other words

```
S*H = [0 1]*[c3 c2 c1 c0] = [0 c3 c2 c1 c0]
```

As we downsample, again keeping the even values we have

```
cD1 = [c3 c1]
```

Note: This "delayed delta" case gives us the same result at this point as if we had kept the *odd* values from the original (**S** = [**1**], **cD1** = [**c2 c0**]) case.

Upsampling the data we have **c3, 0, c1, 0** at **arrow 2**. We need to adjust for the "delayed delta" and so we the result at **arrow 2** as

```
(arrow 2 result) = 0, c3, 0 c1, 0.
```

As we convolve with **H'** we can express this as the *correlation* of

```
0, c1, 0, c3, 0 and H' = (c0, c1, c2, c3).
```

In other words for the "delayed delta' signal we have

---

† Substituting back **–c3** for **b0** and **–c1** for **b2** we have the sum of both paths as **c0c0 + c2c2 +** (**–c3**)(**–c3**) + (**–c1**)(**–c1**) = $\mathbf{c1}^2 + \mathbf{c2}^2 + \mathbf{c3}^2 + \mathbf{c4}^2$ = 1.0, which is another property of Perfect Reconstruction Quadrature Mirror Filters—the sum of the squares is a constant (usually 1.0, if the filters have not been pre-divided).

```
D1 = xcorr({c0 c1 c2 c3], [0 c1 0 c3 0])
```

This is depicted in Figure 12.1–7 at left. Note that the trailing zero is not shown.

For the lowpass path we have the "delayed delta" signal the similar results

```
cA1 = [b3 b1]

A1 = xcorr([b0 b1 b2 b3]), [0 b1 0 b3 0])
```

This process is depicted in the right graphic of Figure 12.1–7. Again, the zero value after **b3** is not shown.

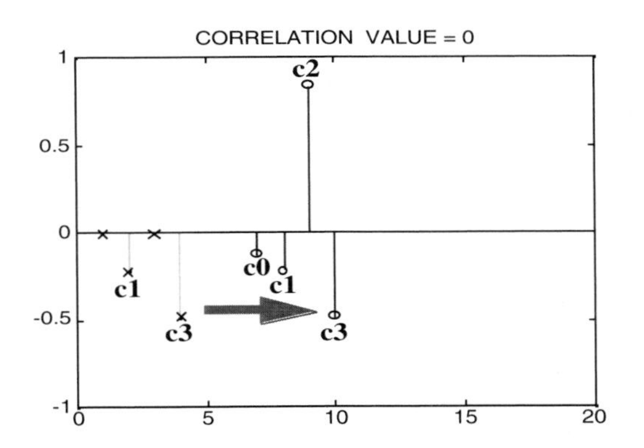

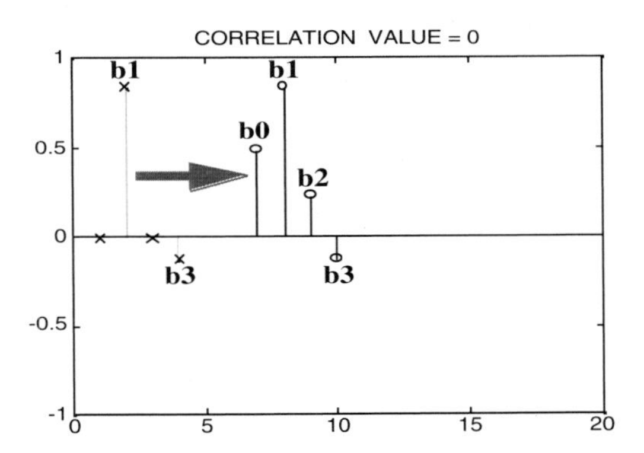

Figure 12.1–7 For "delayed delta signal, **S**, the convolution of **arrow 2** results with **H'** is same as *correlation* of flipped **arrow 2** result (dotted lines with "X") with **H'** (solid lines with "O") as shown in left graph. This produces **D1**. Right graph shows similar convolution for the lowpass side to produce **A1**.

As with the earlier non-delayed case, the first overlap will be zero

```
0 x c0 = 0; 0 x b0 = 0
```

thus the correlation values will be zero for both the highpass and lowpass paths and they will obviously cancel (trivial case not shown—ref. Fig. 12.1–3 for the non-delayed signal).

The first non-trivial overlap (2nd overlap) is shown in Figure 12.1–8 for both the highpass and lowpass paths. Notice that the correlation values cancel. We can also visually observe this cancellation between the 2 graphs (arrows).

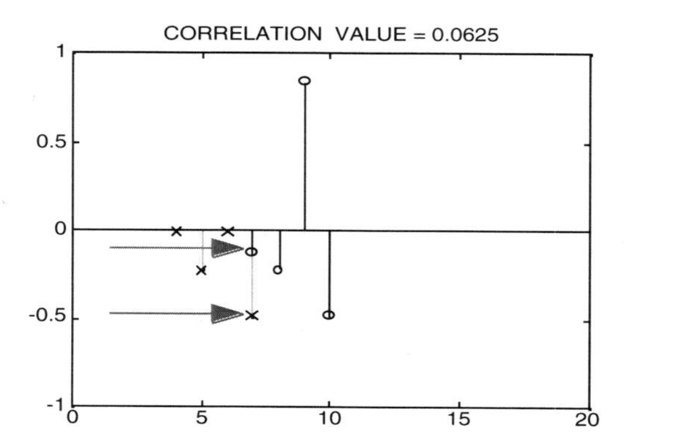

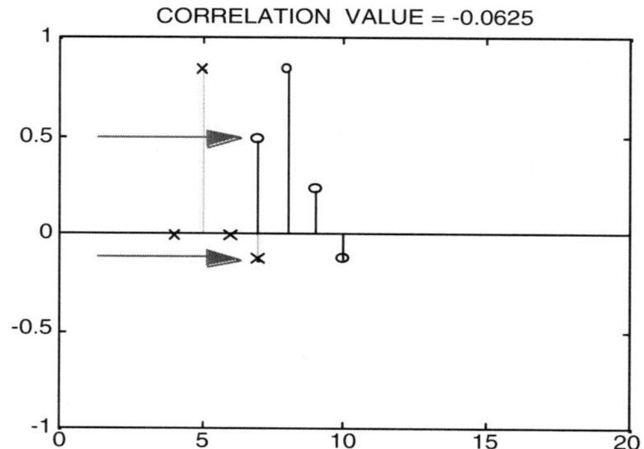

Figure 12.1–8 As we slide the vector **0, c1, 0, c3, 0** (trailing zero not shown) along the upper path (left graph) until we have a non-trivial over lap with **H'** (**c0, c1, c2, c3**) we have **c1c3** which produces +**0.0625**. The right graph shows the same overlap with **b1b3** which produces –**0.0625**. The values cancel for this 2nd overlap.

The 3rd overlap (not shown) produces +**0.1083** and –**0.1083** and thus also cancels. The 4th overlap is shown in Figure 12.1–9

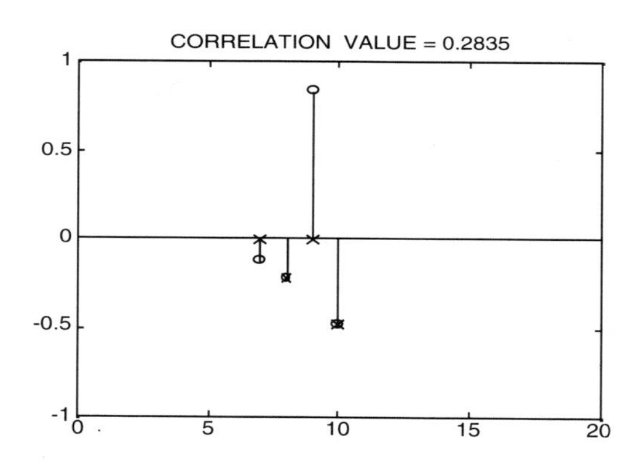

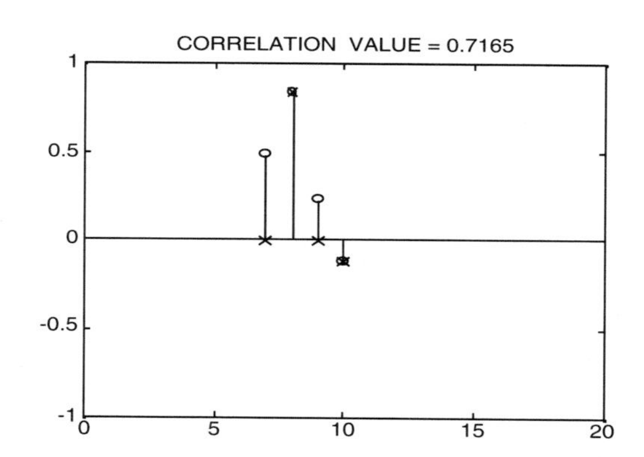

Figure 12.1–9 As we slide the vector (**0 c1, 0, c3, 0**) to where it completely overlaps **H'** (left graph) we have **c1c1** + **c3c3** which produces +**0.2835**. The right graph shows the same overlap with **b1b1** + **b3b3** which produces +**0.7165**. The values now sum to **1.0** for this 4th overlap.

Proceeding with the correlation the remainder of the overlaps cancel. We thus have for our overlap values **0 0 0 1 0 0 0**. However, we recall that we had that extra delay caused by the delayed delta and thus for **S** = [**0 1**] we have **S'** = [**0 0 0 0 1** . . .]. Like the earlier **S** = [**1**] case we have perfect reconstruction, but this time to within a delay of 4, rather than 3. We also note that if we had **S** = [**0 3**] we would have **S'** = [**0 0 0 0 3** . . .]

For a Kronecker delta signal delayed by **2** (**S** = [**0 0 1**]) convolved with **H** (**arrow 1**, Fig. 12.1–1) we have **0, 0, c3, c2, c1, c0**. As we downsample we have **0, c2, c0** and when we upsample by 2 we have at **arrow 2** the values **0, 0, 0, c2, 0, c0**. But this is the same as the original case (**S** = [**1**]) except for a delay of 2! Similarly we will have at **arrow 4** on the lowpass path **0, 0, b2, 0, b0** which is the same as the original case except for the same delay of 2. Thus for **S** = [**0 0 1**] we will have perfect reconstruction with **S'** = [**0 0 0 0 0 1 . . .**].

In a similar fashion we can show that a delta signal delayed by 3 produces the same results as the 2nd case except for an *additional* delay. In other words, for **S** = [**0 0 0 1**] we have **S'** = [**0 0 0 0 0 0 1 . . .**]. It turns out that any delay of the Kronecker delta as the original signal will simplify to either the even or the odd downsample with appropriate delays.

Thus for *any* delayed signal we will have perfect reconstruction. A signal **S** = [**7 8 9**] for example can be re-written as

```
S = 7x{1] + 8x[0 1] + 9x[0 0 1]
```

which will produce

```
S' = 7x[0 0 0 1] + 8x[0 0 0 0 1] + 9x[0 0 0 0 0 1]
```

which is equal to

```
[0 0 0 7 8 9]
```

In other words the signal will be perfectly reconstructed to within a delay of 3. This means that the single-level conventional DWT is a *Linear Time Invariant (LTI) system.*

Jargon Alert: **LTI means Linear, Time-Invariant. Time-Invariant is sometimes called "Shift Invariant". If input x(n) produces output y(n) in an LTI system then c1x1(n) + c2x2(n) produces c1y1(n) + c2y2(n) where c1 and c2 are constants. (Linear property). Also x(n+k) produces y(n+k) (Time or Shift Invariant) where k is a delay in the input. The LTI property is not in general true for downsampling, but *is* true for these DWT filter banks with Perfect Reconstruction Quadrature Mirror Filters.**

To sum up, we have not *proved* linear time invariance of the conventional DWT, but have *demonstrated* it for the Db4 filters and can do so with other filters. A word of caution is in order here—we showed perfect reconstruction and alias cancellation because the upper and lower paths of the conventional DWT canceled point-by-point except when aligned. In other words, for **S** = [**1**] we had

```
D1 = [0 -0.1083 -0.1875  0.7165 -0.375 -0.1083  0.0625]
A1 = [0  0.1083  0.1875  0.2835  0.375  0.1083 -0.0625]
```

and when we added them together we obtained

```
S' = D1 + A1 = [0 0 0 1 0 0 0]
```

Similarly, for **S** = [**0 1**] we had

```
D1 = [0  0.0625  0.1083 -0.375  0.2835 -0.1875  0.1083]
A1 = [0 -0.0625 -0.1083  0.375  0.7165  0.1875 -0.1083]
```

and when we added them together we obtained

```
S' = D1 + A1 = [0 0 0 0 1 0 0]
```

Thus shifting (delaying in time) the signal **S** by one merely produced the same signal with a delay of 4 instead of 3 for **S'**. If we were to set **D1** to zero, however, as we did in the last example in the previous chapter, we would have for **S** = [**1**] the result

```
S' = D1 +A1 = 0 + A1 = A1
   = [0 0.1083  0.1875  0.2835  0.375  0.1083 -0.0625]
```

and for the delayed signal **S** = [**0 1**] the result

```
S' = [0 -0.0625 -0.1083  0.375  0.7165  0.1875 -0.1083]
```

In other words, when we remove **D1**, delaying or shifting (in time) the original signal by one not only does not provide perfect reconstruction, but also does *not* simply delay the outcome by one (**S'** = [**0 0 0 1 0 0 0**] to **S'** = [**0 0 0 0 1 0 0**]). The conventional DWT with de-noising or compression (which removes or thresholds away portions of the subbands) is thus not *shift invariant.* This is why the (conventional, downsampled) DWT is sometimes called a *shift-variant transform* while the Undecimated DWT is sometimes called a *shift-invariant wavelet transform.*

12.2 DWT Alias Cancellation Demonstrated in the Frequency Domain.

We begin with a short tutorial review from Digital Signal Processing. A simple cosine signal sampled at Nyquist is shown in Figure 12.2–1. Here the entire signal is simply

```
S = [1 -1  1 -1  1 -1  1 -1  1 -1  1 -1  1 -1  1 -1]
```

and in the frequency domain (taking the FFT) we have a single point at Nyquist as shown.*

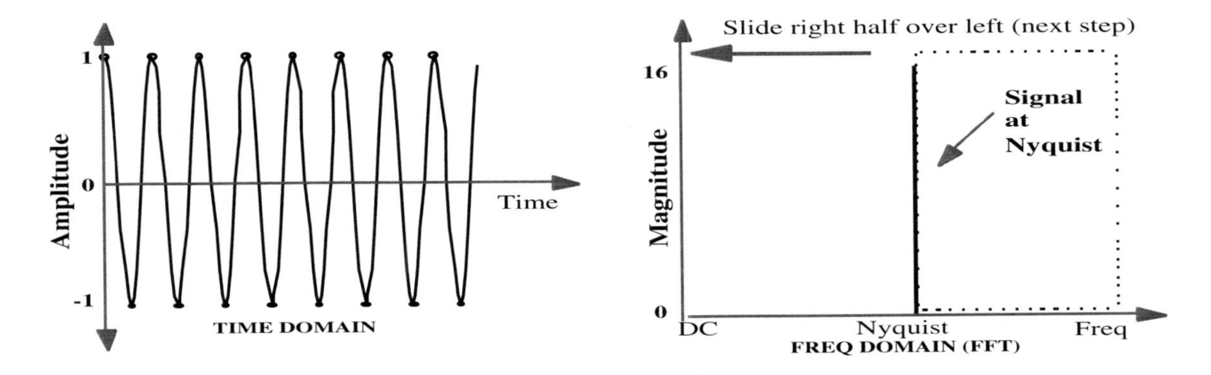

Figure 12.2–1 A signal sampled at Nyquist is shown in both the time and frequency domains. Note the 16 samples (circles) on the left graph are at the top and bottom of each cycle. We can visualize the next step of downsampling by 2 as "sliding" the right half of the frequency chart over the left half.

If we downsample by 2, as we would do in a conventional DWT, we would have the signal as shown in Figure 12.2–2. Our downsampled signal is now a constant value in the time domain In other words we have

```
Sdown = [ 1  1  1  1  1  1  1  1]
```

* This is not generally recommended in practice. Instead of acquiring the values at the peaks we could be out of phase and acquire lesser values—including all zeros if we were 90 degrees out of phase. We should sample more frequently than in this tutorial case.

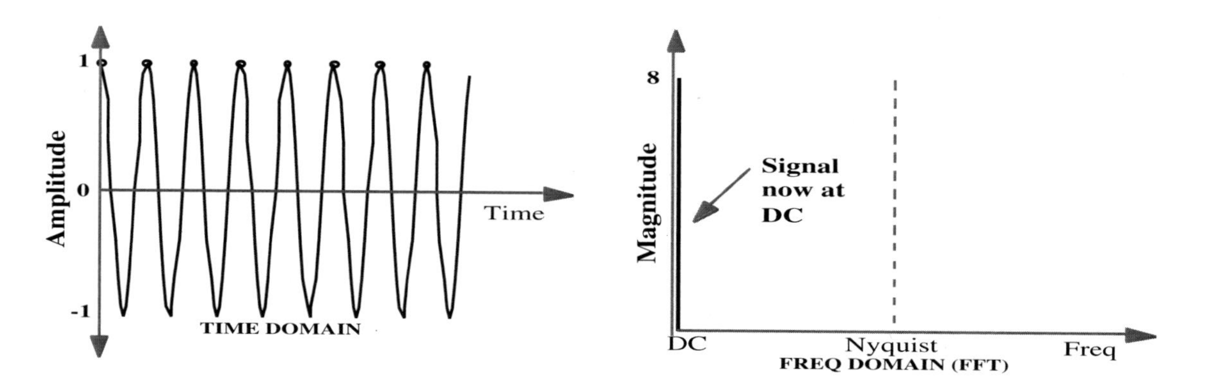

Figure 12.2–2 As we downsample (or "decimate") by 2 we are left with only 8 points on the peaks of the left drawing. These represent a constant value or "D.C.". This can be seen in the frequency domain representation (right graph) with the signal slid all the way to the left. Note that the magnitude of this smaller signal is now 8 instead of 16. Note also that the Nyquist frequency will change for the downsampled signal but the depiction is still valid.

and in the frequency domain the signal has been aliased all the way to DC (zero frequency). For these simple cosine signals the phase is constant at zero and is not shown as part of the tutorial.

The next step in a conventional DWT would be to upsample as shown in Figure 12.2–3. This means placing a zero between the existing points and/or adding a leading or trailing zero. The downsampled-then-upsampled signal now becomes in the time domain

```
Sdownup = [1 0 1 0 1 0 1 0 1 0 1 0 1 0 1 0]
```

and in the frequency domain we have components at both Nyquist *and* at DC. We are now sampling at the same rate as in the original signal and the Nyquist frequency is now the same as the original signal. This makes sense when we consider that the signal looks like the original signal except that it has a DC bias.

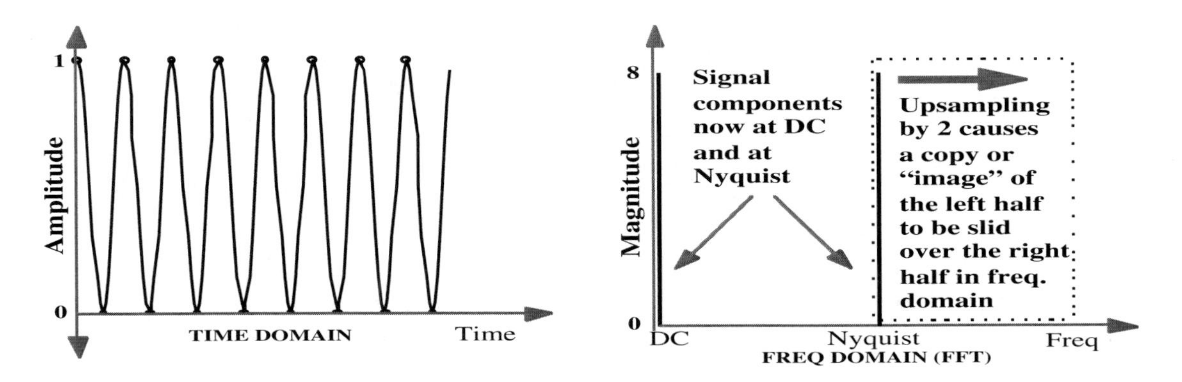

Figure 12.2–3 As we upsample by 2 we have 16 points again but now they alternate between 1 and zero. The FFT at right shows an "image" of the left part of the signal slid over to the right half.

Having completed this tutorial review, we now return to our discussion of alias cancellation demonstrated in the frequency domain. As we did in the last section (ref. Fig. 12.1–1) we begin with a Kronecker delta as our signal. In other words

```
S = [1]
```

We then filter by the (Db4) highpass decomposition filter **H**. In the frequency domain the Kronecker delta is a constant for all frequencies as shown in Figure 12.2–4 at left . The Highpass decomposition filter, **H**, is shown at right in the frequency domain.

We recall from DSP studies that convolution in the time domain means *multiplication* in the frequency domain. Thus when we multiply the highpass filter **H** by the constant **1.0** from the delta we have **H** unchanged and the right half of Fig. 12.2–4 also depicts the result at **arrow 1** in the DWT.

Continuing on the upper path of the single-level DWT we downsample by 2 to obtain **cD1**. As we showed in the tutorial review, downsampling by 2 has the effect of sliding the right half over the left half in the frequency domain.

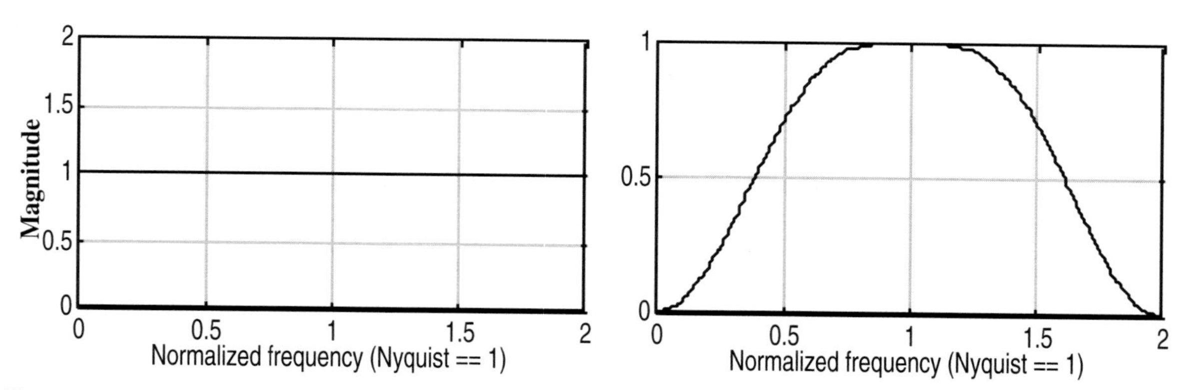

Figure 12.2–4 The Kronecker delta has a constant value of 1 in the frequency domain as shown at left. The highpass decomposition filter, **H**, has the frequency response shown at right.

The first graph of Figure 12.2–5 shows the one-point signal after filtering and downsampling to produce **cA1**. Rather than add the aliased part to the original signal we have drawn the magnitudes of the signal and the alias separately. These are now complex signals and we will shortly deal with the phases.* As we did in the tutorial example, we now upsample by 2. This produces the result shown in right graph of Figure 12.2–5. This is the result in the frequency domain at **arrow 2** in Fig. 12.2–1.

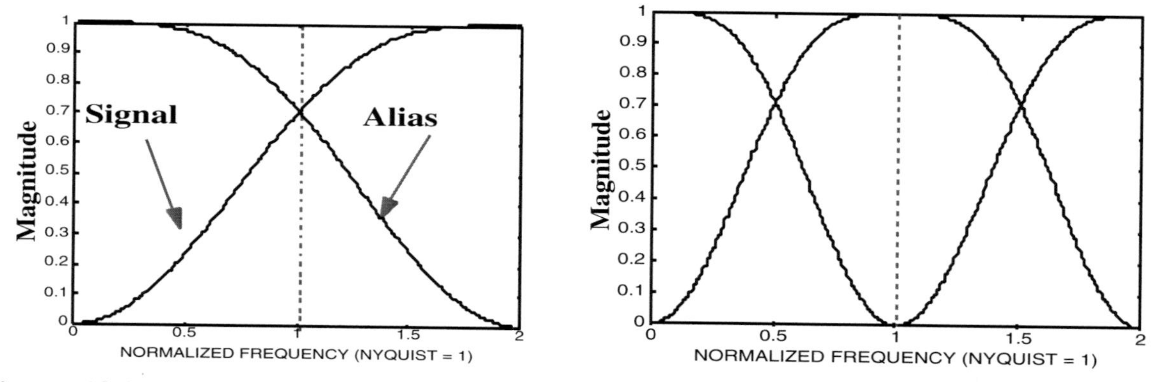

Figure 12.2–5 As we downsample by 2, as shown in the first graph, the right half is slid over the left half and we have aliasing. As we upsample this result by 2 we slide the "image" in the left graph to the right to create the right graph. The magnitudes of the signal and the alias are overplotted on both graphs to keep them separate for now.

* It should be noted that although the aliased points "add back in", with complex numbers they may add, subtract or otherwise change magnitude. For example **1+j** and **1_j** each have a magnitude of sqrt(2) but if we add them together we get a magnitude of **2.0**, not **2sqrt(2)**. Thus, instead of combining the 1st half of the signal with the aliased portion, we keep them separate for now.

The final step to produce **D1** is to multiply in the frequency domain (convolve in the time domain) by **H'**. The magnitude of **H'** is shown below in Figure 12.2–6 at left.* Both the signal portion and the alias portion will be multiplied by **H'**. The result is shown below in the right graph. Notice how both the signal portion and the alias portion are attenuated by this highpass filter.

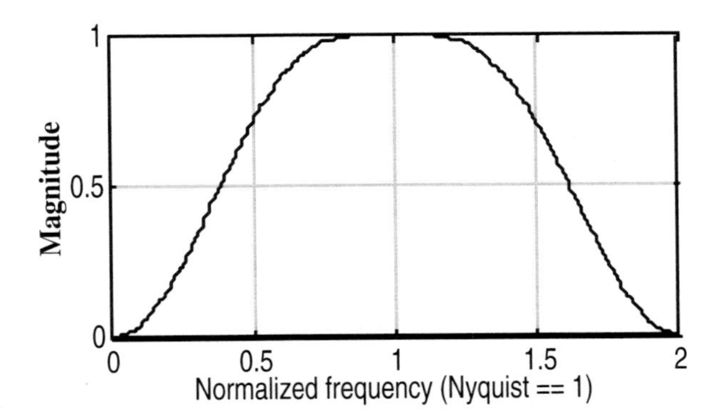

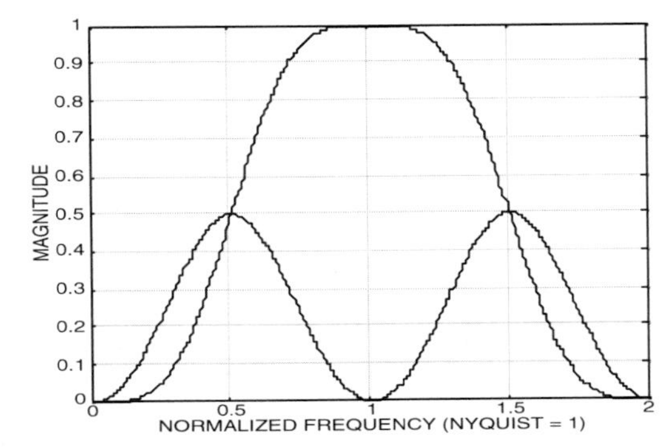

Figure 12.2–6 H' in the frequency domain is shown at left. Multiplying the result at **arrow 2** by H' produces this final value for **D1** as seen at right. Again, only the magnitudes of the signal and alias are shown for now.

We must digress for a moment to recall the single-level Undecimated DWT (ref. Fig. 11.5–6). For our Kronecker delta signal we have for the UDWT **D1** simply the convolution of H and H' in the time domain or the product in the frequency domain. We recall that the convolution of H and H' was the highpass halfband filter (ref. Fig. 8.3–1).

Thus for the single-level UDWT, **D1** is simply the highpass halfband filter as shown in the left of Figure 12.2–7. Comparing with the (conventional, decimated) DWT results in the previous figure we can see that the taller curve is the signal and the lower "humps" are the alias contribution as indicted in the right graph.

* The magnitudes of **H** and **H'** are identical for the Db4, but the phases are different.

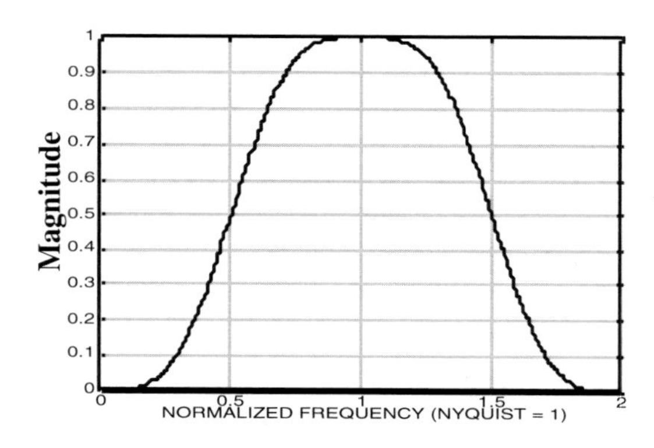

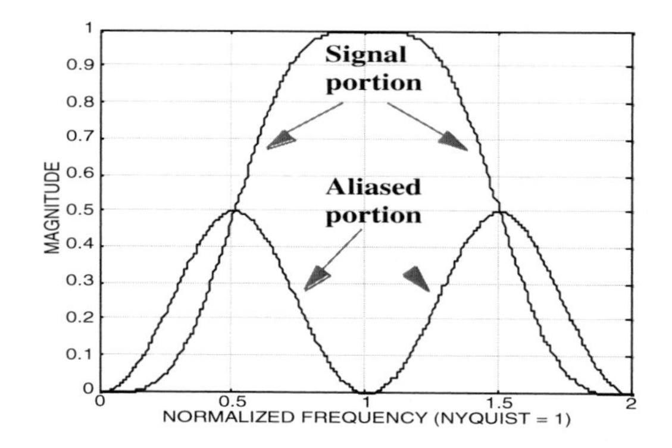

Figure 12.2–7 Db4 highpass halfband filter from a Db4 UDWT is shown at left. Notice the symmetry at Magnitude 0.5. We see the same symmetry in the *signal* portion of the *conventional* DWT result shown again at right.

Following our single-point signal through the *lower (lowpass) portion* of the conventional DWT we obtain the final result at **A1** as shown in the left graph of Figure 12.2–8. The UDWT lowpass path produces a *lowpass* halfband filter for **A1** as shown in the right graph. Note again the signal portion of the conventional DWT matches the lowpass halfband filter produced by the UDWT.

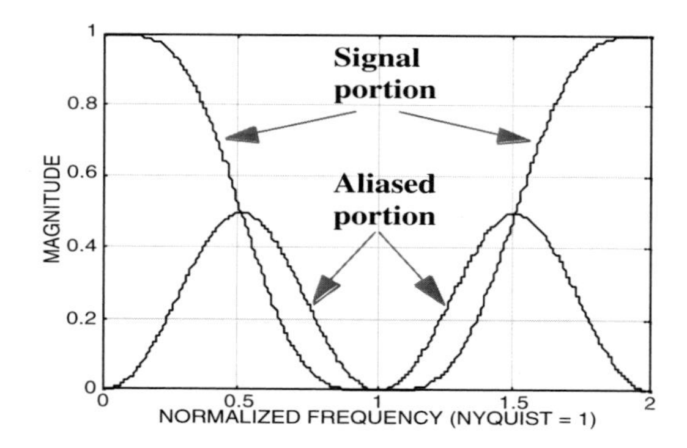

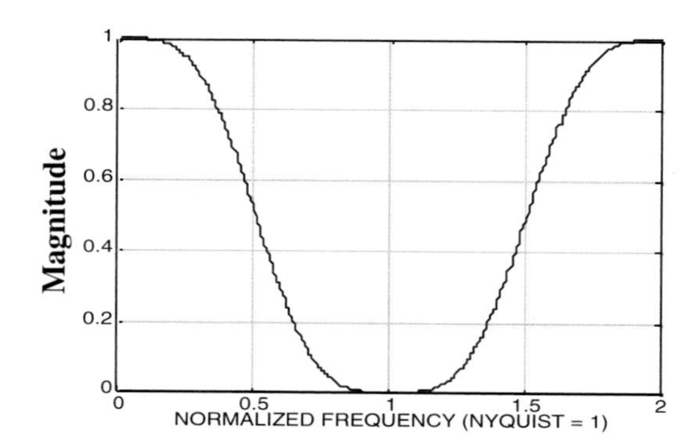

Figure 12.2–8 The end result, **A1**, on the *lowpass path* is shown at left for the conventional DWT. The signal portion replicates a lowpass halfband filter (UDWT **A1**) as shown at right.

Our next step is to show that the *signal* portions of **A1** and **D1** from the conventional DWT add to 1 and that the *aliased* portions cancel. To do this we must look at the *phase*. Actually we are most interested in the *phase difference*.

A very short tutorial review of complex* numbers is in order to clarify the addition of the signals and cancellation of the aliased portions. At the left of Figure 12.2–9 we have 2 complex vectors: [**2** + **2j**] and [–**2** –**2j**]. Because they are 180 degrees (pi radians) apart and are the same magnitude they will cancel.

We can observe that *any* 2 complex numbers having the same magnitude but a phase difference of 180 degrees will cancel. Note that a phase difference of –180 degrees is the same as +180 degrees.

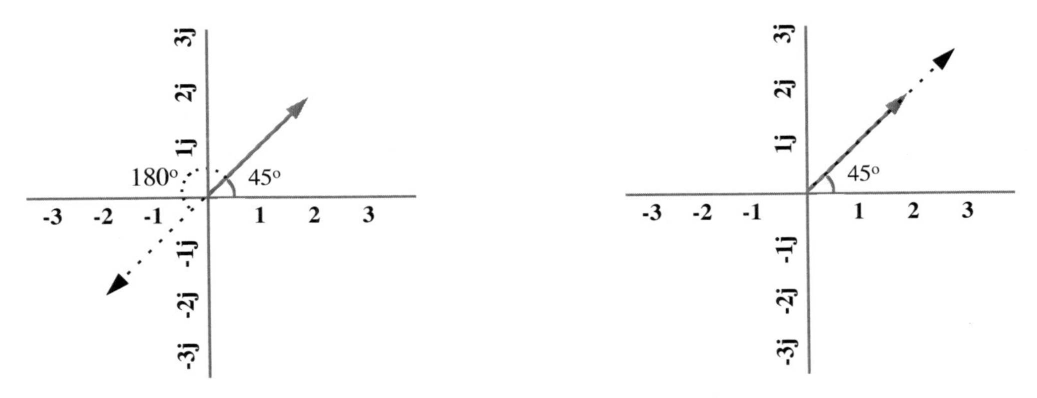

Figure 12.2–9 The left graph illustrates how complex numbers of the same magnitude that are 180 degrees out of phase will cancel. The right graph illustrates how complex numbers that are in phase add their magnitudes. The 45 degree phase angle is used here for simplicity but these principles work for *any* phase angle.

The right graph of Figure 12.2–9 shows 2 vectors "in phase" or with a phase difference of zero. The 2 complex vectors [**2** + **2j**] and [**3** + **3j**] add directly to produce [**5** + **5j**], whose magnitude is the sum of the magnitudes of the 2 vectors.

Returning to our conventional DWT example, we look at the magnitude and the phase angle (degrees) for the alias portion of **D1** as shown below in Figure 12.2–10.

* Complex doesn't necessarily mean *complicated.* The dictionary definition of complex is either (1) hard to analyze, or solve; or (2) composed of two or more parts—such as a *complex* of buildings. We are talking about real and imaginary parts of a vector so we use the latter definition.

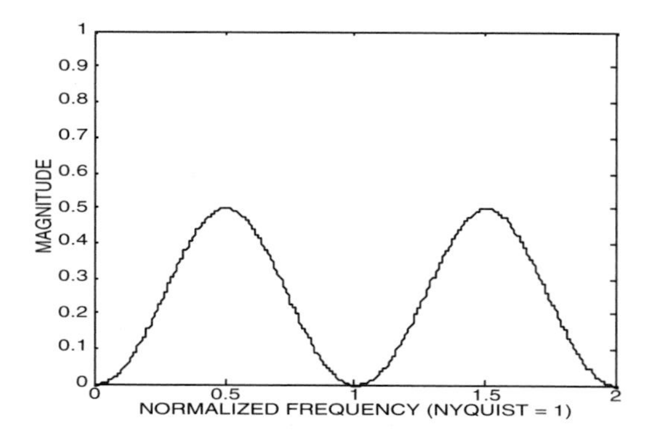

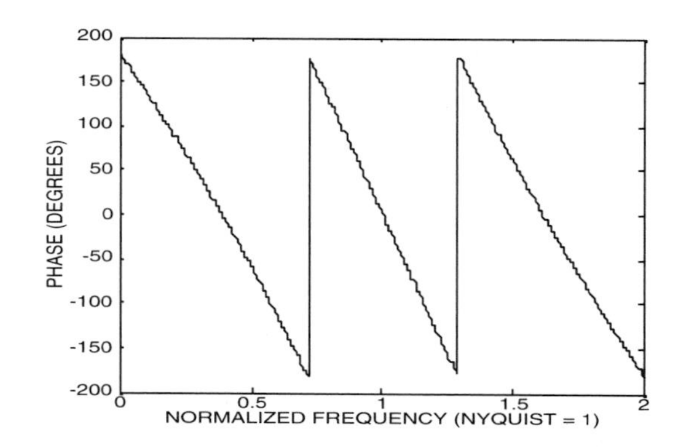

Figure 12.2–10 The left graph shows the magnitude of the aliased portion of **D1**. The right graph shows the phase angle. Note the phase is linear and that as it "jumps" from –180 to +180 degrees it is still linearly decreasing.

We now compare the aliased portion of the lowpass path leading to **A1**. Again the magnitude and phase are shown in Figure 12.2–11 below. We notice that the *magnitudes* of the aliased portions of **D1** and **A1** are identical—they would have to be in order for them to cancel!

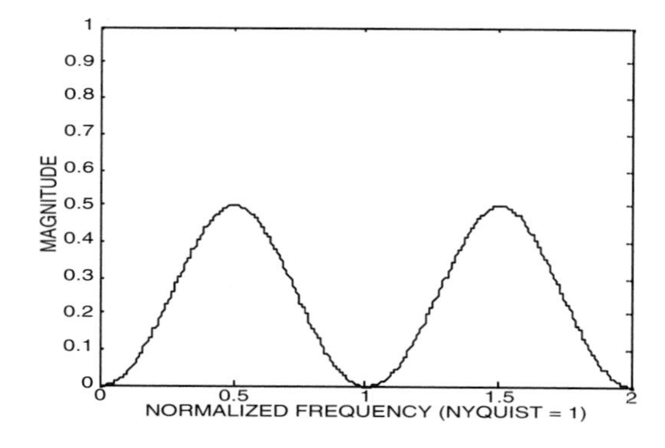

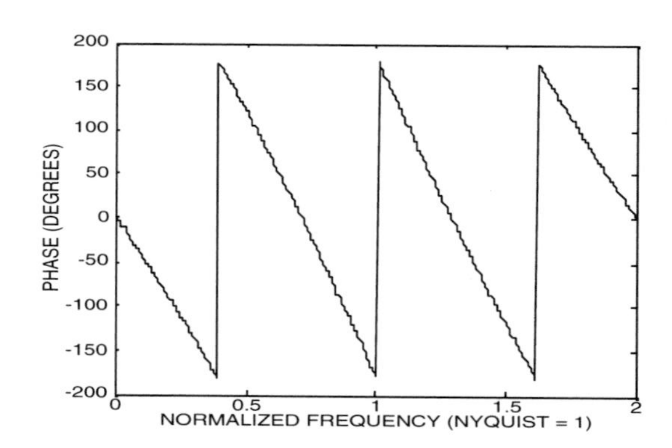

Figure 12.2–11 The left graph shows the magnitude of the aliased portion of **A1** and is identical to the **D1** alias magnitude in the previous figure. The right graph shows the phase angle. and is *not* the same as the **D1** alias phase in the previous figure.

If we now subtract the phase angles of the aliased portions of **A1** from those of **D1** we have the result as shown in the left graph of Figure 12.2–12. Knowing from our tutorial review a moment ago that 2 complex numbers of the same magnitude cancel when they are out of phase by +180 degrees *or* –180 degrees (+π or –π) we see that the aliased portions cancel completely.

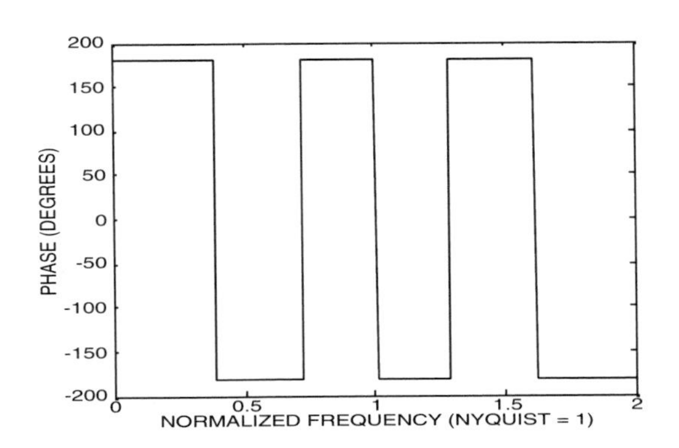

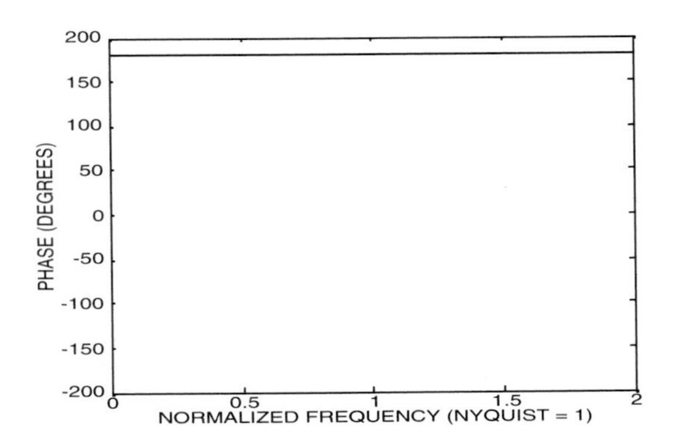

Figure 12.2–12 Left graph shows the phase *difference* between the alias portions of **D1** and **A1**. Right graph shows the absolute value of the phase difference as a constant 180 degrees.

We next look at the signal portions of **D1** and **A1**. We saw earlier that they were highpass and lowpass halfband filters (ref. 12.2–7 and 12.2–9). Figure 12.2–13 below shows the phase of the signal portions of **D1** and **A1**.

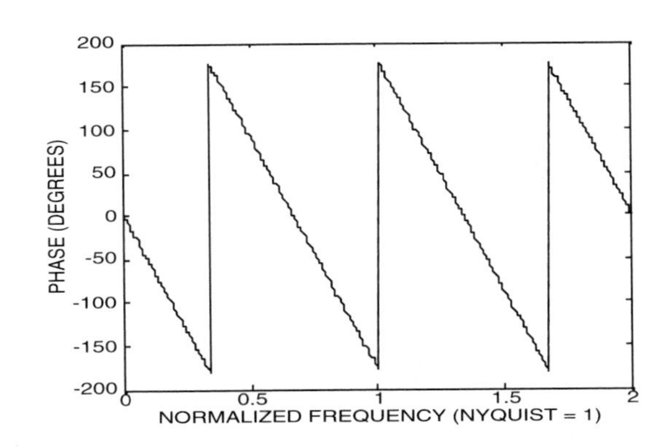

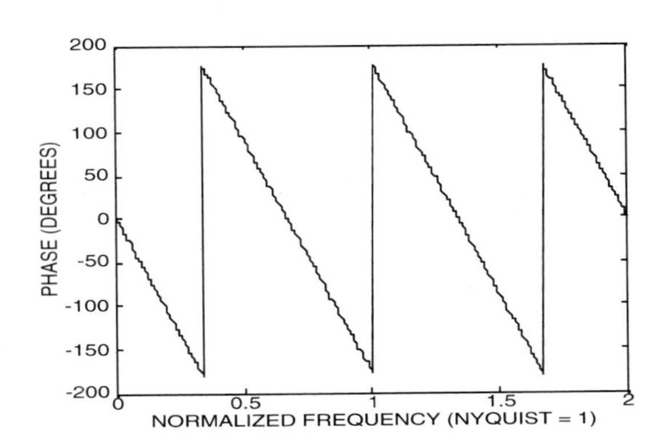

Figure 12.2–13 The left graph shows the phase of the signal portion of **D1**. The (identical) right graph shows the phase of the signal portion of **A1**.

We can see that **A1** and **D1** are in phase. With a zero phase difference the magnitudes add directly. Figure 12.2–14 shows the magnitude of **A1** overplotted with the magnitude of **D1**. It appears they would add to a constant value of 1.0. The right graph shows that this is indeed the case.

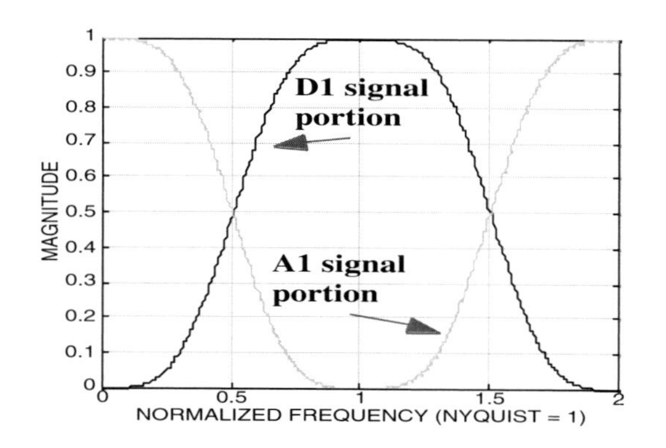

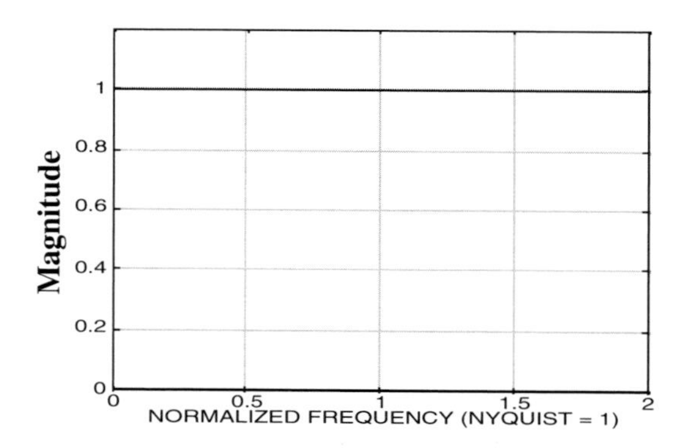

Figure 12.2–14 The left graph shows the magnitude of the signal portion of **D1** (solid line) and the magnitude of the signal portion of **A1** (dotted line). The (right graph shows the summation of these 2 halfband filters.

This is expected with summation of the lowpass and highpass halfband filters. The Kronecker delta has a similar frequency response (ref. Fig. 12.2–4) with a constant magnitude of 1.0 but a constant phase of zero. A *delayed* delta [0 0 0 1] will have the same magnitude in the frequency domain, however the phase will be linear but not constant as shown in Figure 12.2–15.

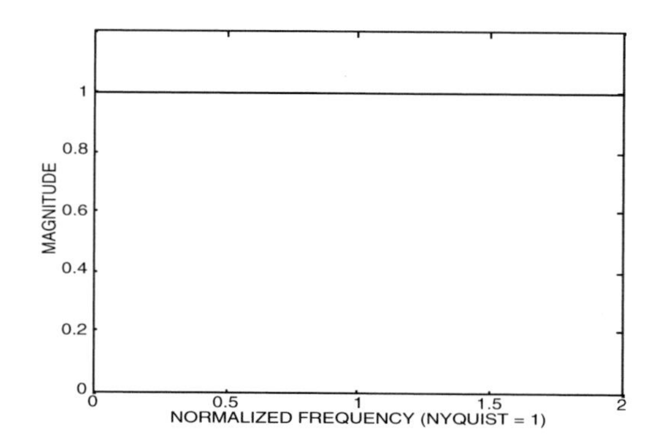

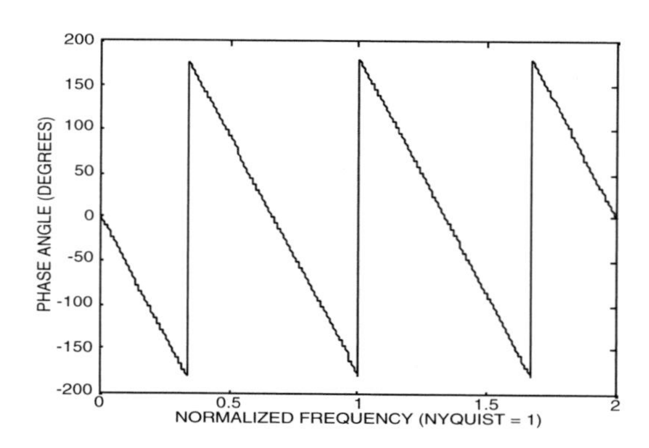

Figure 12.2–15 A Kronecker delta [1]delayed by 3 to become **[0 0 0 1]** has the constant magnitude and linear phase as shown here.

Another look at Figs. 12.2–13 and 12.2–14 will show that we have the identical magnitude and phase. In other words, we have shown again, using the frequency domain, that for the signal

```
S = [1]
```

that we have

```
S' = [0 0 0 1]
```

One more reminder is in order that if we chose to discard **D1** in a conventional DWT, as in the last example of the last chapter, our final result for **S'** will look like the left graph of Fig. 12.2–8. In other words we will have the desired lowpass filtering but will also have the aliasing as shown.

12.3 Relating the Above Concepts to Equations Found in the Traditional Literature

Traditional wavelet literature often shows the "*Alias Cancellation*" and the "*No Distortion*" Equations in the general form

$$F0(z)H0(-z) + F1(z)H1(-z) = 0$$

and

$$F0(z)H0(z) + F1(z)H1(z) =^{*} z^{-t}$$

Substituting our familiar terminology we have

```
L  = Lowpass Decomposition Filter (LOD) for H0
L' = Lowpass Reconstruction Filter (LOR) for F0
H  = Highpass Decomposition Filter (HID) for H1
H' = Highpass Reconstruction Filter (HIR) for F1
```

We recall from DSP studies that

$$z = e^{j\omega} = \cos(\omega) + j\sin(\omega)$$

and that

$$e^{j\pi} = \cos(\pi) + j\sin(\pi) = -1 + j\times 0 = -1$$

thus –z can be expressed as

$$-z = e^{j\omega}e^{j\pi} = e^{j(\omega+\pi)}$$

* The right side of the equation may be $= \mathbf{2z^{-t}}$ depending on the author. Once again, "Caveat Emptor").

We can now rewrite the equations for Alias Cancellation and No Distortion in more familiar terms (note that these equations are in the frequency domain and the terms are multiplied rather than convolved).

$$L'(\omega)L(\omega + \pi) + H'(\omega)H(\omega + \pi) = 0$$

and

$$L'(\omega)L(\omega) + H'(\omega)H(\omega) = e^{-j\omega t}$$

respectively, where **t** is the time delay. Note that the magnitude of $e^{-j\omega t}$ is **1.0** (the delay effects the linear phase but not the magnitude).

Looking at the *no distortion* equation (frequency domain) in more detail we notice that we have

L'(ω)L(ω) = L'L = lowpass halfband filter

and

H'(ω)H(ω) = H'H = highpass halfband filter.

Figure 12.3–1 shows the magnitude and the phase of the complex frequency domain result for the right side of the equation, $e^{-j\omega t}$, with a delay **t** = **3** (for the Db4 we are studying here) plotted below. Comparing with Fig. 12.2–13 through 12.2–15 above we see that the results are identical to those we obtained by adding together the signal portions of **D1** and **A1** (halfband filters) for the conventional single level DWT. Thus the *no distortion equation*

$$F0(z)H0(z) + F1(z)H1(z) = z^{-t}$$

is also telling us that we can have perfect reconstruction to within a time delay and a constant of multiplication. The constant in this case is simply **1.0** ($z^{-t} = 1z^{-t}$).

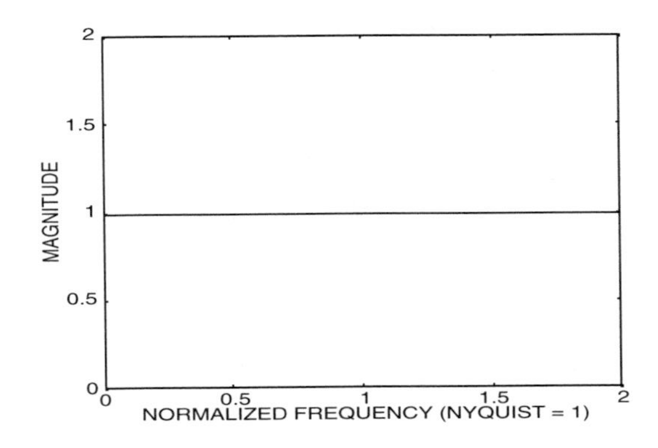

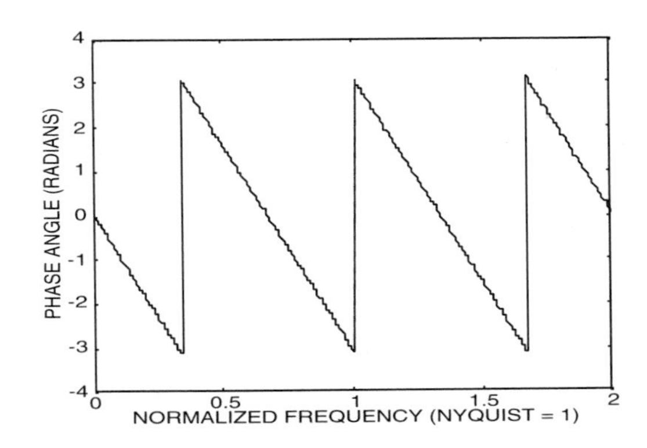

Figure 12.3–1 The expression $e^{-j\omega t}$ is plotted for a delay (t) of 3. The magnitude is shown at left and the phase angle *in radians* is shown at right. Note the phase wraps around π and −π.

We next examine the *alias cancellation* equation (expressed in the frequency domain and in familiar terms)

$$L'(\omega)L(\omega + \pi) + H'(\omega)H(\omega + \pi) = 0$$

First, we look at the magnitude and phase of L with the frequency expressed in radians as shown in Figure 12.3–2. Instead of normalized with Nyquist = 1, we use the raw value π **radians** as the Nyquist frequency.

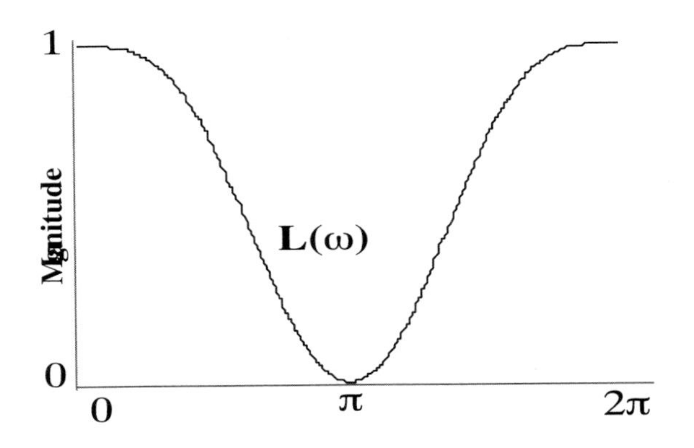

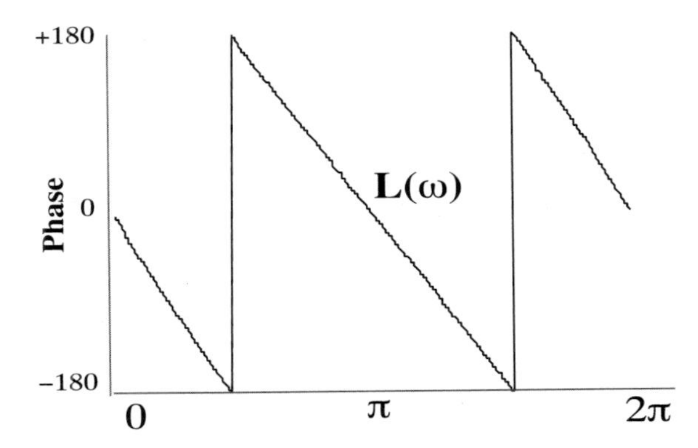

Figure 12.3–2 Magnitude of **L** is shown in the frequency domain with the frequency expressed as 0 to 2π. Nyquist is of course at π.

We recall from DSP studies that the FFT is a fast method of implementing the *Discrete Fourier Transform* (DFT) and that real-life digital computer data is *discrete* in both the time and frequency domains. This means that it can be treated as *periodic* in both domains. The frequency representation for L is now shown in Figure 12.3–3 as repeated (periodic) over *two periods* instead of the usual one period.*

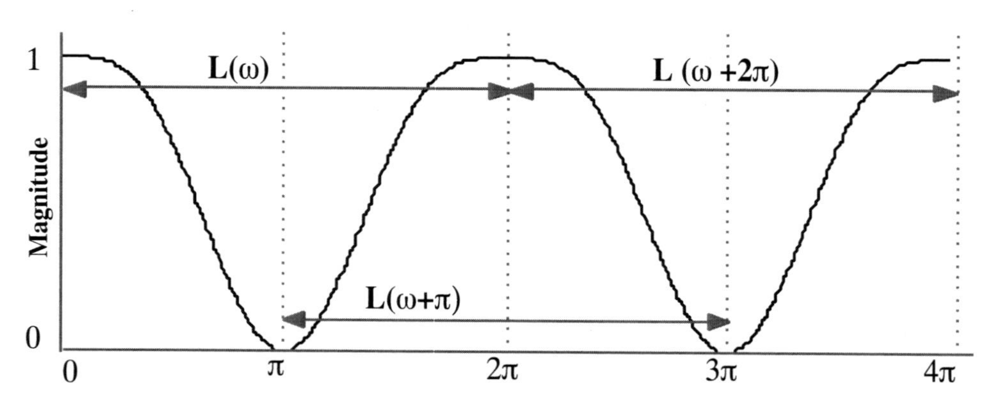

Figure 12.3–3 The magnitude of L is shown with the periodic frequency repeated twice (twice around the unit circle). We can see L(ω) repeated twice. We can also see L(ω+π).

Looking at the phase with L repeated twice (Figure 12.3–4) we notice the phase of L(ω+π) is the same as the phase of L(ω).

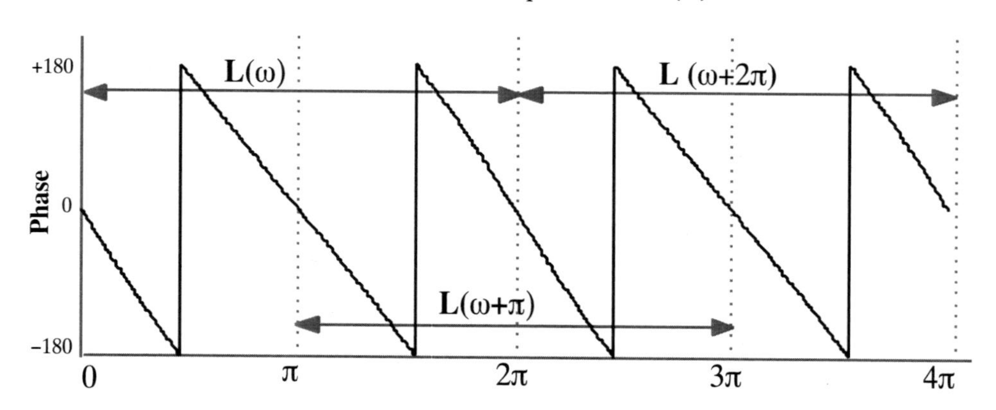

Figure 12.3–4 The phase of L(ω) is shown with the periodic frequency repeated twice. We also can see the phase angle of L(ω+π) and notice that it is in phase with L(ω)

* We could have used any number of periods to show this or we could have shifted within a single period but this method seems the simplest and clearest.

From the above figures we see that the magnitude and phase of L(ω+π) look familiar. Figure 12.3–5 shows the magnitude and phase for H'. We notice that because of the *Quadrature Mirror* attributes of these filters (ref. Fig. 8.1–1) that H'(ω) is identical to L(ω+π) in both magnitude and phase.

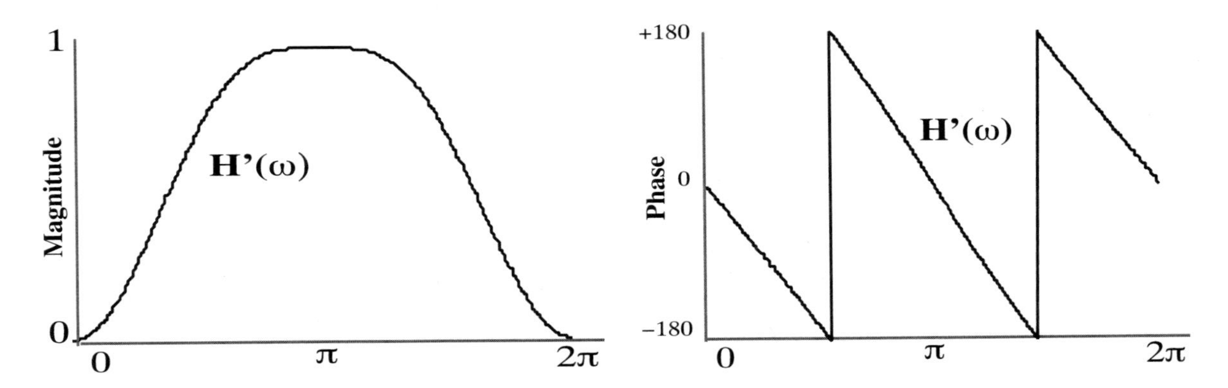

Figure 12.3–5 The magnitude and phase of H'(ω) is shown here.

Thus we can substitute H'(ω) for L(ω + π) in the first part of the alias cancellation equation. In other words

$$L'(\omega)L(\omega + \pi) = L'(\omega)H'(\omega) = L'H'$$

We now (complex) multiply L'(ω) and H'(ω). The magnitude and phase of this complex product are shown in Figure 12.3–6.

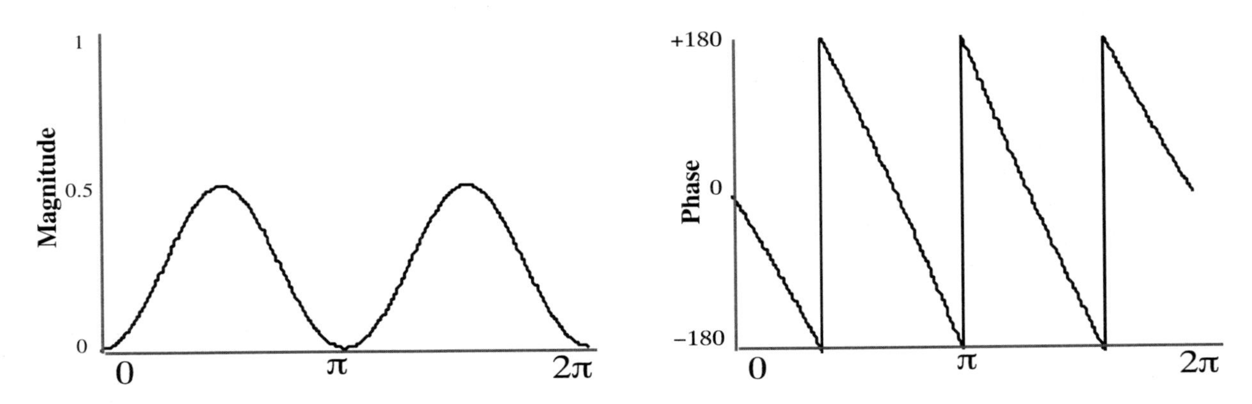

Figure 12.3–6 The magnitude and phase of H'L' (= L'(ω)L(ω + π)) is shown here.

If we look back at Figure 12.2–11, we see the same result in our representation of the alias portion of **A1** in following the lowpass (lower) path of the single-level conventional DWT.

We now look at the 2nd set of terms in the *alias cancellation equation*

$$H'(\omega)H(\omega + \pi)$$

If we look at **H** repeated over 2 periods we have the magnitude as shown in Figure 12.3–7.

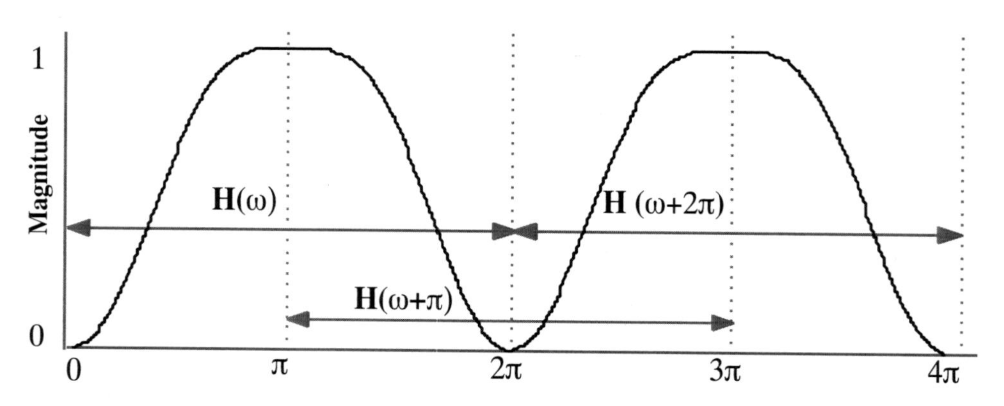

Figure 12.3–7 The magnitude of **H** is shown with the periodic frequency repeated twice (twice around the unit circle).

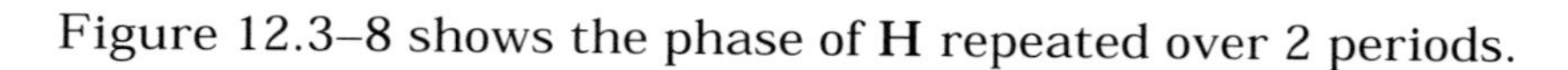

Figure 12.3–8 shows the phase of **H** repeated over 2 periods.

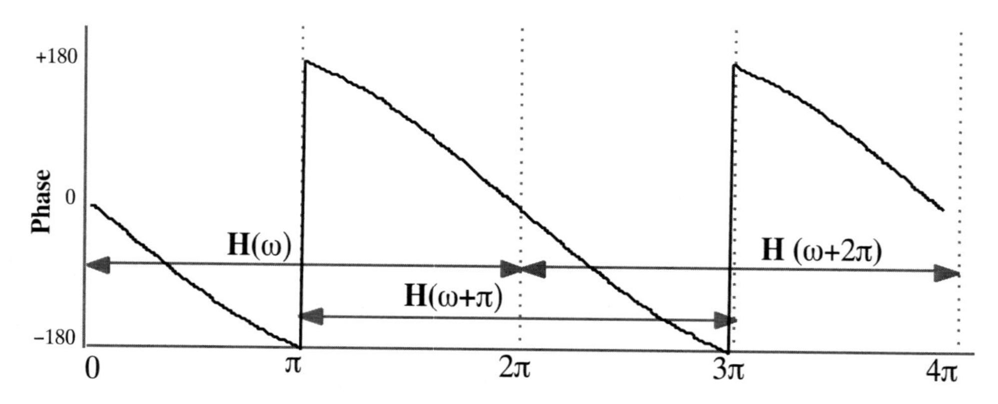

Figure 12.3–8 The phase of **H** is shown with the periodic frequency repeated twice.

We look next at the magnitude and phase of the lowpass reconstruction filter, L', as shown in Figure 12.3–9.

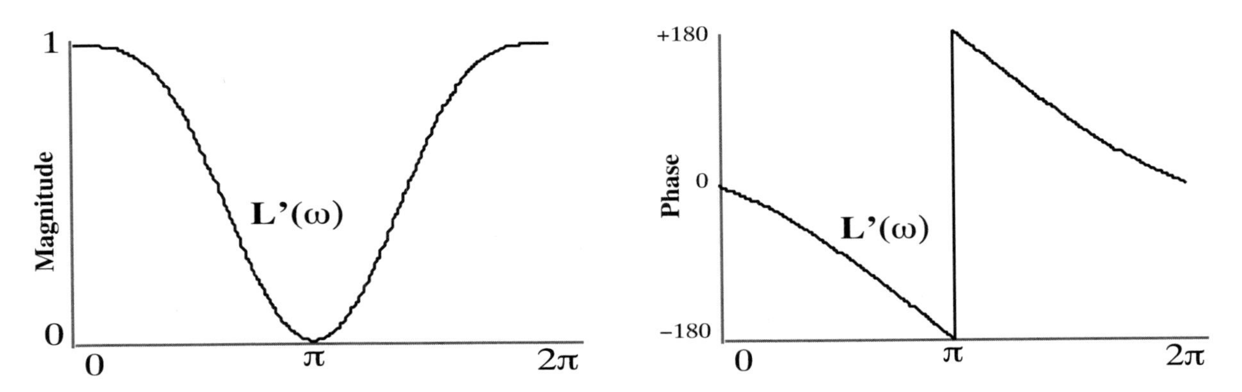

Figure 12.3–9 The magnitude and phase of L' is shown here.

We notice (because of the Quadrature Mirror nature of the filters) that the *magnitude* of L'(ω) is the same as that of H(ω+π) but that the *phase* does *not* match. A closer look, however, shows that L'(ω) and H(ω+π) are *out of phase by 180 degrees.* From our earlier discussion that means that

$$H(\omega + \pi) = -L'(\omega)$$

and that the entire left side of our alias cancellation equation can be rewritten

$$L'(\omega)H'(\omega) - L'(\omega)H'(\omega)$$

which is obviously zero. It is interesting to look at the right part of the above expression

$$-L'(\omega)H'(\omega)$$

Figure 12.3–10 shows this result of the complex multiplication. This is the same result as shown in Figure 12.2–10 for the alias portion of **D1**. Thus we see from the *alias cancellation equation* the same result as from following the highpass and lowpass paths of the conventional DWT—the alias components of **A1** and **D1** cancel. A final reminder that the *no distortion* and the *alias cancellation* equations are strictly true only when we do not discard the components for noise reduction or compression.

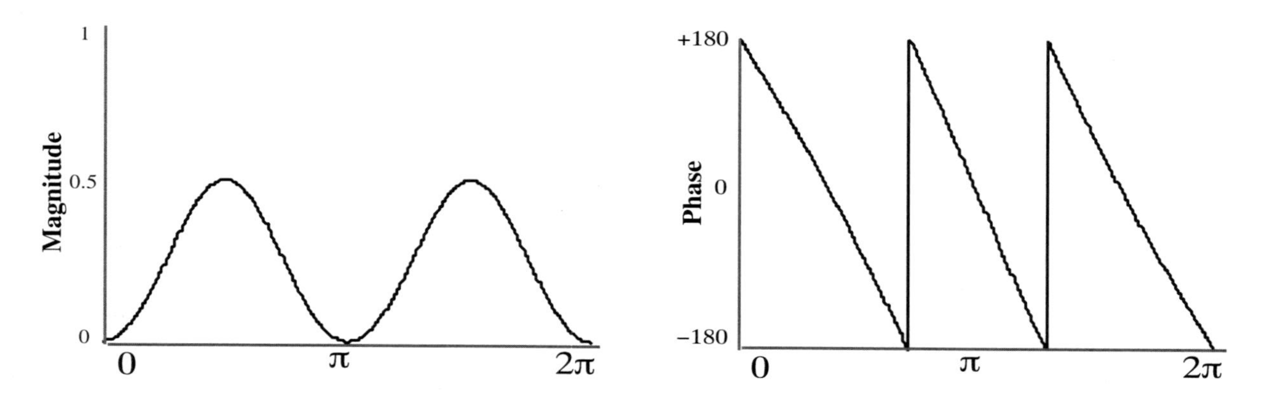

Figure 12.3–10 The magnitude and phase of $-L'(\omega)H'(\omega)$ is shown here.

We have now related the mathematical equations, found in numerous texts and papers, to the concepts of alias cancellation and perfect reconstruction.

12.4 Summary

We demonstrated alias cancellation by using a Kronecker delta as our signal. In the time domain we were then simply looking at the convolution of the downsampled-then-upsampled decomposition filters with the untouched reconstruction filters (**H'** and **L'**).

We were able to literally *see* the sliding convolution and to observe how the highpass and lowpass points canceled—except when aligned and then they summed to unity.

In order to show the LTI capabilities of the downsampled (conventional) DWT it was necessary to also follow the path for a Kronecker delta signal delayed by one. This showed alias cancellation and perfect reconstruction (to within a constant and an *additional* delay) whether we kept the even or the odd samples in "decimation by 2".

We used the Kronecker delta again in our frequency domain demonstration of alias cancellation. The process of downsampling then upsampling the decomposition filters and then *multiplying* by the reconstruction filters (in the frequency domain) gave us a signal and an alias part for the end results **D1** and **A1**. By looking at the phase we saw that the aliased parts were 180 degrees apart. Because they were the same magnitude, they canceled.

The signal portions at **A1** and **D1** were shown to be halfband filters. They were also shown to be *in phase* for all frequencies. As we added them together, we had a constant value of 1. The phase of **S'** was linear, but nonzero. This was shown to be the result of the system delay.

In the last section we related these concepts to some terminology found in the more equation-oriented wavelet literature. We were able to show how the *no distortion equation* was similar to the addition of the halfband filters in the frequency domain. We were also able to show how the *alias cancellation equation* was providing the same results as the addition of our out-of-phase (by π radians or 180 degrees) magnitudes of the alias portions of **A1** and **D1**. We used more mathematics than usual but this was necessary to translate these key equations from math-based treatises into the concepts we have learned from this book.

We by no means *proved* alias cancellation but concentrated on the intuitive *understanding* of the concepts. We used the familiar Db4 filters throughout this chapter but the concepts apply to other wavelet filters as well. In all 3 sections we cautioned that both **A1** and **D1** are necessary for complete alias cancellation and perfect reconstruction. With the intuitive insights gained through the demonstrations in both the time and frequency domains, however, we have a better "feel" for these processes and can learn to remove noise or extraneous data with minimal adverse effects.

We now proceed to look at a few other key equations from the traditional literature and show the concepts behind them and/or demonstrate some alternative methods.

"As far as the laws of mathematics refer to reality, they are not certain; and as far as they are certain, they do not refer to reality."

—Albert Einstein

CHAPTER 13

Relating Key Equations to Conceptual Understanding

Much of the traditional literature treats wavelets as Applied Mathematics. One of the objectives of this book is to relate concepts, once understood and demonstrated, to some key equations found in some of the traditional literature.

In the previous chapter we related 2 equations (the Alias Cancellation Equation and the No Distortion Equation) to the concepts we had acquired in understanding the conventional DWT.

In this chapter we will look at the Dilation Equation and how it relates to the method of interpolation of points using upsampling and convolution. We will find out why artifacts that look like somewhat like the scaling function or the wavelet function often appear. We will also address some terms found in some of the wavelet literature and relate these terms to our acquired conceptual understanding. Additional key equations found in wavelet literature will be addressed in an upcoming appendix.

13.1 Building the Scaling Function from The "Dilation Equation"

We saw earlier (ref. Fig. 6.3–5). how to build an approximation or estimation of a "continuous" Db4 wavelet or "wavelet function",(ψ),* by repeated upsampling and lowpass filtering of the 4-point highpass reconstruction filter (**H'** or **hir**).

We now show how to build an estimation of a "continuous" *scaling function,*(ϕ), from the 4-point lowpass reconstruction filter (**L'** or **lor**). The 4-point H' and L' filters are re-drawn below in Figure 13.1–1

* The theoretical "continuous wavelet function" is often represented in wavelet literature by the symbol *psi* (ψ) while the "continuous scaling function is represented by phi (ϕ).

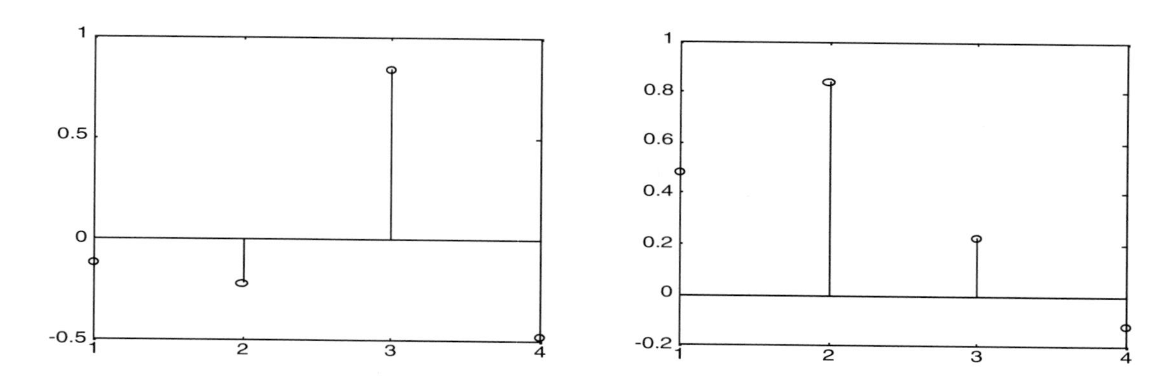

Figure 13.1–1 Four point basic Db4 *wavelet filter* (**H'** at left) and the four point basic Db4 *scaling function filter* (**L'** at right).

As we did with the *wavelet filter*, we first upsample the *scaling function filter* by 2 and then filter by the Lowpass Reconstruction Filter* (**L'** or "**lor**") as shown in Figure 13.1–2 to obtain a stretched 10 point scaling function (filter). We continue to upsample and lowpass filter to produce the 22 point scaling function and then the 46-point scaling function (filter).

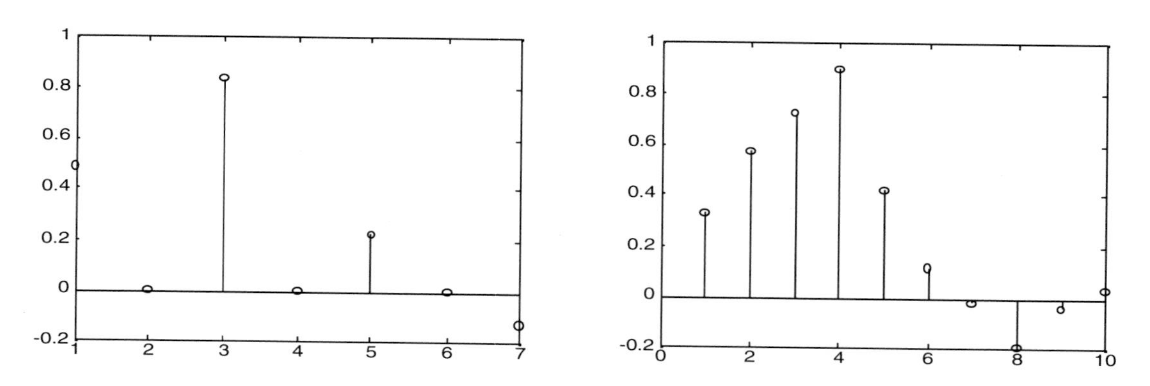

Figure 13.1–2 Basic scaling function filter (**L'**) upsampled by 2 (left graph) and then lowpass filtered by **L'** again to produce the interpolated "stretched" scaling function filter as shown at right.

* When we built the estimate of the wavelet function we started with **H'**, upsampled, then lowpass filtered by **L'**. In building the estimate of the *scaling function* here we start with **L'** but still lowpass filter by **L'** to perform the interpolation.

The interpolation process continues as we upsample and lowpass filter to produce a 94 point stretched filter, then 190,, 382 and finally a 766 point estimation (approximation) of a "continuous" scaling function as shown in Figure 13.1–3 at left. We can map this scaling function to the interval 0 to 3 and overplot the original 4 points at 2/6, 5/6, 8/6, 11/6. Along with these 4 points we plot the 2 trailing zeros at 14/6 and 17/6 that also fit (right graph). Note that as we keep the 2 trailing zeros the stretched filters are all 2 points longer. In other words, keeping the trailing zeros, the filter lengths are 6, 12, 24, 48, 96, 192, 384 and finally 768 points long.

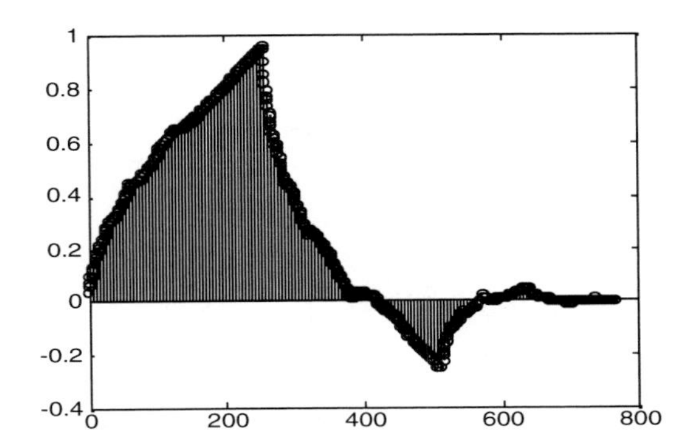

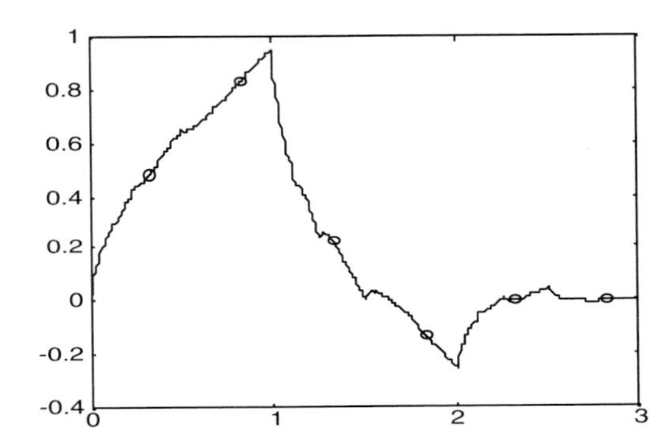

Figure 13.1–3 768 point estimation of a continuous scaling function individually plotted, as we did in the previous graphs, in "stem" format (left). 768 points connected (no stems) and then mapped onto the interval 0 to 3 with the original 4 points from the basic Db4 scaling function filter (**L**') over-plotted along with the 2 trailing zeros.

We find this same process of upsampling and lowpass filtering in the conventional DWT and we will show later how the DWT can "build" what looks like a miniature scaling function from an isolated coefficient.

We will now show how this process is related to the *Dilation Function Equation* found in much of the wavelet literature.

Let us first step back a moment: Mathematics has come a long way from Roman Numerals.* The concepts of irrational numbers and negative num-

* In the last century, motion picture companies were rumored to have used Roman numerals to indicate the production year of the movie to obscure its age. For example a movie made in 1997 would be have a date of MCMXCVII while one made 50 years earlier would have a similar-looking Roman numeral date of MCMXLVII. Things are harder for them in the new millennium with a movie made in 2002 shown as simply MMII.

bers are still relatively new. Algebra is also intimidating when first encountered because an equation such as

```
x = 2x -1
```

seems to say "tell me the twice the unknown value (x) first, next subtract 1, and *then* I'll tell you the unknown value". We've learned how to handle this by subtracting 2x from both sides and then changing the signs to discover

```
x = 1
```

The scaling function "*dilation equation*" encountered in wavelet literature is even more difficult. It is a *2-scale difference equation.* Like the simple algebra equations you often need to know one value to get another but you are now dealing with *functions* of the unknown value. For example

```
"f(x) = f(2x+a)"
```

where **a** is a constant. Below is the well-known *dilation equation* that is found in much of the wavelet literature. This defines, in mathematical terms, a "continuous" scaling function $\phi(t)$ for an infinite number of values of t (recall that in the world of engineering we use a finite number of points). The lowpass reconstruction filter, ***lor*** or ***L'*** is the basic 4-pt Db4 scaling function. Here is the "dilation equation" that leads to an estimation of the scaling function:

$$\phi(t) = K\sum_{m} lor_m \phi(2t-m)$$

Φ
t

Although this equation may appear daunting at first, we can simplify it to a workable form. Instead of the continuous version at *all times* (t), we look instead at **t** = **n/2** (**n** is used in a general sense and may or may not be an integer). Also, instead of an *infinite sum*, we know the Db4 Lowpass Reconstruction filter (**L'** or **lor**) has only 4 values 1/2 integer apart at times 2/6, 5/6, 8/6, 11/6 (and the 2 trailing zeros at 14/6 and 17/6). Thus, the Dilation Equation can be re-written as shown.

$$\phi\left(\frac{n}{2}\right) = K\sum_{m=0}^{3} lor_m \phi(n-m)$$

where **K** is a constant = **sqrt(2)** (usually). From DSP studies you might recognize the right side of the equation as a *convolution*. We will soon show that upsampling combined with convolution *will produce the same values as this dilation equation*, but for now, we continue evaluating this equation.

We will next rewrite the equation slightly for convenience. We will use "c_m" to represent the filter coefficients $\mathbf{lor}_m$. Thus **c0** is the first **lor** coefficient located at **time** = **2/6** and equal to **0.4830** (the first of the 4 "magic numbers" of the Db4 wavelet filters). We then have **c1, c2, c3** as the remaining coefficients located 1/2 integer apart at **time** = **5/6, 8/6, 11/6** on the 0-to-3 mapping and have values **0.8365**, **0.2241** and **–0.1294**. The simplified equation is now

$$\phi(\frac{n}{2}) = K \sum_{m=0}^{3} c_m \phi(n-m)$$

We will start by letting **n** = **2/6**. We then have

$$\phi(\frac{n}{2}) = \phi(\frac{2}{12}) = K \sum_{m=0}^{3} c_m \phi(2/6-m)$$

Expanding the right side of the above equation we have

```
φ(2/12)  = K[c0φ(2/6)+c1 φ(-4/6)+ c2 φ(-10/6)+c3 φ(-16/6)]
```

We recall that we used the 4 **lor** (**L'** or **c** in this equation) filter points to create the "continuous" 768-point scaling function, ϕ. Thus we *know* the value of ϕ at **time** = **2/6** to be simply **c0** and we have

```
φ(2/6)  = c0  = 0.4830
```

We also know that the scaling function, ϕ, is zero outside the interval 0 to 3 (recall that scaling functions, like wavelets, start and stop or in wavelet terminology have "*compact support*"). thus

```
φ(-4/6)  = φ(-10/6)  = φ(-16/6)  = 0
```

Therefore

$$\phi(2/12) = K[c0c0+0+0+0] = sqrt(2) \times 0.4830^2 = 0.3299$$

Figure 13.1–4 shows the 4 values of **L'** (the "scaling function filter") and the 2 trailing zeros starting at **2/6** and spaced 1/2 integer apart. It also shows the new value of **0.3299** at **time** = **2/12**. All these points are overplotted on the 768-point estimation of a continuous scaling function

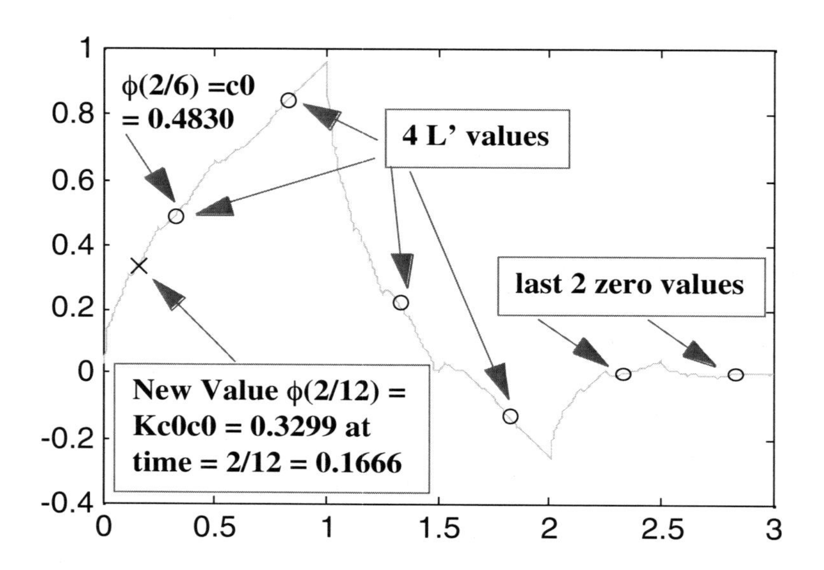

Figure 13.1–4 The 4 lowpass reconstruction filter (**L'** or **lor**) points and the 2 trailing zeros are plotted as "**O**" on the 768-point estimation of a continuous scaling function. The new point we found from the dilation equation is shown as an "**X**" near the left border.

From the same dilation equation we obtain additional points: For **n** = **5/6** we have

```
φ(5/12)=K[c0φ(5/6) + c1φ(-1/6) + c2φ(-7/6) + c3φ(-13/6)]
```

but the scaling function φ at **t** = **5/6** is simply **c1**. Also φ is zero outside the interval from 0 to 3 and we have:

```
φ(5/12)=K[c0c1 + 0+0+0] = K x 0.4830 x 0.8365 = 0.5714
```

Instead of a single term, for **n** = **8/6** we have a result that is the sum of 2 terms as follows:

```
φ(8/12)=K[c0φ(8/6) + c1φ(2/6) + c2φ(-4/6) + c3φ(- 10/6)]
= K[c0c2 + c1c0 + 0 + 0]
= sqrt(2) x [0.1082 + 0.4040]
= 0.7244
```

We repeat this process using the dilation equation and add to our 6 original points (the 4 filter points and 2 end zeros) 12 additional points. These 12 points are spaced *1/4 integer* apart and are located at times

2/12, 5/12, 8/12, 11/12, 14/12, 17/12, 20/12, 23/12, 26/12, 29/12, 32/12, and 35/12

Note that all these new times are still between 0 and 3. The values of ϕ at these 12 times are

0.3299, 0.5714, 0.7244, 0.9012, 0.4183, 0.1121, -0.0173, -0.1941, -0.0410, 0.0237, 0.0000, 0.0000

These 12 additional points are shown along with the original 6 points in Figure 13.1–5 at left.

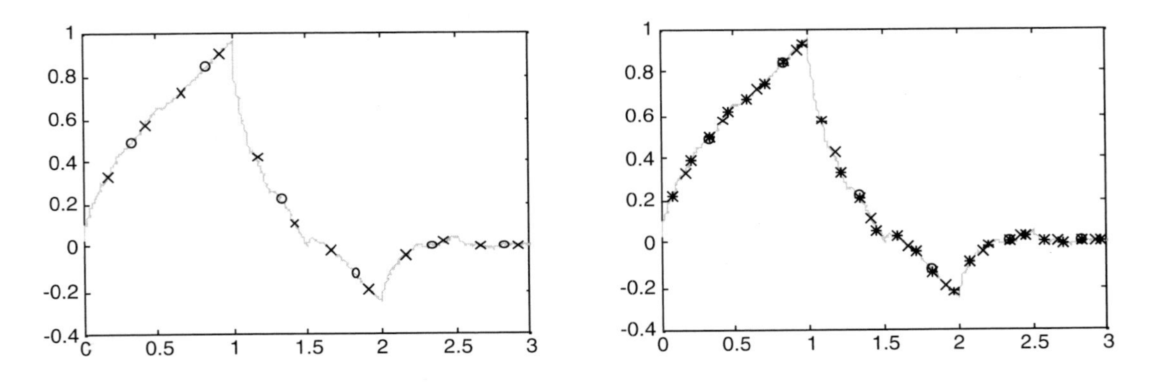

Figure 13.1–5 On the left graph, the 4 reconstruction filter (**L'**) points and the 2 trailing zeros are plotted as "**O**" on the 768-point estimation of a continuous scaling function. The 12 new points starting at 2/12 and spaced 1/4 integer apart are plotted as "**X**". The right graph is identical to the left except it has 24 new points starting at 2/24 and spaced 1/8 integer apart. These are plotted as asterisks ("*").

We can continue to use the above dilation equation to produce 24 points 1/8 integer apart from the 12 points 1/4 integer apart we just found. For example, using $\mathbf{n} = \mathbf{2/12}$, our first point would be at time $\mathbf{n/2} = \mathbf{2/24}$ We would have

$$\phi(2/24) = K[c0\phi(2/12) + c1\phi(-10/12) + c2\phi(-22/12) + c3\phi(-34/12)]$$

All but the first term will be zero (outside $\mathbf{0 \le t \le 3}$). Thus

```
φ(2/24) = sqrt(2) x 0.4830 x 0.3299 = 0.2253
```

Figure 13.1–5 above at right shows the 6 original points at 1/2 integer apart starting at **time** = **2/6**, the 12 points 1/4 integer apart starting at **time** = **2/12** and the 24 points 1/8 integer apart starting at **time** = **2/24**. We of course can continue this process to 48, 96, 192, 384 and to 768 points—each of the 768 points spaced 1/256 integer apart.

Recall that the theoretical *continuous scaling function*, and *continuous wavelet function* are never actually used—only the various filter points. We note that this recursive method is ideal for digital computers and we are seeing another example of why wavelet technology is called "a child of the digital computer age."

Jargon Alert: **Recursion is repetition of a sequence of operations that yields data ever closer to a desired result.**

In review, from the simplified dilation equation

$$\phi(\frac{n}{2}) = K \sum_{m=0}^{3} lor_m \phi(n-m)$$

we used known values of ϕ at specific times (2/6, 5/6, etc.) to produce additional values*. We have also seen, however, that we were able to do this in a simpler and more elegant fashion by upsampling and lowpass filtering the basic filter points. We will now demonstrate why this works.

13.2 Building the Scaling Function Using Upsampling and Simple Convolution

We noted in the previous section that we knew the values of the Db4 scaling function ϕ at times 2/6, 5/6, 8/6, and 11/6 to be those of the basic scaling function filter, the lowpass reconstruction filter, **L'** or **lor**. For convenience, we denoted these values as **c0**, **c1**, **c2**, **and c3**. We also noted that the 2 "end"

* Some wavelet literature uses known values of ϕ at **times** = **0.0**, **1.0**, **2.0**, and **3.0** to produce additional values at **times** = **0.5**, **1.5** and **2.5**. These integer-apart values produce additional values 1/2 integer apart at **times** = **0.25**, **0.75**, **1.25**, **1.75**, **2.25** and **2.75**. This method approximates the same theoretically continuous scaling function as we have produced above.

values at times 14/6 and 17/6 were zero. Also, the values at any time less than 0 or greater than 3 would be zero.

For the first 3 of the 12 additional values we found from the dilation equation that (ignoring the constant K for now) we had

```
φ(2/12)  = c0c0

φ(5/12)  = c0c1

φ(8/12)  = c0c2 + c1c0
```

A direct convolution of **L'** (**lor** or **c**) with itself (presented in sliding graphic form) is simply

```
                c0   c1   c2   c3

 c3   c2   c1   c0 ---->
```

which would yield the results

```
 c0c0, c0c1+c1c0, c0c2+c1c1+c2c0, etc.
```

The first term matches the correct results from the dilation equation method, but the others *do not*! The *secret* is to upsample first, and only *then* convolve*. Doing this we now have

```
                c0   0   c1   0   c2   0   c3

 c3   c2   c1   c0 -------->

 = c0c0, c0c1, c0c2+c1c0, etc.
```

which now agrees perfectly with the results from using the dilation equation. If we were to upsample these latest 12 values and convolve again with **L'** we would obtain the 24 values. We continue this process to obtain 48, 96, 192, 384 and then 768 points. As a further demonstration that this upsample/convolve method works correctly, refer back to Figure 13.1–5. The **X**, **O** and * plot symbols came directly from the dilation equation while the dotted line that they were overplotted upon was built by upsampling and convolution.

If we take a close look at a conventional DWT we see a series of upsampling and lowpass filtering as shown in the oval in Figure 13.2–1.

---

* If the values of ϕ were integer apart in time instead of 1/2 integer it would indeed be a direct convolution.

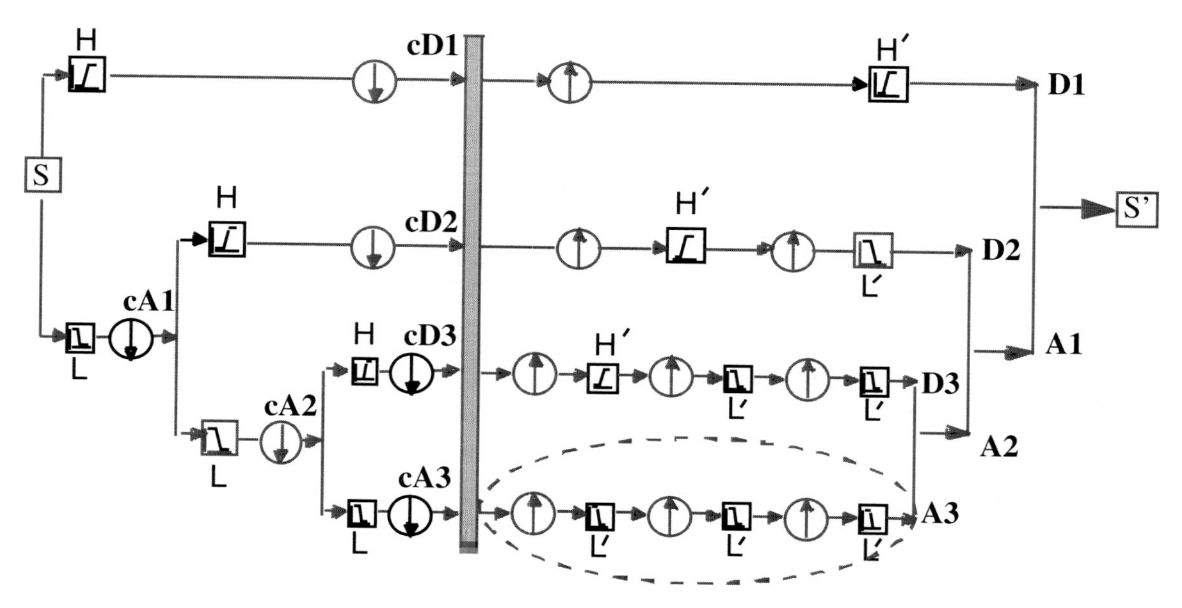

Figure 13.2–1 3-level conventional DWT. Notice the series of upsampling and convolution (filtering in the time domain) in the dotted oval.

Can we see this "building of a scaling function" in a DWT system? Yes! If **cA3** in the above figure were a single point it would first be upsampled by 2 but would still be a single point (with a trailing zero). When convolved by the filter **L'** we would have simply **L'** (with a delay and multiplication constant). Then **L'** would be upsampled and lowpass filtered twice more. **A3** at the end of the path would look a little like a scaling function, in this case the Db4 scaling function.

Figure 13.2–2 (left) shows a signal with a single point artifact at the far end. at **cA3**. When repeatedly upsampled and filtered the artifact begins to take on the shape of a scaling function. The final result at **A3** (right graph) shows a mini-scaling function at the end. Note that the artifact is not a perfect replica of a Db4 scaling function because the spike is not perfectly isolated*.

* The next-to-last point is non-zero and "degrades" the artifact slightly. Notice that with this few points (36) in **cA3** that we can see some of the ragged shapes belonging to a Db4 scaling function in the main portion of the final Approximations **A3**.

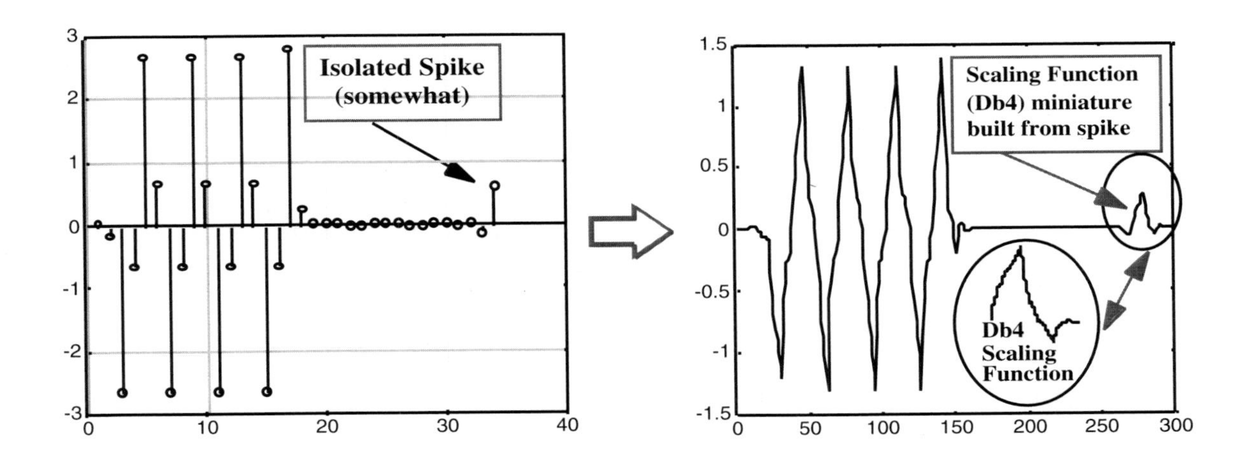

Figure 13.2–2 A (somewhat) isolated spike at **cA3** is "built" into a sort of miniature replica of the Db4 *scaling function* by repeated upsampling and lowpass filtering in the reconstruction portion of the conventional DWT to produce **A3** (right graph)

When we downsample the signal to **cA1**, **cA2**, **cA3** (and **cA4**, **cA5**, etc. for more levels), the "signal" becomes only a few points long and we begin to see these few points built into the shapes of the scaling function for the particular wavelet we are using. In fact, *we can tell we have downsampled enough levels* when we see scaling function patterns in the final results (**A3** in this case). Also, in a multi-level conventional DWT we can often see these artifacts and be able to tell which wavelet or scaling function was used by recognizing the shape!

13.3 Building the Wavelet Function from the Dilation Equation

The *wavelet* dilation equation leading to the *wavelet function* (ψ)

$$\psi(t) = K\sum_{m=-\infty}^{\infty} hir_m \phi(2t-m)$$

is very similar to the *scaling function* dilation equation we saw in the last section

$$\phi(t) = K\sum_{m} lor_m \phi(2t-m)$$

(equation repeated here for convenience) except that instead of the basic *scaling function* filter (lowpass reconstruction filter **L'** or **lor**) we use the basic *wavelet function* filter (highpass reconstruction filter **H'** or **hir**). Note that the *wavelet* values **ψ(t)** come from the *scaling function* **φ(2t–m)**.*

We can simplify the *wavelet function* dilation equation as we did for the *scaling function* dilation equation to get 1/4 integer-apart *wavelet* values from the 1/2 integer-apart *scaling function* values.

$$\psi\left(\frac{n}{2}\right) = K \sum_{m=0}^{3} hir_m \phi(n-m)$$

Similar to our previous scaling function analysis, we now substitute "**d**" for the highpass reconstruction filter, **hir** or **H'**. The *wavelet dilation equation* then becomes

$$\psi\left(\frac{n}{2}\right) = K \sum_{m=0}^{3} d_m \phi(n-m)$$

Thus setting **n** = **2/6**, **ψ(2/12)** would be given by (ignoring K for now)

```
ψ(2/12) =d0φ(2/6) + d1φ(-4/6) + d2φ(-10/6) + d3φ(-16/6)
= d0c0 + 0 + 0 + 0 = d0c0
```

The next value, 1/4 integer later at **t** = **5/12** is given by

```
ψ(5/12) = d0φ(5/6) + d1φ(-1/6) + d2φ(-7/6) + d3φ(-13/6)
= d0c1 + 0 + 0 + 0 = d0c1
```

Next we have

```
ψ(8/12) = d0φ(8/6) + d1φ(2/6) + d2φ(-4/6) + d3φ(-10/6)
= d0c2 + d1c0 + 0 + 0 = d0c2 + d1c0
```

---

* Some texts refer to the *scaling function,* φ, as the *Father Wavelet* and the *wavelet function,* Ψ, as the *Mother Wavelet* because the "mother" needs the "father" to function. Women's Rights activists may have something to say about this!

and so forth.

The results for all 12 of the Psi (ψ) wavelet function values, including the 2 trailing zeros are as follows:

```
d0c0, d0c1, d0c2+d1c0, d0c3+d1c1, d1c2+d2c0,
d1c3+d2c1, d2c2+d3c0, d2c3+d1c1, d3c2, d3c3, 0, 0
```

The values of Ψ at times **2/12, 5/12, 8/12, . . . , 35/12** are thus

```
-0.0884  -0.1531  -0.1941  -0.2415    0.5003    1.0306,
-0.0647  -0.7244  -0.1531   0.0884  0.0 and 0.0
```

These 12 additional points are shown along with the original 6 points in Figure 13.3–1 at left.

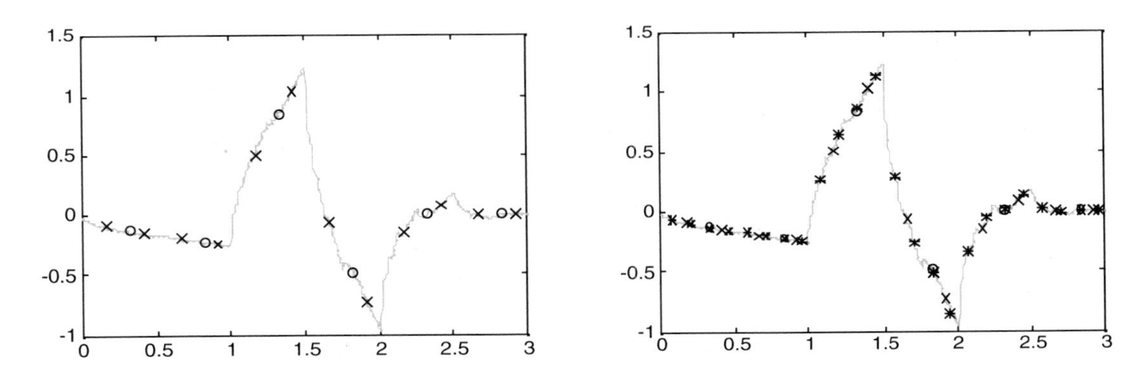

Figure 13.3–1 On the left graph, the 4 *highpass* reconstruction filter (**H'**) points and the 2 trailing zeros are plotted as "**O**" on the 768-point estimation of a continuous *wavelet* function. The 12 new points starting at 2/12 and spaced 1/4 integer apart are plotted as "**X**". The right graph is identical except it has 24 new points starting at 2/24 and spaced 1/8 integer apart. These are plotted as asterisks ("*").

As we did with the *scaling function* values we can find the next 24 *wavelet function* values (now 1/8 integer apart) from the dilation equation. Note that we are using the previous *scaling function* values (ϕ) *and* the highpass reconstruction filter (**H'** = **hir** = **d**). For example, the first value would be

$$\Psi(2/24) = K[d0\phi(2/12) + d1\phi(-10/12) + d2\phi(-22/12) + d3\phi(-34/12)]$$

$$= K[d0\,\phi(2/12) + 0 + 0 + 0] = K\{d0\,\phi(2/12)] = -0.0604$$

The fifth of these 24 values would be

$$\Psi(14/24) = K[d0\phi(14/12) + d1\phi(2/12) + d2\phi(-10/12) + d3\phi(-22/12)]$$

$$= K[d0\,\phi(14/12) + d1\phi(2/12)] = -0.1811$$

All 24 of these wavelet function points are overplotted on the right graph of the above figure. As before, the process continues with 48, 96, 192, 394, and finally 766 points as an estimation of a continuous wavelet function.

13.4 Building the Wavelet Function Using Up-sampling and Simple Convolution

This is a very similar process to that used for the scaling function. We could work with the wavelet dilation equation directly, but we obtain the same results far more efficiently by repeated upsampling and filtering (convolution with the lowpass reconstruction filter). We simply take the latest results, upsample, and then convolve with the *lowpass* filter.* For example, to obtain the values

$\psi(2/12)$, $\psi(5/12)$, $\psi(8/12)$, $\psi(11/12)$, $\psi(14/12)$, etc.

we simply upsample and convolve as follows

```
               d0   0   d1   0   d2   0   d3
c3   c2   c1   c0 -------->
```

For the first 3 of the additional 12 values we have

$$\psi(2/12) = c0d0$$

$$\psi(5/12) = c1d0$$

$$\psi(8/12) = c0d1 + c2d0$$

* It may at first seem counter-intuitive to convolve the upsampled **H'** with ***L'*** instead of **H'** with ***H'*** until we remember that we are in effect *interpolating* points—which involves upsampling and *lowpass* filtering—whether the upsampled filter is highpass or lowpass.

which is identical to the results obtained using the wavelet function dilation equation.

We now take another look at the synthesis portion of the 3-level conventional DWT as shown in Figure 13.4–1. This time we look at the path from **cD3** to **D3** (inside the re-drawn oval). If we have an isolated point at **cD3**, it will be upsampled and then convolved with **H'**. Then it will be upsampled and convolved with **L'** (twice) to produce **D3**. Notice that the first filter in the oval is the highpass reconstruction filter so the shape we will see is that of the *wavelet* function.

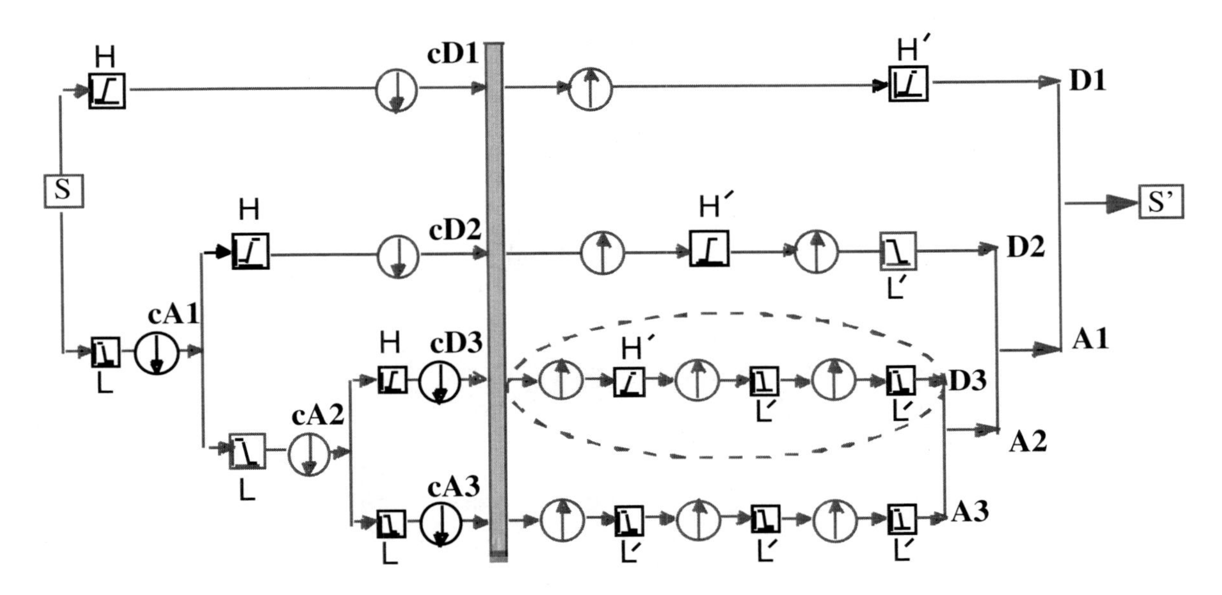

Figure 13.4–1 3-level conventional DWT. Notice again the series of upsampling and convolution in the dotted oval. Note the first convolution is with the highpass reconstruction filter, **H'**, but that the final 2 convolutions are with the *lowpass* reconstruction filter, **L'**, as in the previous oval (ref. Fig. 13.2–2).

Figure 13.4–2 (left) shows a signal with a single point artifact at the far end. at **cD3**. When repeatedly upsampled and filtered the artifact begins to take on the shape of the *wavelet* function. The final result at **D3** (right graph) shows an imperfect but recognizable mini-wavelet function at the end.*

* Again, the point is not perfectly isolated in **cD3** and thus the artifact is influenced slightly by the nearby points. Notice that in the main portion of **D3** that we can see some of the ragged fractal-looking shapes of the Db4 *wavelet.*

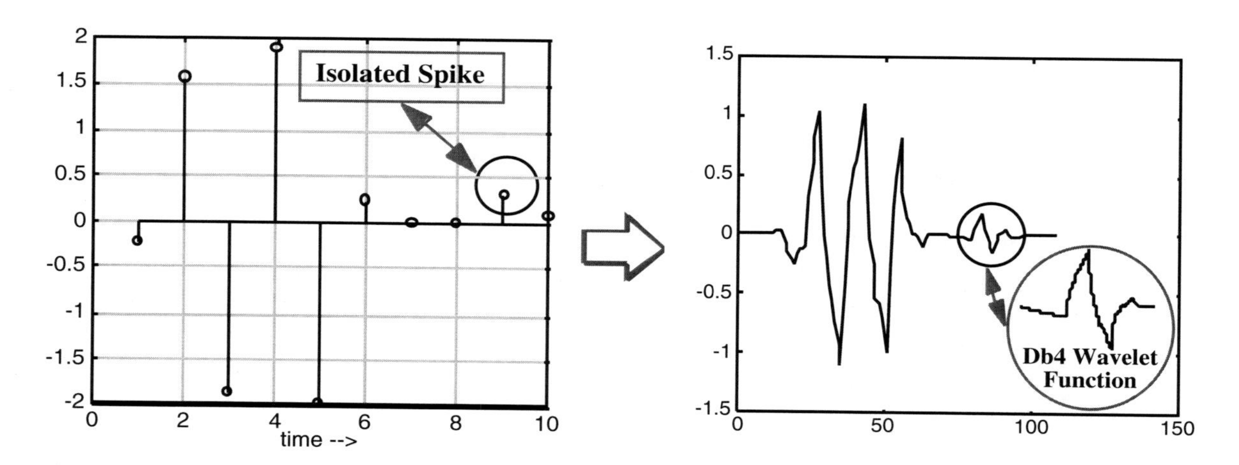

Figure 13.4–2 A (somewhat) isolated spike at **cD3** is "built" into a replica of the Db4 *wavelet function* by repeated upsampling and lowpass filtering in the reconstruction portion of the conventional DWT.

Similar to the scaling function discussion, when we downsample the signal to **cD1, cD2, cD3** (and **cD4, cD5**, etc. for more levels), the "signal" becomes only a few points long and we begin to see these few points built into the shapes of the *wavelet* function for the particular wavelet we are using. Again, *we can tell we have downsampled enough* when we see wavelet function patterns in the final results (**D3** in this case).

We have shown this effect for the Db4 wavelet and scaling function filters. We see this effect, of course, for any of the wavelet filters used in a conventional DWT with downsampling and upsampling. Again, with practice, it is often possible to tell which wavelet (set of 4 wavelet filters) was used by looking at the mini-wavelet function or scaling function artifacts found in the higher levels of the Details and Approximations (where there are fewer points because of the downsampling).

13.5 "Forward DWT", "Inverse DWT" and Other Terms from Wavelet Literature

Let us review some familiar terminology from the *Fast Fourier Transform*. The FFT is best used with stationary signals—those that don't change frequency or amplitude (envelope) as time passes. For these signals, the FFT is able to *transform* data into the *frequency domain.* If we desire, we can later perform an *Inverse FFT* to "perfectly reconstruct" (in wavelet terminology) the original signal. If we are careful, we can also use an FFT for de-noising by removing some of the frequency components (in the frequency domain) and then use an *Inverse* FFT (IFFT) to construct a signal without the noise. The FFT and IFFT also usually work best with signals that are a power of 2 (radix 2) in length (e.g. 16, 32, 256, 4096, etc.). The FFT is functionally equivalent to the *Discrete Fourier Transform* (DFT) but works orders of magnitude faster for large signals when the data is a factor of 2 (or zero-padded to a factor of 2). Hence the term "*fast*" in the Fast Fourier Transform.

We have seen in this book that the *Discrete Wavelet Transform* (DWT) can do many of the same things that the Discrete Fourier Transform (DFT or the functionally equivalent FFT) does and works better for *non-stationary* signals. Because of some similarities between the DFT/*FFT* and the *DWT* some books and papers use very similar terminology. For example, they refer to the decomposition and reconstruction portions of the Discrete *Wavelet* Transform as the "Forward DWT" and the "Inverse DWT", respectively. In a few publications, the results of the "Forward DWT" (the first half of the DWT where the wavelet coefficients **cA1**, **cD1**, etc. are produced) is even referred to as the "*wavelet domain*" Also because the conventional DWT usually works best with radix 2 data it is referred to by some authors as a "Fast Wavelet Transform" or split into 2 parts called the "Fast Forward Wavelet Transform" and the "Fast Inverse Wavelet Transform".

Now that we have gained an intuitive understanding of the DWT process, however, we can see some definite problems in using such terminology. In the first place, the results of a *forward FFT* can stand alone in a usable form. The DWT, on the other hand, requires the "Inverse DWT" (the reconstruction portion) to produce usable results. More importantly, there are some serious problem with a conventional DWT, as we have seen, in that the repeated downsampling causes aliasing after decomposition (the "forward" DWT). We depend heavily on the reconstruction portion (the "inverse" DWT) for alias cancellation.

In other words, with discrete, sampled data we have seen in a conventional DWT that whenever we discard or attenuate the components after downsampling in the so-called "forward DWT' (as we do in de-noising and/or compression), that we lose some of the alias cancellation. This is not a problem in the Discrete *Fourier* Transform (DFT/FFT) as we do not need the Inverse Discrete Fourier Transform (IDFT/IFFT) for alias cancellation.

Thus it is recommended that we stick with the terms "decomposition" and "reconstruction" (or analysis portion and synthesis portion) as the 2 integrated parts of the DWT rather than treat them as stand-alone "forward" and "inverse" transforms.

The term "wavelet domain" for the coefficients produced halfway through the DWT can also lead to misunderstanding. It is recommended that we use the terms "Approximation Coefficients" and "Details Coefficients" to describe the results at the end of the decomposition portion (**cD1, cD2, cA2**, etc.).*

The Fast Wavelet Transform is just the Conventional DWT. To be fair, it is certainly "fast" compared to the *Continuous* Wavelet Transform and the DWT does work better on signals that are a factor of 2 in length. However the term "Fast Wavelet Transform" can also be somewhat misleading in that it does not *replace* the *DWT* like the FFT (using the Cooley-Tukey algorithm) *replaces* the *DFT*. The term DWT is less confusing than "Fast Wavelet Transform". This is especially true with terms such as "Fast Forward Wavelet Transform" and "Fast Inverse Wavelet Transform". The terms "decomposition portion" and "reconstruction portion" of a DWT are more intuitive and lead to a better understanding.

We showed in the previous chapter the *anti-aliasing* and the *no distortion* equations and how they correctly depicted what we have learned about alias cancellation and perfect reconstruction in a more mathematical form.

Serious problems can occur, however, when equations that treat the data as *continuous and infinite* are not at some point adapted to reflect the *discrete sampling* of the infinite data and to allow for the possibility of aliasing when using a digital computer. In other words when data is portrayed as continuous (rather than discrete points) there is no such thing as aliasing† because decimation of an infinite number of points still gives an infinite number of points.

It is unlikely that terms such as *forward DFT, inverse DFT, Fast Wavelet Transform, Fast Inverse Wavelet Transform, Wavelet Domain* and the general notion of wavelets being treated as *continuous functions* will be stricken

* A better use for the term *wavelet domain* might be the insights and overview gained by looking at a CWT display.

† The author once received an e-mail from a student saying that the (continuous time) equations *prove* there is no such thing as aliasing and that this is a scheme cooked up by unscrupulous authors and teachers to make things unnecessarily difficult and thus keep their jobs!

from existing texts and papers.[*] That is not the goal of this section or this book. The objective here is rather to introduce them to you so that when you encounter these terms in further wavelet studies you will be able to understand them and place them in the proper context.

13.6 Summary

We introduced the *dilation equation for the scaling function* and showed how this "*2-scale difference equation*" can be simplified and used with the 6 points (including the 2 trailing zeros) of the basic Db4 *scaling function filter* **L'** to create 12 additional points. This process is then repeated by using the 12 points to produce 24 points and so on until we have a 768 point estimation of the "continuous" scaling function

We noticed that the simplified, discrete form of the dilation equation resembled a convolution. We proceeded to show that our familiar method of upsampling by 2 and then lowpass filtering by **L'** produces the identical result as the *scaling function dilation equation*. We showed how a stray or isolated point at the end of the *decomposition* portion of a conventional DWT produces an artifact at the end of the *reconstruction* portion that can look a lot like the scaling function. This is or course because of the repeated upsampling and lowpass filtering found in the conventional DWT reconstruction portion. With practice, we can recognize the set of filters that was used in the DWT (Db4, Haar, etc.) by these artifacts.

The presence of these artifacts also tells us we have decomposed the signal to enough levels. In other words, when the signal is downsampled by 2 so many times that it becomes a single point or an isolated point, the upsampling and filtering in the reconstruction half of the DWT produces artifacts that look like a rendition of a mini-scaling functions.

We next presented the *wavelet function dilation equation* which is similar to the *scaling function dilation equation* we just discussed except that we use the 6 **H'** filter points rather than the 6 **L'** points. The other major difference is that the *wavelet function* points are derived from *scaling function* points. This means we would want to find the *scaling function points* first and then use those to obtain the *wavelet function* points. We showed how to obtain 12 *wavelet function* points from the 6 *scaling function* points, 24 *wavelet func-*

[*] It far more likely that the *author* be stricken.

tion points from the 12 *scaling function* points and so on until we had enough points for an estimation of the "continuous" wavelet function.

Like the scaling function dilation equation, the *wavelet function* dilation equation takes the form of a convolution and we were able to show that the wavelet dilation equation method is equivalent to upsampling the previous result and then lowpass filtering to interpolate extra points. As with the scaling function, when we have an isolated point in the wavelet coefficients, the reconstruction portion of the DWT will upsample and lowpass filter repeatedly to produce a miniature version of the wavelet function.

As mentioned, we can learn to tell which set of wavelet filters was used by looking at the shapes that are seen in the higher levels (more downsampling) and we can tell when we have downsampled enough. For example a 1024 point signal will be downsampled by 2 10 times in a 10 level conventional DWT and become a single point. This point will then be upsampled and filtered 10 times in the reconstruction portion to produce a replica of the *wavelet function* in the final *Details* and a *scaling function* in the final *Approximations.* Note: If the single point is negative, the wavelet or scaling function appears upside down.

We talked about some other terms found in some of the wavelet literature. We explained the terms *forward DFT, inverse DFT, Fast Wavelet Transform, Fast Inverse Wavelet Transform,* and *Wavelet Domain* and how they relate to the intuitive concepts and terms we have learned in this book.

POSTSCRIPT

It is suggested you read through the various appendices that follow this main body. In addition to providing recommendations of the best of the conventional textbooks, papers, and websites for further study, they provide additional insight and clarification of the concepts we have discussed in this book.

*For those who already have a strong background in mathematics it is hoped that we have provided some insights and intuition to back up some of the equations you are already familiar with. And whatever your mathematical background, it is the author's ardent hope that this conceptual understanding of wavelets and wavelet transforms will help you figure out what is really going on when you have a problem.**

A final word: The amount of wavelet books and papers is increasing exponentially in almost every engineering discipline. It is fervently hoped that this book will allow you, the reader, to have gained an intuitive, conceptual understanding of wavelets and wavelet transforms along with a familiarity with the various terminology used in this literature. Then as you proceed to learn more about wavelets in your particular area of interest you will have already seen much of this new information demonstrated in a simpler form in this book and you can proceed to use wavelets confidently, effectively and profitably.

Best Wishes in your Wavelet Studies and Applications!

D. Lee Fugal
San Diego, California, USA

* "We can't solve problems by using the same kind of thinking we used when we created them." — Albert Einstein

Be sure to visit our website

www.ConceptualWavelets.com

- Free downloads including audio and/or video clips and selected color slides from presentations
- Post-publication information and updates
- Additional case studies and examples
- Suggestions and corrections
- Ordering information
- Additional recommended wavelet books, articles, papers, and websites (including links)
- Information on upcoming lectures, seminars and short courses
- Consulting and contact information
- Frequently asked questions (FAQs)

APPENDIX

A

Relating Wavelet Transforms to Fourier Transforms

In the introductory chapter (Chapter One) and throughout the book we saw how the various wavelet transforms are related to the more familiar Fourier transforms. In this appendix we show some further comparisons to facilitate conceptual understanding.

A.1 Example of a Pathological Case Using the Fast Fourier Transform

To further illustrate the need for being able to use the wavelet transforms, we look at a pathological case using the FFT. Figure A.1–1 shows a high frequency sinusoid added to a slower frequency sinusoid. The FFT of this signal is displayed at right.

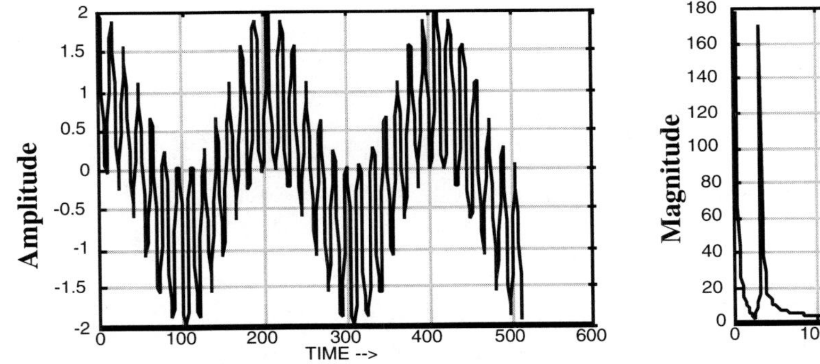

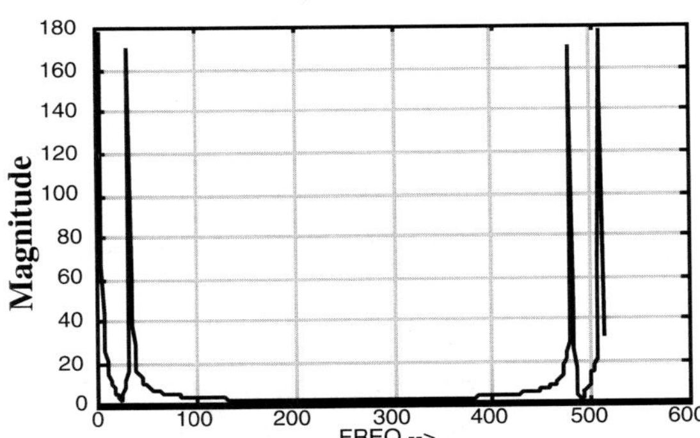

Figure A.1–1 Low frequency sinusoid with higher frequency sinusoid superimposed is shown at left. The Fast Fourier Transform of this signal is shown at right.

Now consider a much different signal as shown in Figure A.1–2. This time we have the low frequency sinusoid by itself for the first half of the signal followed by the higher frequency sinusoid for the second half of the signal. The FFT is shown at right. Comparing this FFT with that in the previous graph shows the two FFTs to look very similar. In other words, looking at an FFT by itself gives very little information about the shape of the signal.

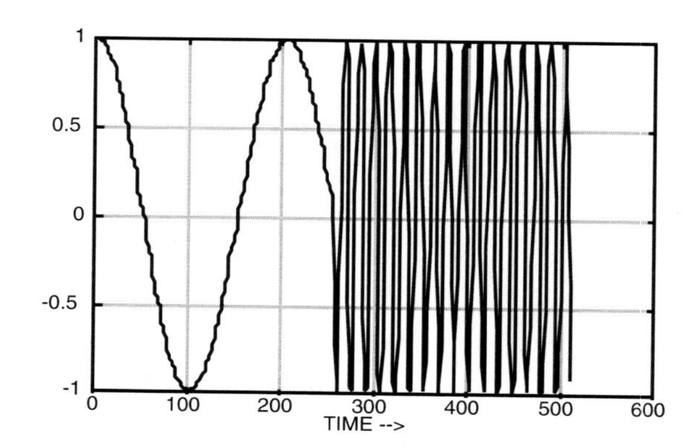

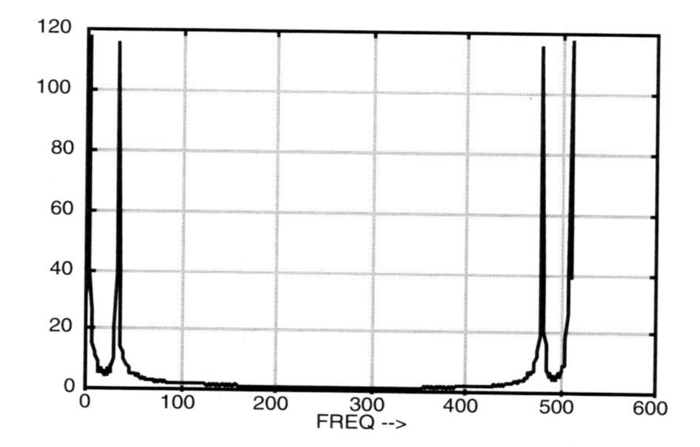

Figure A.1–2 In this example the low frequency sinusoid is followed by the higher frequency sinusoid. The FFT is shown at right and is disturbingly similar to the FFT in Fig. A.1–1.

Knowing the shape of the signal in advance, we can more intelligently process the data. The signal shown in Fig. A.1–1 is *stationary* and is an excellent candidate for a conventional FFT. For the signal in Fig. A.1–2 we can perform two Short-Time FFTs (STFT)—one for the first half and one for the 2nd half. But we do not usually know *a priori* the shape of the signal and even if we can discern start and stop times for the STFTs, the process can become tedious. Wavelets, on the other hand, can produce better results and do it automatically as we have described in the various chapters.

A.2 FFT and STFT Results Shown In Continuous Wavelet Transform Format

We can better understand the display of the Continuous Wavelet Transform (CWT) by starting with an FFT display. Figure A.2–1 shows an arbitrary FFT display in the familiar form at left and then rotated clockwise 90 degrees (center graph). At right, the rotated form is shown with the magnitude indicated by *brightness* (or color) as is done in the CWT. Notice that although frequency decreases in the y direction, the *scale*, as we have discussed in Chapter One, now *increases* in the y direction.

Notice also that the above FFT was performed for a specific integration interval and can be assigned only one specific time (usually the start of the interval).

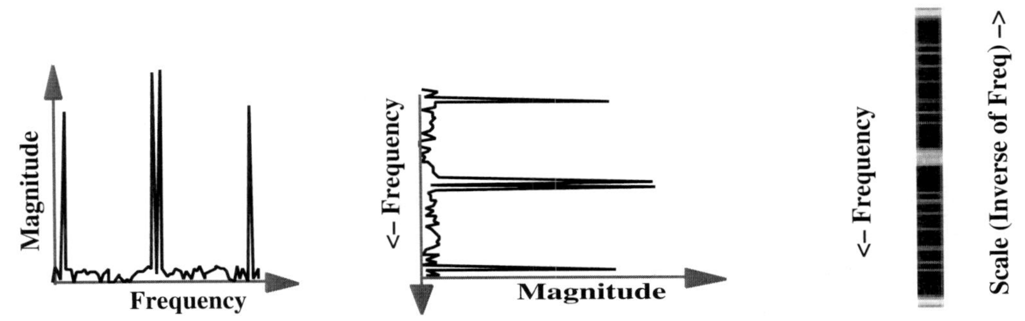

Figure A.2–1 FFT or "Frequency Domain" of an arbitrary signal with large components near DC and near Nyquist Frequency is shown in the familiar form at left. The same FFT is next shown rotated clockwise by 90 degrees. Note that frequency now *decreases* in the y direction. The rotated middle graph is now plotted at right with magnitude indicated by brightness. The bright spots at the top, middle, and bottom of the right graph indicate the peaks (large magnitude).

Suppose we now wished to look at the FFT of the data for the *next consecutive* integration interval. In the wavelet format of the right graph above we would have a different graph with a slightly later time assigned. Figure A.2–2 shows this process repeated 9 times with our first graph and 8 more graphs. Notice that we now have *time* (9 distinct times) increasing in the **x** direction, *scale* (inverse of frequency) increasing in the **y** direction, and magnitude (at a particular time and frequency) indicated by brightness ("z" direction).

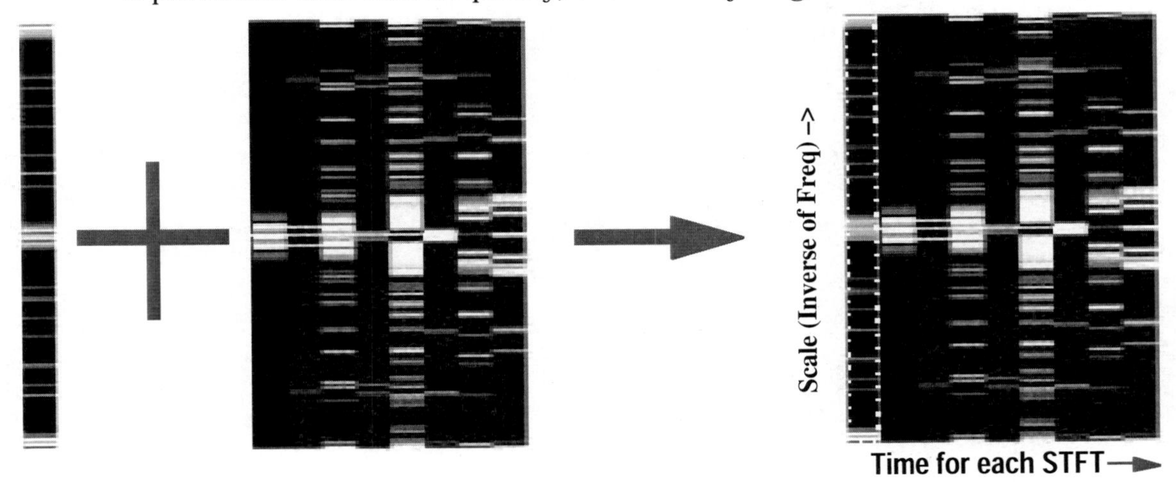

Figure A.2–2 The FFT result (for a short time interval) as displayed in the previous figure are now added to 8 additional FFT results from subsequent time intervals to form an **x-y-z** 3-D representation of time, frequency, and magnitude with **x** representing *increasing time*, **y** representing *decreasing frequency* (but increasing scale), and brightness representing *magnitude* (z). The first FFT results are indicated by the dotted line.

Note that all 9 time intervals are equal and that this figure is a depiction of 9 Short-Time (Discrete) Fourier Transforms (STFT). A Continuous *Wavelet* Transform (CWT) will have longer time intervals as needed for the lower frequencies and shorter time intervals for the higher frequencies (less stretching o *scaling).* This general 3-D form (x, y, and intensity or color), however, is used for the CWT displays and can now be better related to the familiar FFT displays.

A.3 The Wavelet Terms "Approximation" and "Details" Shown in FFT Format.

Although we touched on the meaning of the terms "Approximation" and "Details" as used in wavelet terminology in the book, we now show graphically a comparison with the results of an FFT. Consider a 2 MHz signal with a smaller 60 MHz noise added as shown in Figure A.3–1 in the time and frequency domains. *.

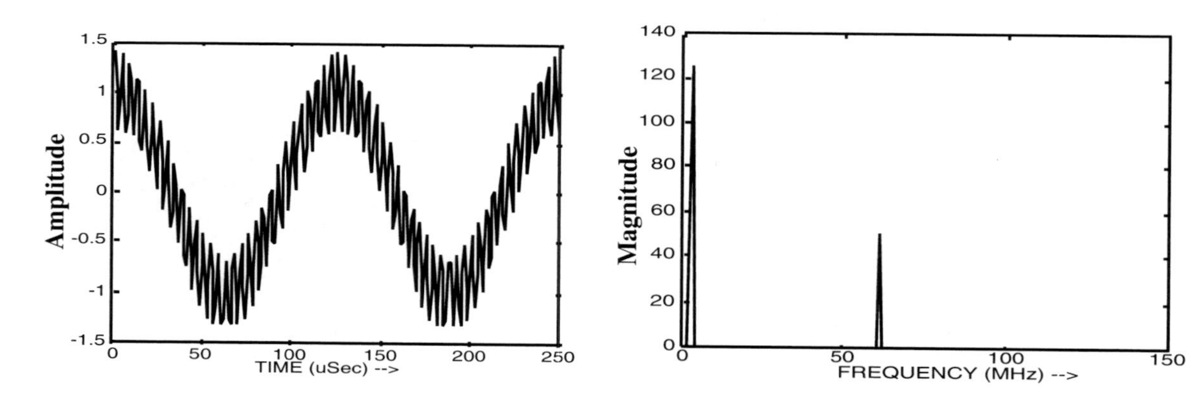

Figure A.3–1 Noisy signal shown in both time and frequency domains. Note this combined signal is *stationary* and a good candidate for an FFT rather than the wavelet transforms.

De-noising of this signal is accomplished by lowpass filtering in the time domain. After we have de-noised the signal we are left with the results shown in Figure A.3–2 at left. Notice this follows the general pattern of the original noisy signal and we could call it an "Approximation" in wavelet terminology.

* It will be recalled from Digital Signal Processing (DSP) that because time and frequency have an inverse relationship, this example could be presented using time in Seconds and frequency in Hertz, in MilliSeconds/KiloHertz, MicroSeconds/MegaHertz (as shown here) or any other equivalent inverse relationship.

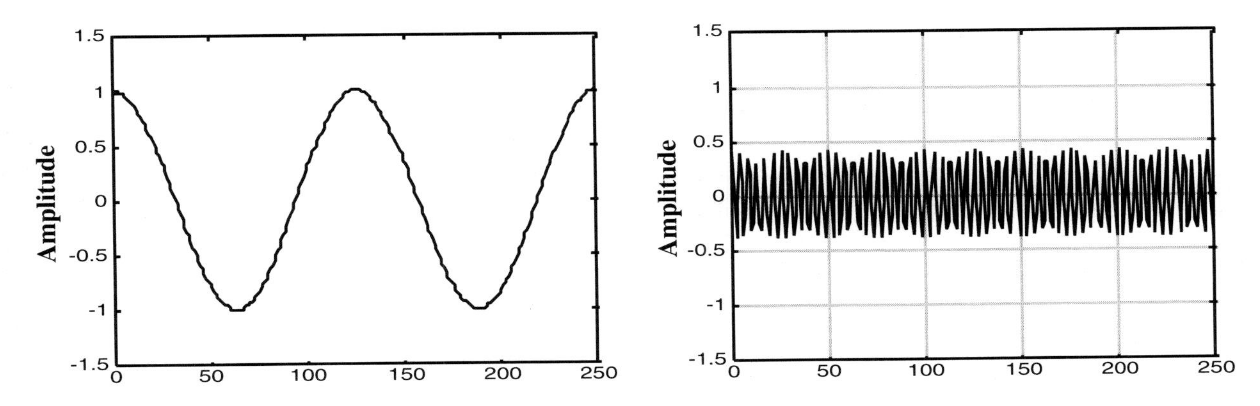

Figure A.3–2 The low frequency part of the signal *approximates* the shape of the original noisy signal and could be considered an "*Approximation*" as shown at left. While the high frequency part (the noise) could be considered the "*Details*" as shown at right.

If we were to instead *highpass filter* the noisy signal we would have the results as shown in the right graph. This result could be thought of as the smaller "Details" of the signal in wavelet terminology.*

As is done in wavelets, if we combine the Details and the Approximation we get the original noisy signal back. In other words if we were to call the lowpass portion "A" and the highpass portion "D" we would have for the signal

S = A + D

which is very similar to

S' = A1 + D1

which we encountered in the DWT and UDWT.

* It is unfortunate that such common English words such as "*approximation*", "*details*", "scaling", and "*translation*" are appropriated for specific wavelet terminology in most texts. For example, describing a 768 point stretched Db4 wavelet filter as an *approximation* of a continuous scaling function would be more descriptive than calling it an "estimation" as we do in this book. Engineers usually use the term "scaling" to indicate changing the amplitude (multiplication in the **y** direction) instead of *stretching* in the **x** direction and thus lowering the frequency as is done with wavelets. "Translation" usually means changing languages (e.g. from MATLAB to C++). Capitalizing the terms helps a little. For example, looking at the "*Details*" indicates a specific end result in wavelets such as **D1** or **D2** rather than the "fine points" such as the *details* of a contract. The use of these terms is so widespread in wavelets, however, that boycotting them would be a disservice in an introductory book such as this one. In Borg terminology "We will comply. Resistance is futile." ;-)

A.4 The FFT Presented as a Sinusoid Correlation (Similar to Wavelet Correlation)

We portrayed the DFT/FFT. as a correlation with sinusoids (sines and cosines) in our discussion of wavelet orthogonality (ref. Fig. 8.5–2). We now look in a little more detail and then compare with the wavelet correlation process in a Continuous Wavelet Transform (CWT). We will draw the discrete sinusoids as continuous in this section for simplicity and conformity

Figure A.4–1 shows a DC (Direct Current or zero Hz) "sinusoid" and a stationary sinusoidal signal. If we were to compare the 2 using a dot or inner product we would have a result of zero. We can visually discern this because for every positive value in the stationary signal there is a negative value. Multiplying these values by a constant (DC) will still leave values that will cancel out.

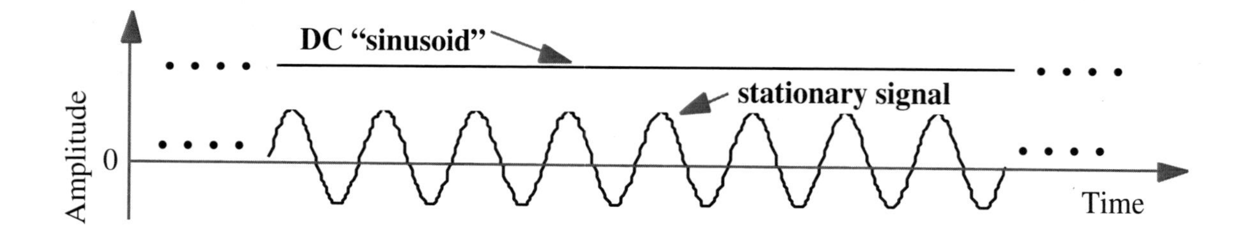

Figure A.4–1 A stationary signal, as would be best processed using an FFT, is compared with a DC sinusoid. The result of the dot product of these 2 would be zero. Note that this could be considered a single-point correlation.

We next look at the (single-point) correlation or dot product of a low frequency sinusoid with our same signal. From Figure A.4–2 we can see that the multiplied values in time interval #1 will cancel those in time interval #2. and, with this cancellation repeated, we have a dot product of zero again.

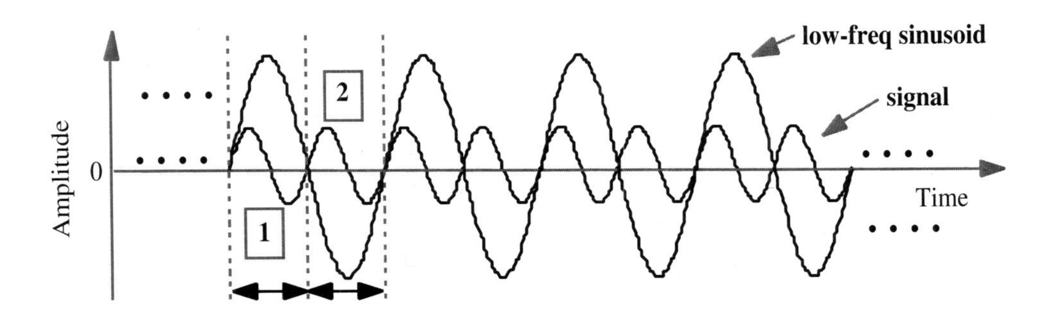

Figure A.4–2 The same signal is now multiplied by a sinusoid at half the frequency of the signal. The values in the first 2 time intervals (first 2 cycles of the signal) will cancel. In fact, looking closer we can see that the values in *either* of the first 2 time intervals will cancel.

We next look at the dot product when the sinusoid is at the same frequency as the signal as shown in Figure A.4–3. Now in time intervals #1 and #2 we have positive values times positive values and then negative values times negative values—each yielding a positive contribution. Thus when the frequencies are the same we have a large positive value for the dot product.

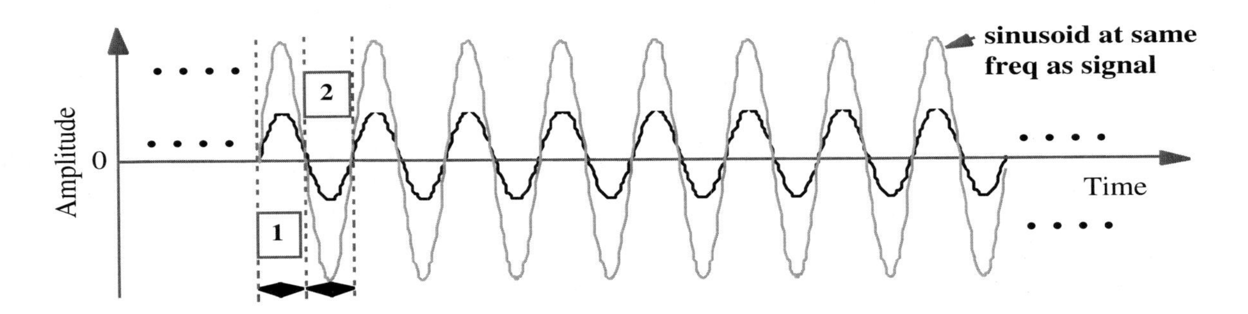

Figure A.4–3 The signal is now multiplied by a sinusoid having the same frequency as the signal. The values in the first time interval will be **positive x positive = positive**. The values in the second time interval will be **negative x negative = positive**. All other time intervals yield positive results.

We look at one final case where the sinusoid is twice the frequency of the signal. As can be seen from Figure A.4–4, the values in time interval #1 cancel those of time interval #2. In fact there is cancellation in each of the time intervals.

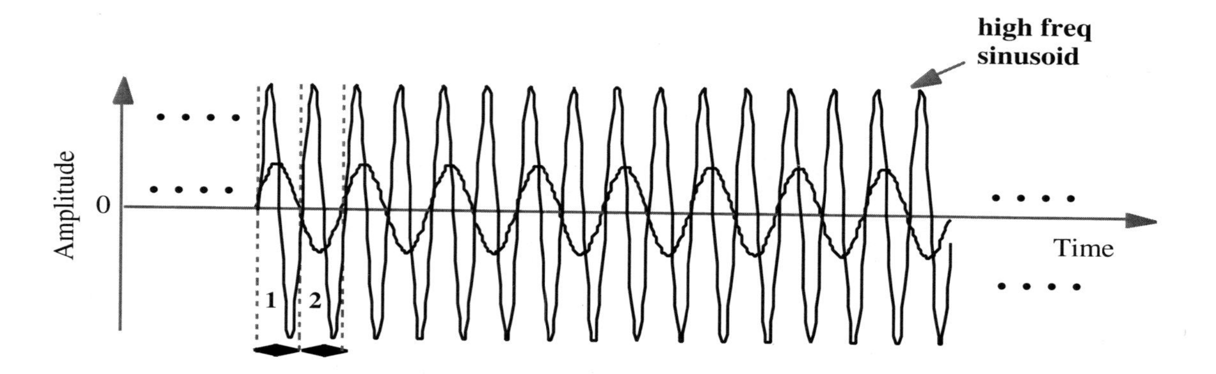

Figure A.4–4 The same signal is now multiplied by a high frequency sinusoid. The values cancel giving a value for the total dot product of zero.

We can plot these 4 dot products as shown in Figure A.4–5. The results are somewhat like the familiar FFT (although we did this for only 4 data points).

The FFT is comparable to a *prizm* that breaks up the signal into various frequencies (colors). In this case we had a monochromatic or single color. The "prism" showed only one "spectral line". We can do the inverse and reconstruct the signal by a "weighted sum" of these sinusoids (only 1 in this case).

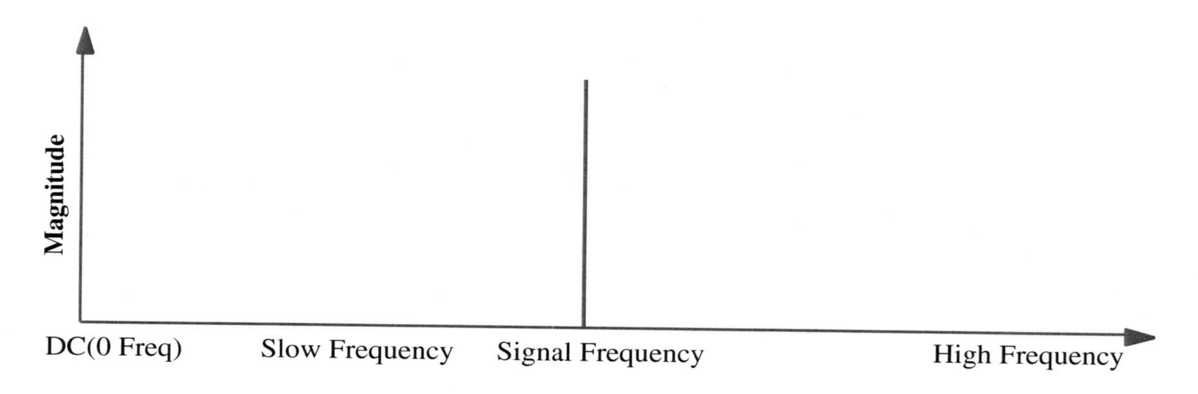

Figure A.4–5 A simple FFT-type plot showing magnitude versus frequency. For the 4 scenarios we saw we should have zero dot products except when the comparison sinusoid was at the same frequency as the signal.

An FFT/DFT is basically a set of dot products similar to what we have shown here except the set can be much larger and includes imaginary sines as well as real cosines.* The principles of *comparison, correlation* (single point) and *dot product* are still the same, however.

A continuous wavelet transform, shown in terms of the simplified depiction of the DFT/FFT presented above can be seen as follows. Figure A.4–6 shows an arbitrary signal. The finite-length wavelet is then compared to only the very first part of the signal (recall the wavelet is zero at all other times). This produces a single value for this dot product. In other words, we

- Compare the wavelet to the section at the start of the signal.
- Calculate a number, **C**, that represents how closely correlated the wavelet is with this section of the signal. The higher **C** is, the more the similarity.
- If the signal energy and the wavelet energy are equal to one, C = the correlation coefficient which can be thought of as a "Resemblance Index". If the wavelet looks like that portion of the signal, we will have a strong correlation and large dot product or correlation coefficient.

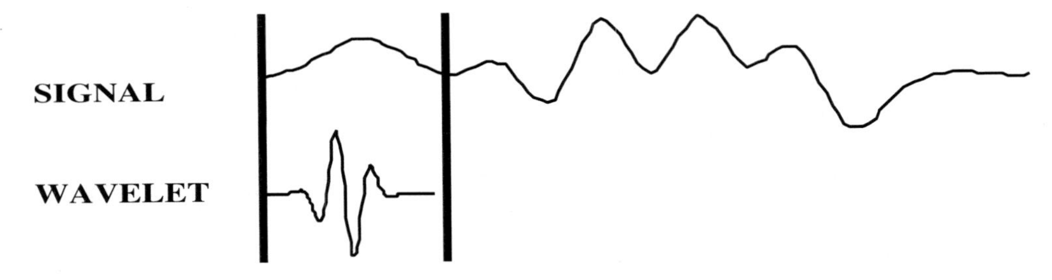

Figure A.4–6 The wavelet is correlated with the first part of the signal. The dot product (or the correlation coefficient if the energies are one) will be small here because we can visually discern there will be much cancellation of positive and negative values.

As shown in Figure A.4–7 we next shift (*translate* in wavelet terms) the wavelet (in time) slightly to the right. We then calculate another correlation coefficient or dot product.

* The DFT equation repeated for convenience here defines **X(*k*)** as

$$\sum_{n=0}^{N-1} x(n)\cos(2\pi nk/N) - j\sum_{n=0}^{N-1} x(n)\sin(2\pi nk/N)$$

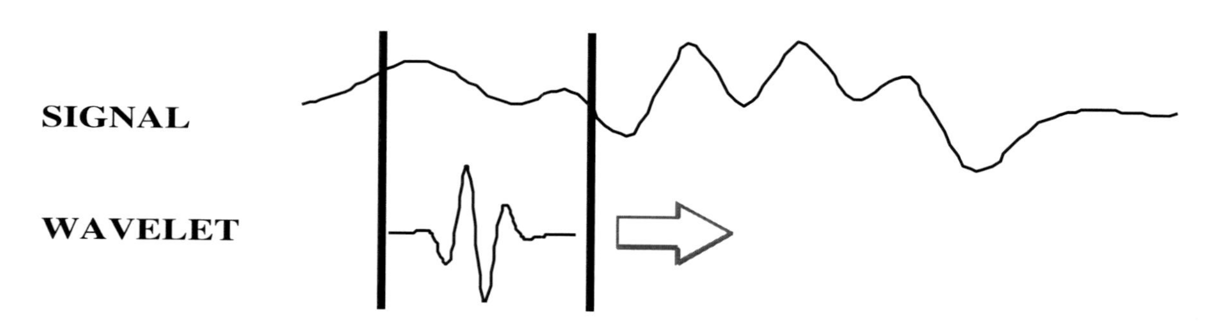

Figure A.4–7 The shifted wavelet is now correlated with the next part of the signal (recall that the wavelet values are zero before the start and after the end of the wavelet so we can take the dot product of the wavelet with the entire signal). Again, we can visually discern there will be much cancellation of positive and negative values and this correlation coefficient will be small.

This process is repeated as the wavelet is shifted along the entire signal. Note we have just performed a cross-correlation of the signal with the basic wavelet. These results will be plotted on the CWT display at a very low scale (unstretched or *unscaled* wavelet means high frequency).

We now proceed, in the CWT, to stretch the wavelet slightly as shown in Figure A.4–8. We calculate a correlation coefficient at beginning of signal as we did before, but this time with the stretched wavelet. We next shift the stretched wavelet slightly to the right (by the same amount as for the original wavelet) and calculate another correlation coefficient for this stretched or "dilated" wavelet

This process is repeated for the entire signal. We have now done a 2nd cross-correlation—that of the signal with the slightly stretched wavelet. These results will appear at the next larger scale on CWT display.* Recall that the CWT display shows time in the **x** direction, scale (inverse of frequency) in the **y** direction and correlation intensity as brightness or color for the **z** direction (ref. Fig. 1.6–3).

* The author, D. Lee Fugal, uses a child's "slinky" toy to demonstrate the inverse relationship between *stretching* and *frequency* in his courses and lectures. A slightly stretched (or scaled in wavelet terminology) slinky looks a lot like a sinusoid (if considered as 2-Dimensional). As the slinky is stretched further, it looks like the frequency has been lowered. As it is relaxed (less stretching or scaling) it appears to have a higher frequency.

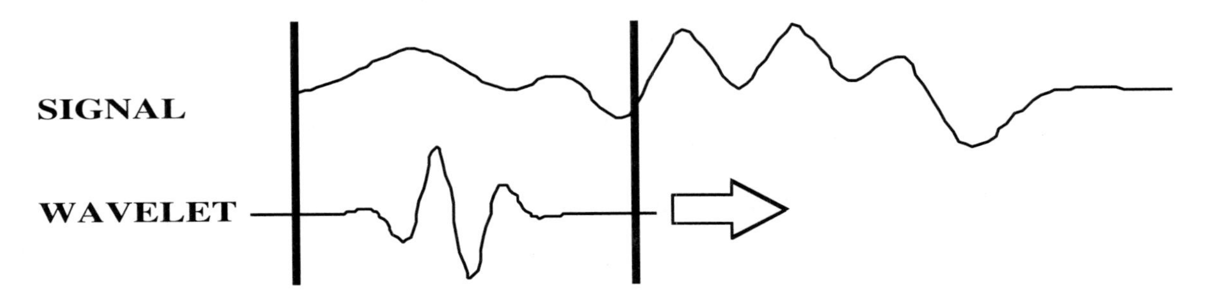

Figure A.4–8 The basic or "mother" wavelet is stretched slightly and we begin again the correlation process of the stretched wavelet with the signal. We can see that the single-point correlation will still be weak.

We continue the process by stretching the wavelet, starting at the beginning of the signal and performing cross correlations.

If we did this at 8 levels (the original "mother" wavelet and 7 increasingly stretched versions) and for 300 steps in time we would have a 300 by 8 set of correlation coefficients.

Recall that in the above DFT/FFT simplification the sinusoids at the various frequencies were compared to the full length of the signal giving a single value for the dot product for each frequency.

Figure A.4–9 shows a particular stretching and shifting (*dilation* and *translation*) in which the wavelet will correlate well with the signal. This will be indicated by a bright area on the CWT display.

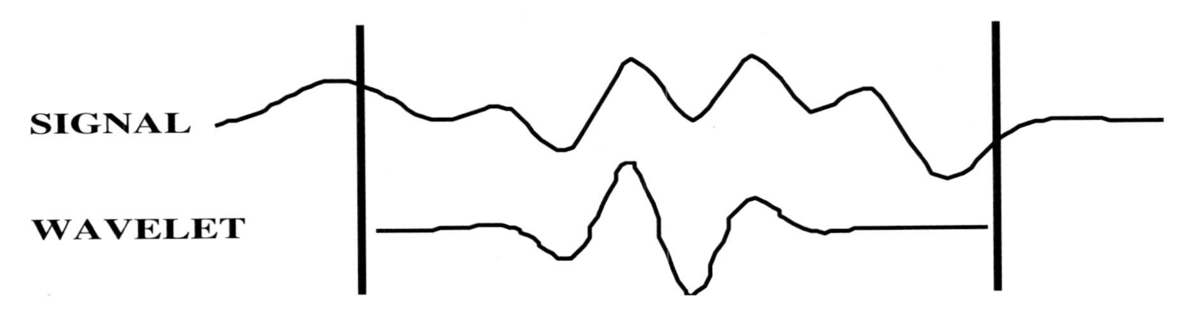

Figure A.4–9 The wavelet is now stretched and shifted to where it "matches" a portion of the signal fairly well. The dot product at this stretching and shifting should be fairly large.

Here is a "*tip*": If you know what the waveform you are trying to find looks like, choose or build a wavelet that resembles that waveform and the CWT display will indicate where it found it in time and frequency (scale). If you do *not* know what the waveform looks like, try a series of wavelets in software (very quick and easy) until you get a strong correlation in the form of a very bright area. Then you will know the time, frequency (scale) and the general form of the event or waveform (because it "matches" the wavelet that gave you the strong correlation and bright spot.

A.5 The Ordinary Acoustic Piano: An Audio Fourier Transform

The 88 strings* of an ordinary acoustic piano each resonate to a different frequency. In his lectures and courses, the author, D. L. Fugal, plays a recording of the strings resonating after he has said the word "Hello". You can clearly hear the frequencies of "Hello" but you have no time information. This can be thought of as an audio version of the Discrete Fourier Transform.

The author next splits the word "Hello" into 2 words "Heh" and "Low". When the 2 "audio DFTs" are played one after another, the results start to sound a little like "Hello". This is a simple demonstration of an (audio) Short Time (Discrete) Fourier Transform or STFT.

We can continue this audio STFT process with shorter and shorter recording (integration) times until the word "Hello" becomes recognizable. If we make our recording time too short, however, we will not be able to record full cycles of the lower notes. For example the lowest string on a piano vibrates 55 times a second (55 Hz) so very short recording times would not pick it up. Wavelets would not have this problem.

Recordings of these audio DFTs and STFTs can be found on the author's website at **www.ConceptualWavelets.com**.

Another demonstration, mentioned in a footnote in Chapter Eight, is repeated here.

---

*Pianos have additional strings for the higher notes. For example, the 3 strings on high "C" are all at the same frequency but as the hammer strikes them they produce 3 times the volume to compete with the large single strings on the low notes.

"In his short courses and seminars the author demonstrates this concept by playing his "Fugal Bugle" along with several other musical instruments in a rendition of Yankee Doodle—but with one of the several Middle C notes played wrong (an easy task for him, but for this demonstration done on purpose).

Denoising using Fourier techniques gets rid of the wrong note but also all other "C notes" at that **particular frequency**. *Denoising using time-series techniques gets rid of the wrong note but also all other notes at that* **particular time**. *Denoising using wavelet techniques, however, gets rid of the wrong note but leaves all the other notes untouched."*

This demonstration can also be downloaded from the author's website.

Outside of a dog, a book is a man's best friend. Inside of a dog it's too dark to read.

—Groucho Marx

APPENDIX B

Heisenberg Boxes and the Heisenberg Uncertainty Principle

We touched on the Heisenberg Uncertainty Principle as applied to wavelets in Chapter Ten. Despite the sophisticated-sounding name, this principle is actually very intuitive when applied to wavelets and signal processing in general—as we will now demonstrate.

B.1 Natural Order of Time and Frequency

In the introductory Chapter One (Section 1.4) we showed how ordinary sheet music is very much like a *discrete* wavelet transform. This figure is reproduced here as Figure B.1–1. We know, for example that in signal processing lower frequencies require longer integration times to determine how *frequent* these slower varying cycles occur and that the Tuba cannot change frequencies (musical notes} as quickly as a Piccolo. This is a "natural order" of time and frequency of waveforms and can be seen in signals, music, or watching water waves from a pier.

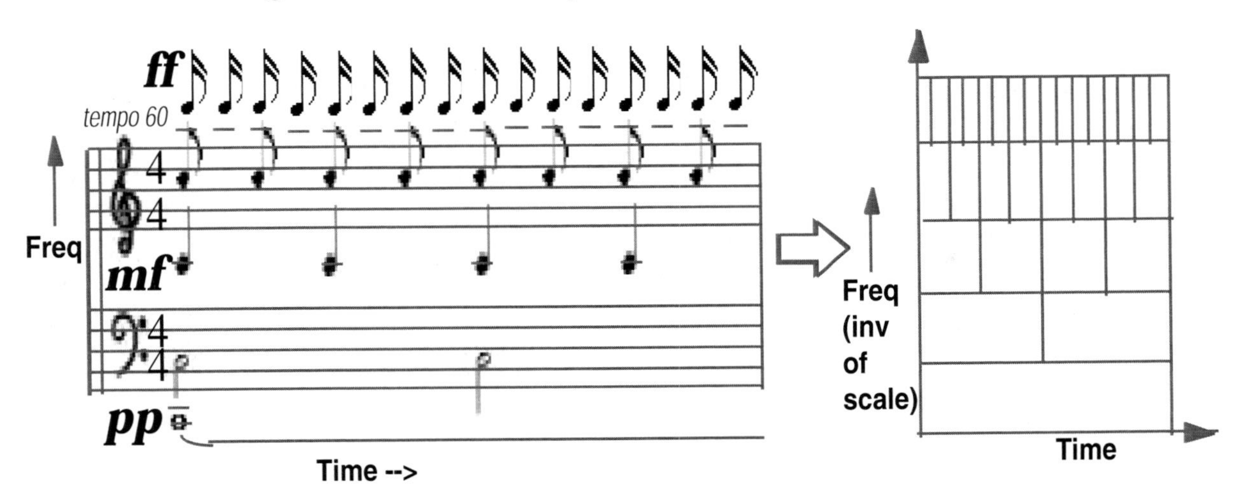

Figure B.1–1 Comparison of ordinary sheet music to the time/frequency map of the *Discrete* Wavelet Transform (DWT)

We also mentioned in this earlier discussion that sheet music uses a logarithmic (base 2) scale—each "C note" in Figure C.1–1 is twice the frequency of the one below it.

B.2 Heisenberg Boxes (Cells) and the Uncertainty Principle

With middle C on a piano being approximately 250 Hz we have for the notes depicted 62.5, 125, 250 , 500 , and 1000 Hz*. We now redraw the log base 2 graph from Figure B.1–1 with some notes indicated by frequency as shown below in Figure B.2–1.

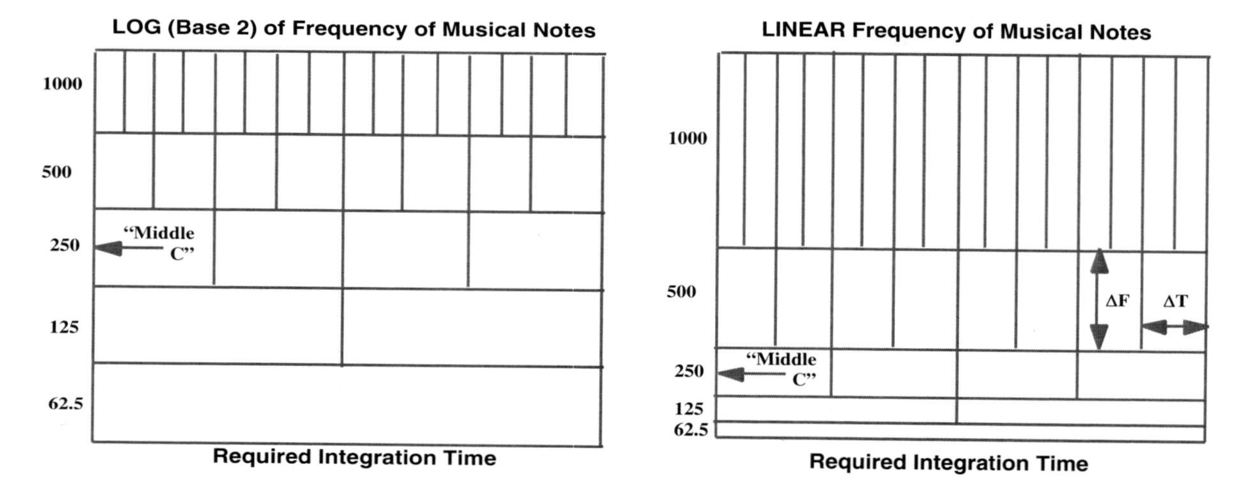

Figure B.2–1 Sketch of *sheet music* format, first in the familiar *logarithmic* (base 2) form and then in *linear* form showing where 5 octaves of the note "C" on a piano, from Deep C to Soprano C, might appear (integration time is linear in both cases).

In the linear graph (right) we notice that the areas in the various boxes are the same (this is not surprising since time and frequency are inverses). Those at the bottom are short and fat while those at the top are tall and thin.

* These approximate frequencies of Middle C and the 2 octaves above and below it are used here for simplicity and clarity in instruction. The exact frequencies (in Hz) for all 8 of the "C" notes on a piano are 32.7032, 65.4064, 130.813, 261.626, 523.251, 1046.50, 2093.00, and 4186.01 on a tuned piano. My own piano needs tuning up, so it's probably closer to the "62.5, 125, 250, 500, 1000" values used above.

These are called *Heisenberg Cells* or *Heisenberg Boxes* and the boxes being the same size is related to the *Heisenberg Uncertainty Principle.*

Stated again here, the Heisenberg Uncertainty Principle for Quantum Physics says that you can't know the exact position and the exact momentum of a particle simultaneously ($\Delta X \Delta P \geq h/2$ where h is Planck's Constant). In time/frequency analysis this principle refers to the fact that you can't know the exact time and the exact frequency of a signal simultaneously ($\Delta T \Delta F \geq$ non-zero constant).

These cells or boxes with Δ Integration Time as the width and Δ Frequency as the height remain at a constant value. One of the powerful capabilities of wavelets is that the wavelets also follow this "natural order". The basic or unstretched wavelet has a short time and a high frequency. As it is stretched is has a longer time and a lower frequency. Identifying pulses, "glitches", events, or other signals that vary over time depends upon matching the analyzing waveform to data. Wavelets with their amoeba-like behavior are ideally suited to this task.

B.3 Short Time Fourier Transforms are Constrained to Fixed Heisenberg Boxes

Until the introduction of wavelet transforms, the Short-Time Fourier Transform was the method of choice for dealing with both time and frequency. Instead of taking the FFT of the entire length of the signal—which gives us the frequency information but no information about time—the signal would be divided into shorter time segments (Δ Time). Taking the FFT of each of the time segments then gives us both discrete frequencies (Δ Frequency) and the time associated with that particular segment as shown in the simplified sketch in Figure B.3–1.

As can be observed from this diagram (see also sheet music in Fig. B.1–1) the lowest "Low C whole-note" at 62.5 Hz may be poorly defined due to the integration interval being so short. Meanwhile, the 16 very short Soprano C notes above the Treble Clef would have only 4 times to represent them.*

* At tempo = 60, each "quarter note" plays for 1 second and each of the 4 STFTs would be 1 second long. With 400 STFTs each 0.01 seconds long we would have 10 cycles of the highest notes in integration time but less than 1 cycle (0.625) for the lowest note. With the frequency range of a piano going from 27.5 Hz (lowest "A") to 4186 Hz (highest "C") this becomes a sticky problem trying to use STFTs.

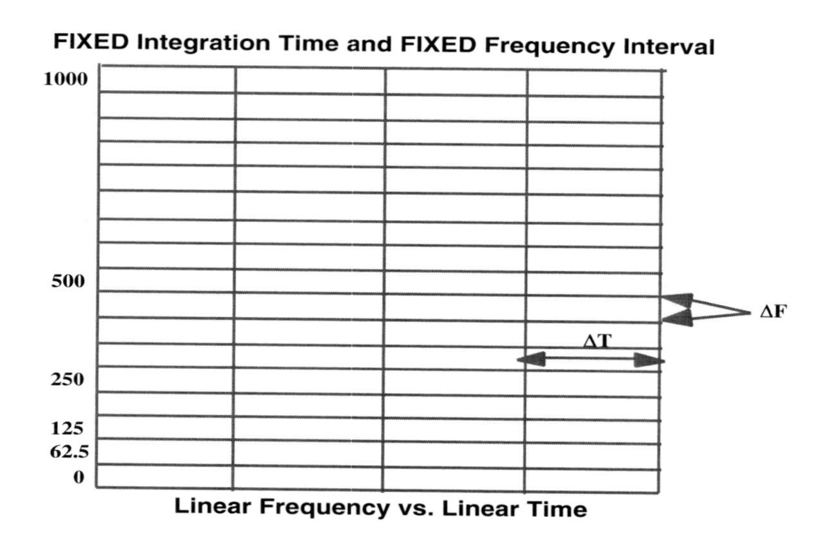

Figure B.3–1 Sketch of the time/frequency representation of an STFT. A fixed-length FFT is taken for each of the 4 time intervals in this simplified diagram. Δ Time and Δ Frequency are fixed.

Another example of the difficulty in using an STFT is portrayed in Figure B.3–2 below. Here we have the quadratic chirp signal that was used to generate the cover of this book. If we use 2 windows (left) and take the FFT of the signal in each window we will possibly have enough integration time to identify at least an average of the low frequencies of the first half of the chirp signal. On the other hand, the high frequencies at the end have no time information associated with them. other than "they are in the 2nd half of the signal".

If we use 4 windows (right graph) and take the FFT for each of the 4 time periods we will still have plenty of time to identify the higher frequencies at the end of the signal and be able to assign those frequencies to one of the 4 windows. However, the first window is not long enough to provide much information about the low frequency signal. A closer look will reveal that the signal has not even completed one cycle in this STFT configuration.

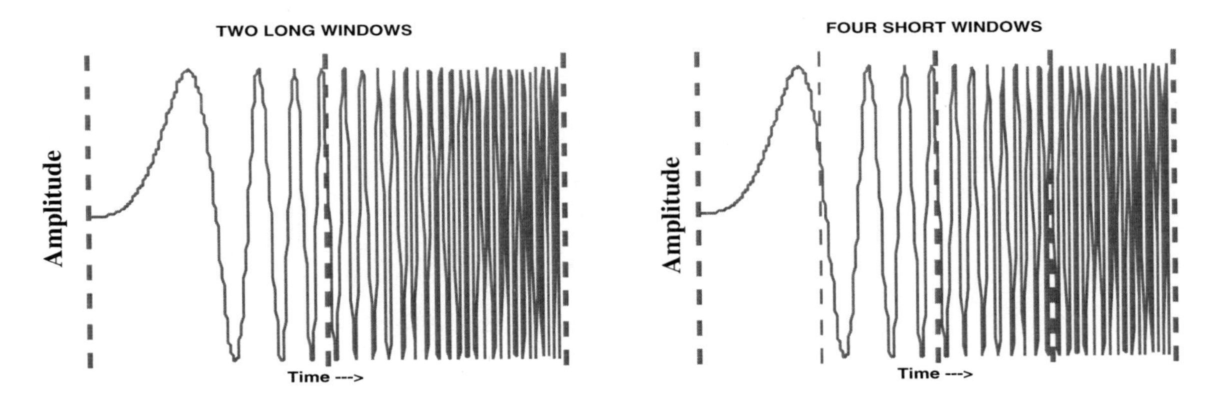

Figure B.3–2 Amplitude vs. Time plot is shown at left for a quadratic chirp signal. The time interval is split into 2 windows so an FFT can be performed on the data in each window. This produces two Short-Time Fourier Transforms. The same signal is shown on the right but with four shorter windows leading to four STFTs.

If we know the signal very well we can perhaps custom-tailor a series of the STFT Δ Time and Δ Frequency intervals for good results. But if we don't know the signal or if the signal has variable frequencies or finite "*events*" during its duration, then the wavelet transforms are a far better method. Wavelets have been referred to as a *mathematical microscope* for their uncanny ability to mimic the behavior of signals and waves found in nature.

In conclusion, although FFTs and even STFTs are the method of choice for certain pre-known signals, the CWT and DWT are also the method of choice for the unpredictable and/or non-stationary signals we encounter each day. By following the Heisenberg Uncertainty Principle as applied to Time/Frequency analyses, wavelet transforms are a natural and powerful method of Digital Signal Processing.

"Since the mathematicians have invaded the theory of relativity, I do not understand it myself anymore."

—Albert Einstein

APPENDIX C

Reprint of Article "Wavelets: Beyond Comparison"

WAVELETS: "BEYOND COMPARISON" - D. L. FUGAL

Wavelets are used extensively in Signal and Image Processing, Medicine, Finance, Radar, Sonar, Geology and many other varied fields. They are usually presented in mathematical formulae, but can actually be understood in terms of simple comparisons with your data.

As a background, we first look at the Discrete Fourier Transform (DFT) or it's faster and more famous cousin, the Fast Fourier Transform (FFT). These transforms can be thought of as a series of comparisons with your data, which we will call for now a "signal" for consistency. Signals that are simple waves of constant frequencies can be processed with ordinary DFT/FFT methods.

Real-world signals, however, often have frequencies that can change over time or have pulses, anomalies, or other "events" at certain specific times. This type of signal can tell us where something is located on the planet, the health of a human heart, the position and velocity of a "blip" on a Radar screen, stock market behavior, or the location of underground oil deposits. For these signals, we will often do better with wavelets. We now demonstrate both the Fourier and Wavelet Transforms of a simple pulse signal.

The Discrete Fourier Transform/Fast Fourier Transform (DFT/FFT)

We start with a point-by-point comparison of the pulse signal (D) with a high frequency wave or "sinusoid" of constant frequency (A) as shown in Figure 1. We obtain a single "goodness" value from this comparison (a correlation value) which indicates how much of that particular sinusoid is found in our own pulse signal.

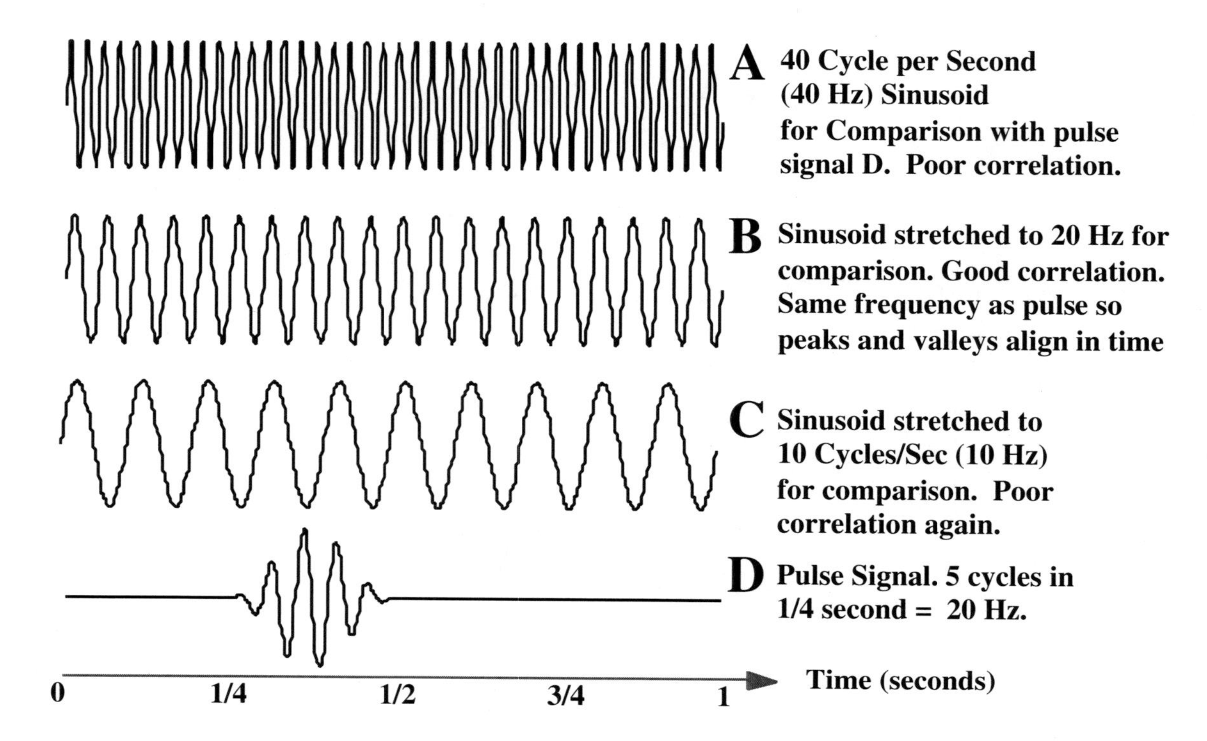

Figure 1 FFT-type comparison of Pulse Signal with several stretched sinusoids.

We can observe that the pulse has 5 cycles in 1/4 of a second. This means that it has a frequency of 20 cycles in one second or “20 Hz”. The comparison sinusoid, A, has twice the frequency or 40 Hz. Even in the area where the signal is non-zero (the pulse) the comparison is not very good.

By lowering the frequency of A from 40 to 20 Hz (waveform B) we are effectively “stretching” the sinusoid (A) by 2 so it has only 20 cycles in 1 second. We compare point-by-point again over the 1-second interval with the pulse (D) This next correlation gives us another value indicating how much of this lower frequency sinusoid (now the same frequency as our pulse) is contained in our signal. This time the correlation of the pulse with the comparison sinusoid is very good. The peaks and valleys of B and the pulse portion of D align (or can be easily shifted to align) and thus we have a large correlation value.

Figure 1 above shows us one more comparison of our original sinusoid (A) stretched by 4 and trimmed so it has only 10 cycles in the 1 second interval (C). This comparison with D is poor again. We could continue stretching and trimming until the sinusoid becomes a straight line having zero frequency or "DC" (named for the zero frequency of Direct Current) but all these comparisons will be increasingly poor.

An actual DFT (or functionally equivalent FFT) compares many "stretched" sinusoids ("analysis signals") to the pulse rather than just the 3 shown here. The best correlation is found when the sinusoid frequency best matches that of the pulse. Figure 2 shows the first part of an actual FFT of our pulse signal D. The locations of our sample comparison sinusoids A, B, and C are indicated. Notice that the FFT tells us correctly that the pulse has primarily a frequency of 20 Hz, but does NOT tell us where the pulse is located in time!

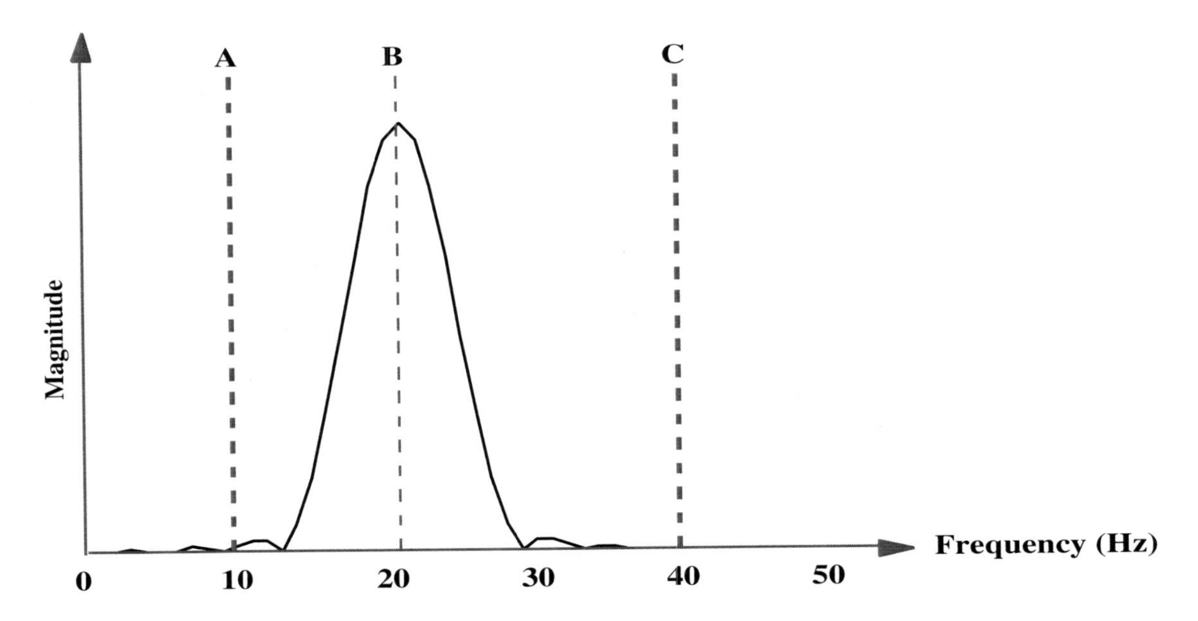

Figure 2 Actual FFT plot of above pulse signal with the three sinusoids indicated

The Continuous Wavelet Transform (CWT)

Wavelets are exciting because they too are comparisons, but instead of correlating with various stretched, infinite length unchanging sinusoids, they use smaller or shorter waveforms ("wave–lets").that can start and stop where we wish.

By stretching and shifting the wavelet numerous times we get numerous correlations. If our signal has some interesting events embedded, we will get the best correlation when the stretched wavelet is similar in frequency to the event and is shifted to line up with it in time. Knowing the amounts of stretching and shifting we can determine both location and frequency.

Figure 3 demonstrates the process. Instead of sinusoids for our comparisons, we will use wavelets. Waveform A shows a Daubechies 20 (Db20) wavelet about 1/8 second long that starts at the beginning (t = 0) and effectively ends well before 1/4 second. The zero values are extended to the full 1 second. The point-by-point comparison with our pulse signal D will be very poor and we will obtain a very small correlation value.

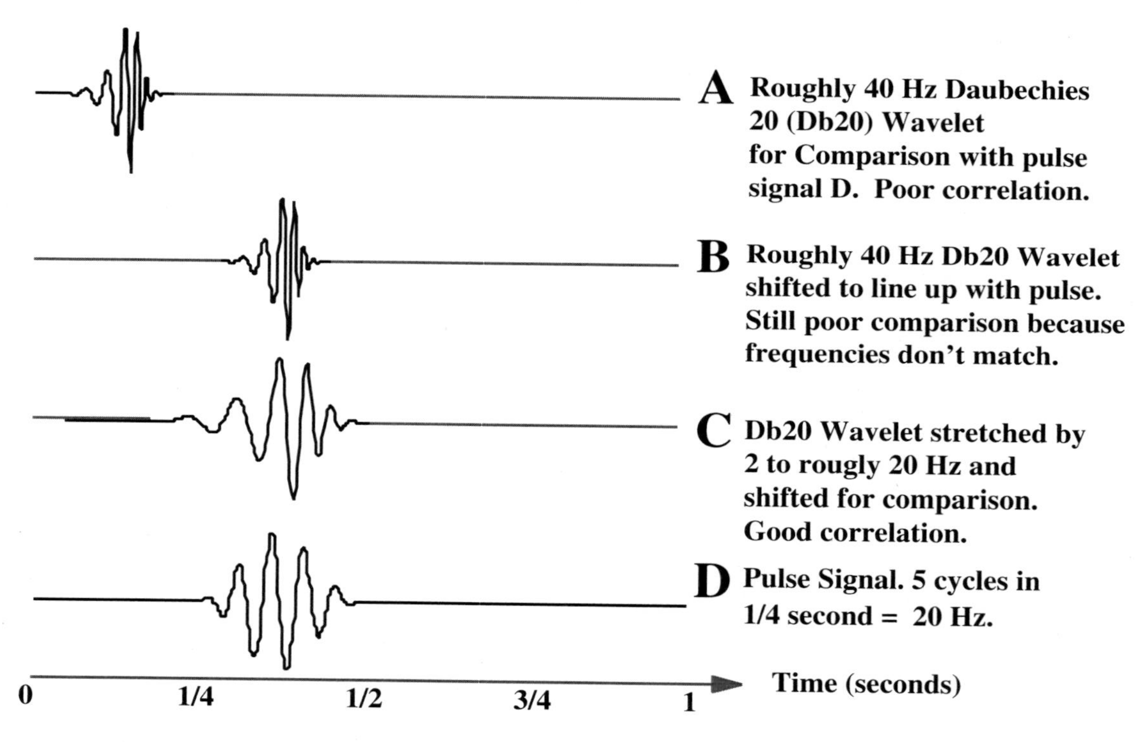

Figure 3 CWT-type comparison of Pulse Signal with several stretched and shifted wavelets

In the previous FFT/DFT discussion we proceeded directly to stretching. In the Wavelet Transforms we shift the wavelet slightly to the right and perform another comparison with this new waveform to get another correlation value. We continue to shift until the Db20 wavelet is in the position shown in

B. We get a little better comparison than A, but still very poor because B and D are different frequencies.

After we have shifted the wavelet all the way to the end of the 1 second time interval we start over with a slightly stretched wavelet at the beginning and repeatedly shift to the right to obtain another full set of these correlation values. C shows the Db20 wavelet stretched to where the frequency is roughly the same as the pulse (D) and shifted to the right until the peaks and valleys line up fairly well. At this particular shifting and stretching we should obtain a very good comparison and large correlation value. Further shifting to the right, however, even at this same stretching will yield increasingly poor correlations.

In the CWT we thus have one correlation value for every shift of every stretched wavelet. To show the data for all these stretches and shifts, we use a 3-D display with the stretching (roughly inverse of frequency) as the vertical axis, the shifting in time as the horizontal axis, and brightness (or color) to indicate the strength of the correlation. Figure 4 shows a Continuous Wavelet Transform (CWT) display for this particular pulse signal (D). Note the strong correlation of the 3 larger peaks and valleys of the pulse with the Db20 wavelet, the strongest being where all the peaks and valleys best align.

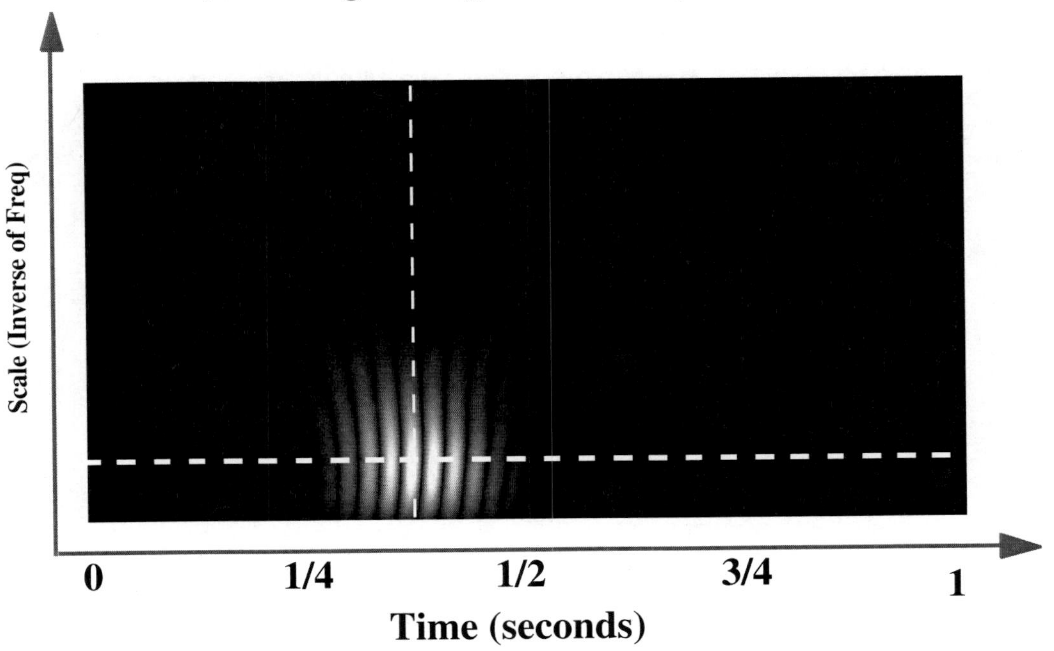

Figure 4 Actual CWT display indicating the time and frequency of the Pulse Signal

The display shows that the best correlation occurs at the brightest point or at about 3/8 second. This agrees with what we already know about the pulse, D. The display also tells us how much the wavelet had to be stretched (or "scaled") and this indicates the approximate frequency of the pulse. Thus we know not only the frequency of the pulse, but also the time of it's occurrence!

We run into this simultaneous time/frequency concept in everyday life. For example, a bar of sheet music may tell the pianist to play a C-chord of three different frequencies at exactly the same time on the first beat of the measure.

For the simple example above we could have just looked at the pulse (D) to see its location and frequency. The next example is more representative of wavelets in the real world.

Figure 5 shows a signal with a very small, very short discontinuity at time 180. The Amplitude vs. Time plot of the signal is shown at the upper left but does not show the tiny "event". The Magnitude vs. Frequency FFT plot tells what frequencies are present but does not indicate the time associated with those frequencies.

With the wavelet display, however, we can clearly see a vertical line at 180 at low scales when the wavelet has very little stretching, indicating a very high frequency. The CWT display also "finds" the large oscillating wave at the higher scales where the wavelet has been stretched and compares well with the lower frequencies. For this short discontinuity we used a short wavelet (a Db4) for best comparison.

This is an example of why wavelets have been referred to as a "mathematical microscope" for their ability to find interesting events of various lengths and frequencies hidden in data.

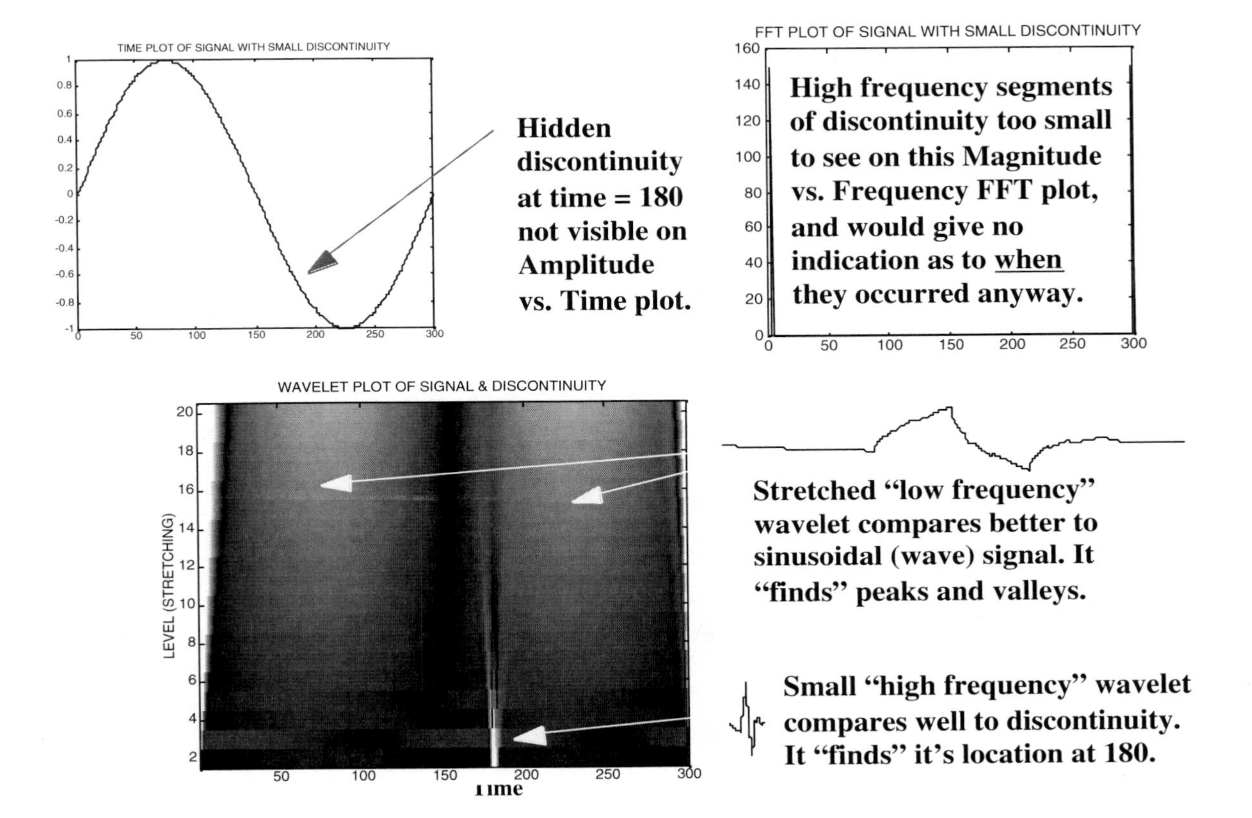

Figure 5 Comparison of Time Plot, Frequency Plot (FFT), and Wavelet Plot (CWT) of a signal with a hidden discontinuity

Discrete Wavelet Transforms Overview

Besides acting as a "microscope" to find hidden events in our data, wavelets can also separate the data into various frequency components, as does the FFT. The FFT/DFT is used extensively to remove unwanted noise that is prevalent throughout the entire signal such as a 60 Hz hum. Unlike the FFT, however, the wavelet transform allows us to remove frequency components at specific times in the data. This allows us a powerful capability to throw out the "bad" and keep the "good" part of the data in that frequency range.

These types of transforms are called "Discrete Wavelet Transforms" (DWT). They also have easily computed inverse transforms (IDWT) that allow us to reconstruct the signal after we have identified and removed the noise or superfluous data for denoising or compression.

Undecimated or "Redundant" Discrete Wavelet Transforms (UDWT/RDWT)

In one type of DWT, the Redundant Discrete Wavelet Transform or RDWT, we first compare (correlate) the Wavelet "filter" with itself. This produces a "Highpass Halfband Filter" or "superfilter". When we compare or correlate our signal with this superfilter we extract the highest half of the frequencies. For a very simple denoising, we could just discard these high frequencies (for whatever time period we choose) and then reconstruct a denoised signal.

Multi-level RDWT's allow us to stretch the wavelet, similar to what we did in the CWT, except that it is done by factors of 2 (twice as long, 4 times as long, etc.). This allows us stretched superfilters that can be halfband, quarter-band, eighth-band and so forth.

Conventional (Decimated) Discrete Conventional Transforms (DWT)

We stretched the wavelet in the CWT and the RDWT. In the conventional DWT, we shrink the signal instead and compare it to the unchanged wavelet. We do this by "downsampling by 2". Every other point in the signal is discarded. We have to deal with "aliasing" (not having enough samples left to represent the high frequency components and thus producing a false signal). We must also be concerned with "shift invariance" (do we throw away the odd or the even values?—it matters!).

If we are careful, we can deal with these concerns. One amazing capability of the filters in the conventional DWT is alias cancellation where the basic wavelet and 3 similar "filters" combine to allow us to reconstruct the original signal perfectly. The stringent requirements on the wavelets to be able to do this is part of why they often look so strange (see Figure 8).

As with the RDWT, we can denoise our signal by discarding portions of the frequency spectrum—as long as we careful not to throw away vital parts of the alias cancellation capability. Correct and careful downsampling also aids with compression of the signal. Modern JPEG compression uses wavelets. Figure 6 shows JPEG image compression. The image on the right was compressed by a ratio of 157:1 using a Biorthogonal 9/7 set of wavelets.

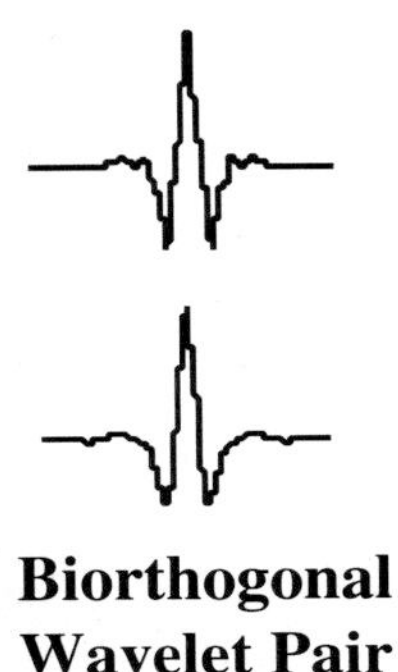

Biorthogonal Wavelet Pair

Figure 6 JPEG image compression of 157:1 using Biorthogonal 9/7 set of symmetrical Wavelets.

There are many types of Wavelets. Some come from mathematical expressions. Others are built from basic Wavelet Filters having as little as 2 points. The Db4 , Db20, and Biorthogonal wavelets shown earlier are examples of this 2nd type. Figure 7 shows a 768 point approximation of a continuous Db4 wavelet with the 4 filter points (plus 2 zeros) superimposed.

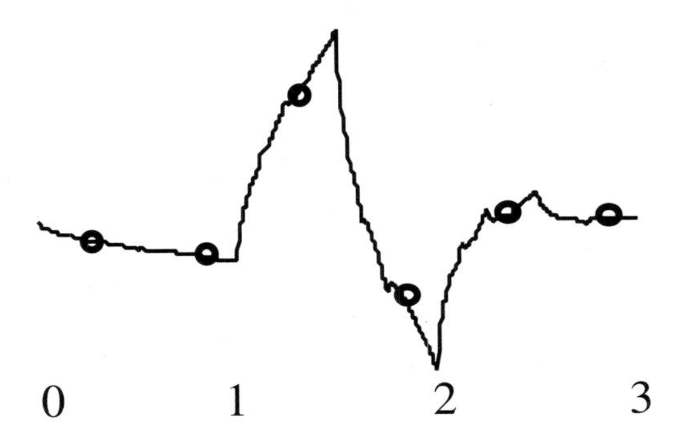

0 1 2 3

Figure 7 Daubechies 4 wavelet with 4 original filter points and 2 end zeros

Some wavelets have symmetry (valuable in human vision perception) such as the Biorthogonal Wavelet pairs. Shannon or "Sinc" Wavelets can find events with specific frequencies (these are similar to the Sinc Function filters found in traditional DSP). Haar Wavelets (the shortest) are good for edge detection and reconstructing binary pulses. Coiflets Wavelets are good for data with self-similarities (fractals) such as financial trends. Some of the wavelet families are shown in Figure 8.

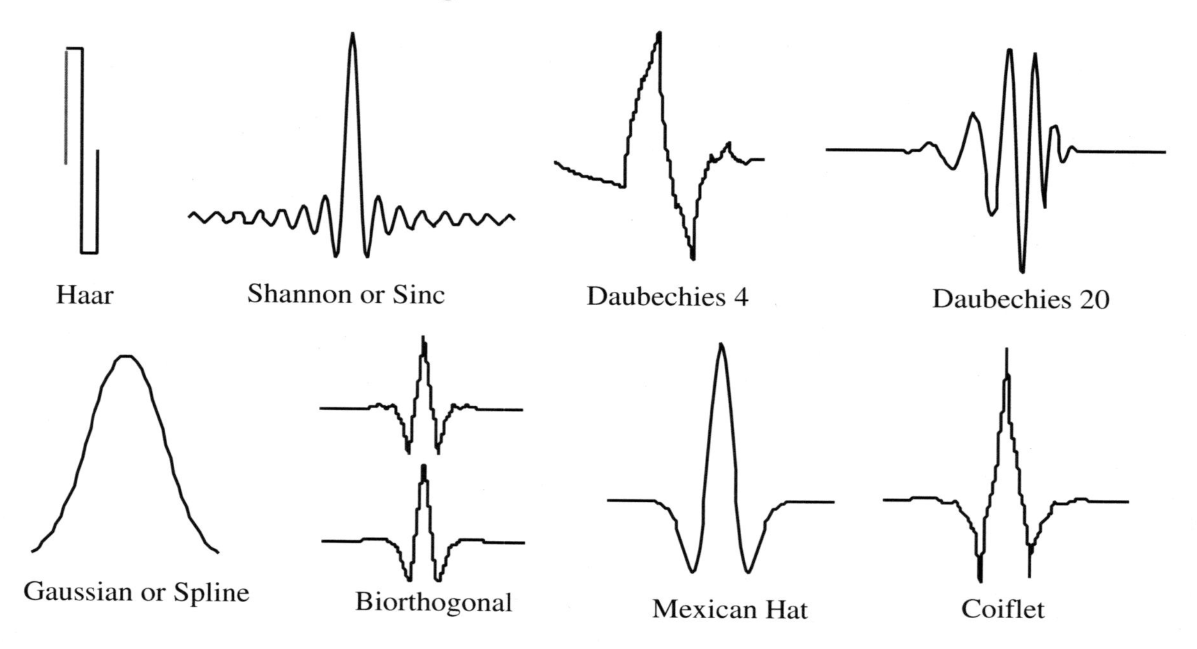

Figure 8 Examples of types of wavelets

You can even create your own wavelets, if needed. However there is "an embarrassment of riches" in the many wavelets that are already out there and ready to go. We have already seen that with their ability to stretch and shift that wavelets are extremely adaptable. You can usually get by very nicely with choosing a less-than-perfect Wavelet. The only "wrong" choice is to avoid wavelets due to an abundant selection.

There is much more to discover than can be presented in this short overview. The time spent, however, in learning, understanding and correctly using wavelets for these "non-stationary" signals with anomalies at specific times or changing frequencies (the fascinating, real-world kind!) will be repaid handsomely.

APPENDIX

D

Further Resources for the Study of Wavelets

As we have seen, wavelets add literally another dimension to Digital Signal Processing—instead of looking at a signal in the time domain or in the frequency domain we can simultaneously process the signal in time and frequency (or equivalently scale or level). Events within the signal that are time and/or frequency limited (the most interesting kind!) can be now isolated and analyzed using the various wavelet transforms. The price to be paid for this powerful capability is of course to learn how to use it—and this means an additional "dimension" of difficulty in the learning process.

While this book fills the need for a fairly in-depth intuitive understanding of wavelets with a minimum of mathematical equations, further study is not only possible but recommended. Familiarity with the concepts and uses of wavelets (and having seen how some key equations are related to these concepts) should (hopefully) make this further study easier and more palatable having seen wavelets "in action".

*There is no shortage of books, papers, websites, and software that deal with wavelets. Some are highly specialized and highlight some specific applications. Others deal with wavelets as Applied Mathematics and are very equation-oriented. A few tell the history of wavelets and/or provide a different perspective from a less sterile environment. There are even a small number of authors who will not only extol the virtues of wavelets but will point out the pitfalls and even share some experiences where things went awry from the improper use of wavelets.**

While it is impossible to review all of the literature (especially with the exponential rise in new publications), there are some particularly good examples that can be recommended. These books are listed alphabetically by author. Which books are "best" depends on your level of comfort with mathematics. The level of math used in each book, however, is stated in these concise reviews, along with some additional comments to help establish the "flavor". MATLAB's Wavelet Toolbox Instruction Guide, although not really a "book" in the classical sense, is included here because of its chapter by chapter layout and its great value as a an illustrated (full color) tutorial. Three excellent general DSP books (Smith, Lyons, Strum et. al.) must also be included for their outstanding readability, clarity, and background support of wavelet study.

---

* My father, D. J. Fugal, used to say that the young often mistake *learning* for *experience* while the old mistake *experience* for *learning*. At any age we must be careful not to mistake learning *or* experience for true *wisdom*.

In this age of internet availability, articles are most easily obtained through a website and are thus presented in this form. Wavelet websites have the URL website reference shown in **bold***. With the URLs of both the websites and the articles changing, however, enough information is given to allow a search engine to find them.*

There has been a virtual explosion of wavelet papers in recent years. For this reason, and because so many papers are application-oriented I have not attempted to include them in this appendix. Learning wavelets from papers is like trying to get a drink of water from a fire hydrant! Also, most (but not all) papers tend to become outdated and difficult to find and retrieve. It is thus recommended that first you gain an intuitive understanding (a major goal of this book) and then use a good search engine to find up-to-date papers in the application or area of further study that you desire.

D.1 Wavelet Books

- Addison, Paul. (2003) "The Illustrated Wavelet Transform Handbook". (IOP Publishing) One of the best introductions to the Continuous Wavelet Transform of any book (Matlab User's Guide is also very good for the CWT). The treatment of the Discrete Wavelet Transform is more equation-oriented. Many diagrams, some in full color. The wavelet transform is correctly and intuitively depicted as a cross-correlation of a group of wavelet filters of various lengths with the signal.

- Boggess, A, Narcowich, F. J., (2001) "A First Course in Wavelets with Fourier Analysis" (Prentice-Hall). Mathematically-oriented with theorems and proofs. As the title implies, relates wavelets to Fourier analysis. Unlike some books that portray wavelets as the method of choice for all signals, this book is enlightened and shows instances where the FFT is the better choice for stationary signals (those without rapid changes and/or having a repetitive nature).

- Burrus, et. al. (1998) "Introduction to Wavelets and Wavelet Transforms – a Primer" (Prentice Hall). Excellent book with lots of words to go with the many equations. One of the few to show the capabilities of the UDWT/RDWT (see page 211). Burrus also substantiates the half-integer method we have used (see page 71) and mentions upsampling/downsampling and convolution as the core of wavelet determination and filter banks.

- Daubechies, Ingrid. (1991) "Ten lectures on Wavelets", (CBMS-NSF Regional Conference Series in Applied Mathematics). This one started it all. The text is concise, highly mathematical, and referenced by the majority of wavelet books. It advances (and provides proofs for) many theorems. As such, it requires of the reader a mathematical background.

- Farhang-Boroujeny, Behrouz., (1999) "Adaptive Filters: Theory and Applications", (Wiley). With his excellent insights and his strong mathematical skills, Behrouz helped D. L. Fugal prepare his first wavelets course and was a co-presenter. His highly mathematical but readable book provides valuable information.

- Hubbard, B. B. (1998) "The World According to Wavelets", (A. K. Peters). Fascinating history of wavelets that adds excitement and a human element. Wonderful quotes are found throughout the book Fluent in both French and English, Ms. Hubbard was able to interview such giants as Daubechies, Meyer, Donoho, Coifman, Strang, Mallat, Wickerhauser, and others. Math-based explanations are clearly separated in sections called "Beyond Plain English".

- Jensen, A., LaCoor-Harbo, A. (2001) "Ripples in Mathematics" (Springer-Verlag). These talented professors from Aalborg University in Denmark introduce a way to understand wavelets using the *lifting scheme* we have talked about earlier. A knowledge of concepts such as Laurent polynomials, invertable matrices, and polyphase representations is helpful.

- Kaiser, G., (1999) "A Friendly Guide to Wavelets", (Birkhauser). Good book with practical examples. Those with a strong math background will find it the most "friendly". An excellent treatment of Spectral Factorization expands on the "square root" analogy to include z-plane technology (poles, zeros, unit circles, and minimum/maximum phase filters).

- Lyons, R. G. (2004) "Understanding Digital Signal Processing – Second Edition" (Prentice-Hall). This is Amazon.com's top-selling DSP book for 5 years in a row. Lyons is an Associate Editor for *IEEE Signal Processing* magazine where he created and edits the "DSP Tips and Tricks" column. Extremely well written and almost conversational in style, Lyons uses liberal diagrams and walks you through examples step-by-step. Lyons explains very well what the equations mean from a pragmatic standpoint for the engineer who will be using them.

- Mallat, S. (1990) :"A Wavelet Tour of Signal Processing" (Academic Press). Stephane Mallat 's pioneering work in wavelets is well known worldwide. This book is mathematically rigorous but is so comprehensive that it has been used as a basis for many other texts. One of the earliest treatments of the UDWT.

- Misiti M. et. al. (1996-2008) "MATLAB's Wavelet Toolbox Instruction Guide" (The Mathworks). All or parts of the *guide* itself can be downloaded for free from mathworks.com. The introductory chapters require almost no math background while the later ones are more mathematically intensive. With or without the actual software, an excellent resource, especially for demonstrating the Continuous Wavelet Transform. Add the software and you can speed the learning process by working through examples.

- Mix, D. W., Olejniczak, J. K. (2003). "Elements of Wavelets for Engineers and Scientists" (Wiley-Interscience). A good book whose goal (like that of the book you are reading) is to provide comprehension and an introduction to the more rigorous texts. By using diagrams, examples and worked problems, Mix's book is thus less math-intensive than most, but does require understanding of calculus, power series, set theory, and matrix manipulation.

- Moon, T. K.; Stirling, W. C. (1999) "Mathematical Methods and Algorithms for Signal Processing" (Prentice-Hall). Math-based treatment of Digital Signal Processing which places wavelets in context (see Chapter Three).

- Nievergelt, Y. (2000) "Wavelets Made Easy" (Birkhauser). It could be argued that although wavelets can be *understood* and (correctly) *used* they deal simultaneously with time, frequency, and magnitude and are thus inherently not really "easy". Although using some sophisticated mathematics (including proofs and lemmas), this book is easier to understand than most and provides some valuable background and insights.

- Proakis, J. G.; Manolakis, D. G., (2006) "Digital Signal Processing – Seventh Edition", (Prentice-Hall). Venerable equation-based textbook on DSP in wide use for upper and graduate level DSP courses worldwide (the author used this textbook in teaching advanced DSP at the University of Utah). Chapter Ten on Multirate Digital Signal Processing is particularly applicable to wavelets.

- Rao, R. M.; Bopardikar, A. S., (1998) "Wavelet Transforms: Introduction to Theory and Applications", (Addison Wesley). One of the better introductory books. Very well laid out and the equations are explained rather than simply presented.

- Smith, S. W. (1997) "The Scientist and Engineer's Guide to Digital Signal Processing". (California Technical Publishing). Written with the practicing engineer in mind, Smith's book addresses everyday DSP problems in a very clear and readable manner. The math is kept as simple as possible and even the complex numbers are explained. Quote from page xii: "The goal (of this book) is to present practical techniques while avoiding the barriers of detailed and rigorous mathematics." The book you are reading has used Smith's book (with his permission) as a model of not only the physical layout, but as a model of clarity and practicality.

- Strang, G.; T. Nguyen (1996), "Wavelets and filter banks", (Wellesley-Cambridge Press). *Very* mathematical but very well done. His tongue-in-cheek reference to a very low signal-to-noise ratio in many wavelet texts when it comes to obtaining useful information is right on the money.

- Strum, R. D., Kirk, D. E. (1989) "First Principles of Discrete Systems and Digital Signal Processing", (Addison-Wesley). Designed as a first course in DSP, this classic textbook was one of the first to work with signals and systems in the discrete, digital domain. You will find summations instead of integrals and methods to design digital filters directly instead of always adapting analog filters. Basic calculus is the only math requirement.

- Teolis, A. (1998) "Computational Signal Processing with Wavelets", (Birkhauser). Fairly mathematical treatment using frame-based theory, Hilbert spaces, Fourier dual spaces, etc. Dr. Teolis is a gifted instructor and lecturer and, if comfortable with this level of applied mathematics, you will enjoy this treatment of wavelets, Gabor transforms, overcomplete wavelet transforms and much more.

- Vaidyanathan, P. P. (1993). "Multirate Systems and Filter Banks" (Prentice-Hall). This mathematically rigorous, well-written book addresses the lifting (raising) of filters and how this applies to spectral factorization. He presents the general principles and leaves specific application to other texts.

- Vetterli, M.; Kovacevic, J. (1995) "Wavelets and Subband Coding", (Prentice-Hall PTR). Very mathematical, but very well-written. Addresses spectral factorization in Chapter Two. He underscores the concept that wavelets provide both time and frequency resolution. His comment about being able to see the forest *and* the trees using wavelets is most insightful.

- Walker, J. S. (1999) "A Primer on Wavelets and their Scientific Applications", (CRC Press). Good verbal explanations and helpful graphics. Many equations, but at the undergraduate level. The step-by-step walk-through on the Daubechies 4 wavelet filter (4 filter points) is especially good.

- Wickerhauser, M. V. (1994) "Adapted Wavelet Analysis from Theory to Software" (A. K. Peters). The title pretty much says it. Very rigorous treatment of the theory. This pioneering work also discusses the technical aspects of implementation. "Best Basis" (referenced earlier in the book you are reading) is also discussed here.

D.2 Wavelet Articles

- ***Barry Cipra*** wrote this very readable article "Wavelet Applications Come to the Fore" for the SIAM NEWS about the use of wavelets by the FBI. (**siam.org/www.siam.org/siamnews/mtc/mtc1193.htm**).

- ***D. L. Fugal*** (the author) wrote an article entitled "Wavelets: Beyond Comparison" hosted on the Applied Technology Institute website (**www.aticourses.com/ati_tutorials.htm**) as a "Staff Tutorial". This seven-page article uses examples, displays, diagrams and figures but no math whatsoever. The familiar FFT/DFT is first reviewed and then compared to the Continuos Wavelet Transform (CWT). The conventional (decimated) Discrete Wavelet Transform and the Undecimated DWT (DWT) are introduced and compared. Examples of the capabilities of these transforms are shown, along with a short overview of the various types of wavelets.

- ***Dana Mackenzie*** writes of how wavelets were used in "A Bug's Life". (**www.beyonddiscovery.org/content/view.article.asp?a=1952**) Tells in plain English about many other applications of wavelets. Another reference to being able with wavelets to see *both* the forest and the trees (theory and practicality).

- ***Don Morgan*** wrote "Why Wavelets?" for *Embedded* (**embedded.com**). Very readable article on wavelets. Discusses the musical octave nature of wavelets, relates to Fourier transforms, and talks of methods of noise reduction. He shows how the FFT/DFT multiplies orthogonal frequency vectors by the signal and adds the results together.

- ***Mark J. Jensen*** wrote an article for *Financial Engineering News* called "Making Wavelets in Finance" (**fenews.com/fen1/wavelets.html**) Interesting to see wavelets in action in the stock market.

- ***Paul Addison*** wrote a very interesting article in Physics World, March 2004, (**physicsweb.org**) on the use of wavelets in monitoring breathing in premature infants. Discussed in "The Little Wave with a Big Future" this is a very good plain English overview of wavelets with good color diagrams.

- ***Peter Schroeder*** published an article in *Wired* magazine (www.**wired.com/wired/archive/3.05/geek.html**). called "Beating the Bandwidth Bottleneck" dealing with wavelet compression. Very readable.

- ***Selesnick, Baraniuk, and Kingsbury*** wrote and article for the *IEEE Signal Processing Magazine*, Volume 22, Number 6, November 2006. (**ieeexplore.ieee.org/iel5/79/33042/01550194.pdf?arnumber=1550194**) This article deals with the Dual-Tree Complex Waveform. This type of wavelet transform is a compromise between the simple UDWT (undecimated) and the conventional (decimated) DWT. It requires twice as much data storage/handling as the conventional DWT, but less than the UDWT. In the Dual-Tree Complex DWT we both downsample (like the conventional DWT) *and* use different filters at each stage (like the UDWT). The method is complicated and still theoretical in nature and requires new designs with new constraints on the filters (e.g. there are *two* lowpass decomposition filters at each stage). The article is groundbreaking in that it acknowledges that the conventional DWT has problems with aliasing when you throw out some of the coefficients. It also acknowledges shift-variance.

- ***Starck and Murtagh*** have an article in PDF form on using a Haar wavelet with an a' trous method to predict the fractal nature of the Stock Market. (**http://strule.cs.qub.ac.uk/~fmurtagh/mr/**). They are aware (unlike many other authors) that the conventional DWT can cause aliasing and thus they use the UDWT in this article.

- ***Torrence and Compo*** "A Practical Guide to Wavelet Analysis". (**paos.colorado.edu/research/wavelets/bams_abstract.html**) Fascinating diagrams and explanations of the El Nino weather phenomenon using wavelets.

D.3 Wavelet Websites

- (**amara.com/current/wavelet.html**) Amara Graps Internet Site. A list of pointers, covering theory, Papers, books, implementations, resources. Amara has done a great service in this overview of wavelets. She has also published several wavelet papers. This is a great guide to other websites. She tells of 39 types of Wavelet Software. She illustrates the fractal nature of wavelets and even lets you listen to a chirp. Good depiction of the fractal nature of wavelets. Her writing is clear and concise. for example, she reminds us that wavelets are *localized* in both *time* and *frequency*.

- (**bearcave.com/misl/misl_tech/wavelets/index.html**) Ian Kaplan "Wavelets and Signal Processing" covers several topics. Ian has done a great job in reviewing the various books and websites out there. This person must be absolutely brilliant because his opinions are mostly the same as the author's ;-).

- (**cas.ensmp.fr/~chaplais/Wavetour_presentation_US.html**). Francois Chaplais has a tutorial website titled "A short presentation" This is based/inspired by "A Wavelet Tour of Signal Processing" by Stephane. Mallat. Chaplais covers: Fourier analysis, time-frequency analysis, frames, singularity analysis and reconstruction, and wavelet bases and filter banks. This is one of a valuable minority of resources that correctly addresses the A' Trous or UDWT as having "stretched wavelets". He also shows that we lose shift invariance and alias cancellation when we discard or modify the coefficients in the conventional DWT.

- (**cmis.csiro.au/Hugues.Talbot/dicta2003/cdrom/pdf/0029.pdf**) Andrew P. Bradley's "Shift-invariance in the Discrete Wavelet Transform" This is one of the rare places where the "soft underbelly" of the conventional DWT is exposed. It tells what happens (loss of full alias cancellation) when you do not use ALL the coefficients in the reconstruction.

- (**ecs.syr.edu/faculty/lewalle/tutor/tutor.html**) Jaques Lewall has written a thorough "introduction" using a Mexican Hat wavelet to analyze a cosine wave. It is not simple, but it has lots of diagrams. He explains energy well and shows the fractal nature. Also good explanation of the Morlet wavelet.

- (**home.od.ua/~relayer/algo/dsp/surfingw/wavelet.htm** or try **wavelet.org/phpBB2/gallery.php?c=Tutorial&t=6312**) Center of Machine Condition Monitoring (CMCM) at Monash University. "Surfing the Wavelets" gives figures of the "scaling and shifting" along with a comparison of a DWT and CWT.

- (**http://users.rowan.edu/~polikar/WAVELETS/WTtutorial.html**). The Engineer's Ultimate Guide to Wavelet Analysis by Robi Polikar One of the better introductions to wavelets.

- (**MathWorks.com**) Matlab Wavelet Toolbox User's Guide. Free PDF files that do an excellent job as an introduction to wavelets, especially the CWT. Valuable even if you use other software.

- (**pagesperso-orange.fr/polyvalens/clemens/lifting/lifting.html**). "The Fast Lifting Wavelet Transform" by Clement Valens Goes into detail on the Lifting Scheme. A valuable adjunct to the "Ripples in Mathematics" text.

- (**pagesperso-orange.fr/polyvalens/clemens/wavelets/wavelets**). "A *Really* Friendly Guide to Wavelets" by Clement Valens A response to Kaiser's "Friendly Guide to Wavelets". Still fairly mathematical but readable. He talks about the last Approximation as a "cork" in the frequency subbands. For example a signal can be broken into **D1** + **A1** and **A1** will be the "cork" that contains all the lowest frequencies. If we decompose it further into **D1** + **D2** + **D3** + **A3**, then **A3** will now be the "cork" that contains all the lowest frequencies.

- (**star.stanford.edu/projects/sswrg/basics.html**) This is a very nice, concise introduction to the Continuous Wavelet Transform. The tutorial stops short of the Discrete Wavelet Transform, however.

- (**williamcalvin.com/bk7/bk7ch13.htm**) Interesting portion of the book "*Conversations with Neil's Brain: The Neural Nature of Thought and Language*" by William H. Calvin and George A. Ojemann dealing with how retinal neurons deal with bright spots on a dark background in a manner similar to the Mexican Hat wavelet. Refer to Chapter Ten (in the book you are reading) in the "Mexican Hat" wavelet description for an interesting experiment described in the footnote.

- (**wavelet.org**). Wavelet Digest Internet Site. Anything going on in the wavelet community will be found here. Books, papers, lectures, seminars, course announcements (including those of the author) are given. Originally founded by Wim Sweldens and edited by the extremely knowledgeable Dimitri Van De Ville and Michael Unser, the Digest is a valuable resource.

- (www2.**cs.uregina.ca/~gerhard/courses/Audio/WaveletsDummies.pdf**) "Wavelets for Dummies" is a *very* high level overview of wavelets (4 slides!). It provides a concise review of some of the main concepts.

- (www.**conceptualwavelets.com**) Website by the author, D. L. Fugal, has some free downloads of book chapters, information on Conceptual Wavelet Courses, and other goodies. Regularly updated. Check periodically. All you ever wanted to know (and more) about my wavelet work can be found here and at **www,SpaceAndSignals.com**.

INDEX

A

B

J

K

L

M

T